Lecture Notes in Earth Sciences

Vol. 1: Sedimentary and Evolutionary Cycles. Edited by U. Bayer and A. Seilacher. VI, 465 pages. 1985

Vol. 2: U. Bayer, Pattern Recognition Problems in Geology and Paleontology. VII, 229 pages. 1985.

Vol. 3: Th. Aigner, Storm Depositional Systems. VIII, 174 pages. 1985.

Vol. 4: Aspects of Fluvial Sedimentation in the Lower Triassic Buntsandstein of Europe. Edited by D. Mader. VIII, 626 pages. 1985

Vol. 5: Paleogeothermics. Edited by G. Buntebarth and L. Stegena. II, 234 pages. 1986.

Vol. 6: W. Ricken, Diagenetic Bedding. X, 210 pages. 1986

Vol. 7: Mathematical and Numerical Techniques in Physical Geodesy. Edited by H. Sünkel. IX, 548 pages. 1986.

Vol. 8: Global Bio-Events. Edited by O.H. Walliser. IX, 442 pages. 1986.

Vol. 9: G. Gerdes, W.E. Krumbein, Biolaminated Deposits. IX, 183 pages. 1987.

Vol. 10: T.M. Peryt (Ed.), The Zechstein Facies in Europe. V, 272 pages. 1987.

Lecture Notes in Earth Sciences

Edited by Somdev Bhattacharji, Gerald M. Friedman, Horst J. Neugebauer and Adolf Seilacher

10

Tadeusz M. Peryt (Ed.)

The Zechstein Facies in Europe

Springer-Verlag Berlin Heidelberg GmbH

Editor

Dr. Tadeusz M. Peryt
Instytut Geologiczny
ul. Rakowiecka 4, PL-00-975 Warszawa, Poland

ISBN 978-3-540-17710-4 ISBN 978-3-540-47781-5 (eBook)
DOI 10.1007/978-3-540-47781-5

Originally published by Springer-Verlag Berlin Heidelberg New York in 1987

Preface

During the last few years, evaporites have increasingly been regarded as sediments and not only as chemical precipitates. Especially the intensive study of the Zechstein facies has resulted in a vast amount of observations and interpretations which are of general significance, offering important information to all sedimentologists interested in carbonates and evaporites. It seems therefore useful to introduce the sedimentological approach in a basin where various chemical concepts have been developed. This is the aim of the present volume, and this approach will be recognized by the reader in most of the chapters.

The idea of publishing a collection of papers on the Zechstein facies and related rocks found an enthusiastic response, although later some contributors were, for various reasons, unable to meet the deadline. However, the papers submitted cover all major fields and will certainly stimulate further research.

The resulting volume is dedicated by the editor and contributors to Professor Dr. Gerhard Richter-Bernburg, the father of modern Zechstein research, on the occasion of his 80th birthday. The help of Krzysztof Görlich and Szczepan Porebski during work on the volume is gratefully acknowledged.

Tadeusz Marek Peryt

Table of Contents

INTRODUCTION

The extent of the Permian Zechstein basin, contoured in Fig. 1, is well known. The attempt has been made to gather information from a wide distribution of localities (see numbers in Fig. 1), in order to give an almost complete overview of the whole area. Since for the understanding of basin development the rocks underlying the evaporites are equally important, investigations of both Kupferschiefer and carbonate facies have been included in this volume.

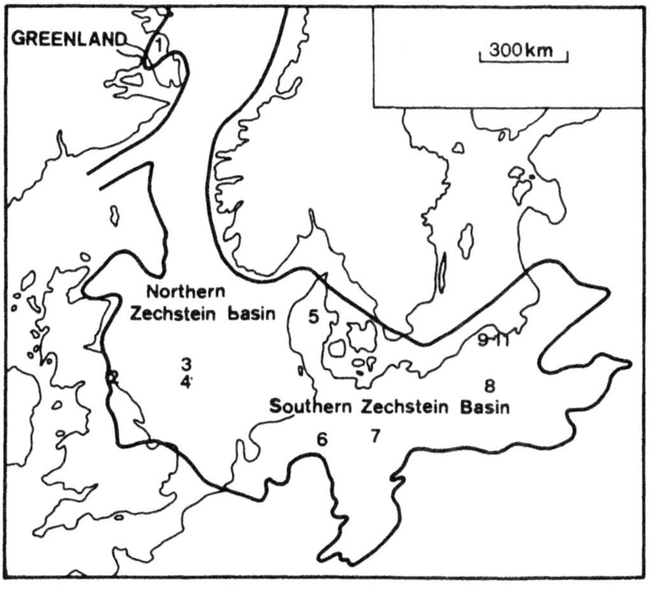

Fig.1. Approximate location of areas discussed in the papers of this volume. 1 - Stemmerik, 2 - Hollingworth & Tucker, 3 - Jenyon & Taylor, 4 - Jenyon, 5 - Sønderholm, 6 - Mausfeld & Zankl, 7 - Langbein, 8 - Oszczepalski & Rydzewski, 9 - Czapowski, 10 - Peryt, 11 - Gasiewicz, Gerdes & Krumbein

Oszczepalski and Rydzewski summarize the data on Kupferschiefer lithofacies distribution throughout Poland, and conclude that the Kupferschiefer was deposited in a

Lecture Notes in Earth Sciences, Vol. 10
T. M. Peryt (Ed.), The Zechstein Facies in Europe

relatively shallow, mud-dominated stratified shelf sea. The deposition generally occurred below wave base, and the variability of sequences is related to the supposed fluctuating redoxcline. Hollingworth and Tucker reconstruct the reef palaeocommunity structure and the evolution of palaeocommunities through time and space. They stress the importance of aragonite cementation in the formation of the Tunstall Reef, as has also been recorded in the Polish and Thuringian reefs, although without being so convincingly documented. Mausfeld and Zankl present the history of deposition and the facies analysis of the Main dolomite in NW Germany. According to them, the deposition took place on a prograding distally steepened ramp with an overall regressive facies sequence. They distinguish two distinct phases of deposition, probably separated by a slight drop in sea level. They conclude that coated grains in high-energy environments are mainly formed by microbiological activity. Such activity is well documented by Gasiewicz, Gerdes and Krumbein in the Platy Dolomite of northern Poland. They describe sabkha-type stromatolites. Stromatolites of the lower part of the Platy Dolomite are dominated by filamentous cyanobacteria and those of the upper part by coccoid ones. The authors suggest an increase in salinity of seawater toward the end of Platy Dolomite deposition. Stemmerik describes four different types of cyclically interbedded shallow marine limestones, often algal, and sulphates in the Upper Permian of Greenland. The cyclicity resulted from rapid fluctuations of sea level and therefore the different cycles are very localized. Peryt presents a sequential model of the early diagenetic history of the Main Dolomite in northern Poland which may be used as a standard for comparison with other intra-evaporitic carbonate systems, and concludes that the early diagenetic imprint was decisive for later diagenetic history.

However, when compared to the earlier three volumes on Zechstein facies (edited by Füchtbauer and Peryt, Depowski et al., and Harwood and Smith), it may be noticed that in the present volume new fields are covered (or developed), both areal (Fig. 1; Greenland, Jutland, the Leba elevation in northern Poland; the latter area is discussed in three papers) and thematical. Of the latter, two papers using seismic surveys of recent date contain important implications. Jenyon presents the seismic zonation of the Zechstein. His data indicate that movement of the Zechstein salt in the English Southern Zechstein Basin took the form of lateral salt flow centrifugally from the basin toward the margin. Jenyon and Taylor present evidence that many features across the Mid North Sea High are due to dissolution and removal of salt rock. Other structures recorded may be carbonate buildups or anhydrite pods.

The growing interest in evaporites and evaporite-related carbonates is reflected in half of this volume being devoted to them. In his essay on Zechstein anhydrites, Langbein summarizes his own research and literature data and concludes that the features observed in the anhydrites depend exclusively on their diagenesis, cementation being the main factor which governed the route of later compaction. Considering cementation and compaction, a new classification of anhydrites is proposed. The breccias are related to different processes, and specifically the giant breccias occurring in the Upper Werra anhydrite, considered by several workers to be olisthostromes, are thought by the author to be compaction or collapse breccias. This somewhat provocative essay will surely avoke a response in the evaporitic world. Czapowski presents a sedimentological analysis of the Werra halites in northern Poland and concludes that medium and high dynamic facies prevailed in the basin at that time. The low dynamic sequences have been found only in the central parts of larger bottom depressions. Maximum water depth was estimated to be from several to tens of metres. Sønderholm describes the Z1-Z2 transition in northern Jutland. Detailed sedimentological and geochemical investigation made it possible to distinguish three facies associations which reflect major events in the evolution of the basin and are closely related to the evolution seen in the Southern Zechstein Basin.

Tadeusz Marek Peryt

CYCLIC CARBONATE AND SULPHATE FROM THE UPPER PERMIAN KARSTRYGGEN FORMATION, EAST GREENLAND

Lars Stemmerik
Institute of Historical Geology and Paleontology
Øster Voldgade 10
DK-1350 Copenhagen K
Denmark

Abstract: Four different types of cyclically interbedded shallow-marine limestone and nodular-mosaic sulphate have been recognised within the Upper Permian Karstryggen Formation in central East Greenland. They are composed of: 1) lime mudstone, algal laminated limestone, and nodular-mosaic sulphate; 2) algal laminated limestone and nodular-mosaic sulphate; 3) intraclast carbonate conglomerate and nodular-mosaic sulphate; and 4) oolitic grainstone and nodular-mosaic sulphate. Subaerial aeolian sediments do not occur within the Upper Permian cycles. Type 1 and 2 cycles were formed as the result of repeated subaerial exposure of lagoonal sediments. Type 3 and 4 cycles were formed as shallow, high-energy shoal sediments became exposed. These cycles result from rapid fluctuations of sea-level. The fluctuations were too rapid to allow facies progradation of any significance and the different cycles aggrade and are therefore very localized.

Introduction

Cyclic sedimentation comparable to the Zechstein cycles (Z1-Z5) of NW Europe has not been recognised in the East Greenland basin (Fig. 1). Here, deposition of the Late Permian Foldvik Creek Group reflects an overall transgressive event dominated by deposition of limestones and black shales followed by progradation of clastic material filling up the basin in the latest Permian (Fig. 2).

The only deviation from this pattern occurs in the southern part of the basin. Here, an apparently structurally controlled regression occurred in late Karstryggen Formation times, and before deposition of the Wegener Halvø and Ravnefjeld Formations (Fig. 2) (Stemmerik, 1985).

Lecture Notes in Earth Sciences, Vol. 10
T. M. Peryt (Ed.), The Zechstein Facies in Europe

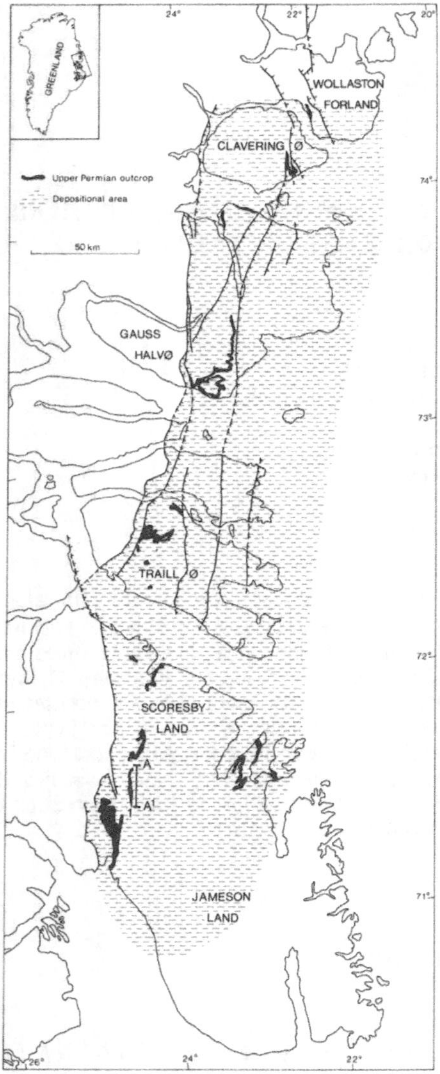

Fig. 1. Locality map of East Greenland showing the outcrop of the Upper Permian Foldvik Creek Group and the proposed outline of the depositional basin. A–A' indicates the position of the section shown in Fig. 3.

Cyclic deposition of limestones and nodular sulphates is, however, common in a small scale within the Karstryggen Formation. The significance of cyclic alternation of nodular anhydrite and shallow marine carbonate became evident from observations of recent sabkhas along the Persian Gulf (Shearman, 1966). The sabkha cycle consists of shallow-marine lagoonal limestone at the bottom, intertidal algal carbonate and supratidal aeolian sediments with diagenetically formed evaporite at the top (Shearman, 1966). The ideal cycle represents progradation of a subaerial sabkha surface into a shallow marine lagoon (Shearman, 1966). Sequence of repeated cycles separated by erosion surfaces represent rapid relative rises in sea-level followed by renewed progradation of the sabkha. The dynamic processes involved appear to allow correlation of individual cycles over long distances, exceeding 100 km along strike in the Ordovician of Arctic Canada (Mossop, 1979).

In East Greenland, four different types of cyclically interbedded shallow marine limestone and nodular sulphate occur within the Upper Permian Karstryggen Formation. The Late Permian cycles include algal laminated limestone and lime mudstone of the protected lagoonal and intertidal environments as well as intraclast conglomerates and oolitic grainstone formed in shallow, high-energy environments. Diagenetic sulphate was formed during repeated exposure of the carbonate. The Late Permian cycles include no aeolian deposits in contrast to the classical sabkha cycle.

In this paper the different types of cycles are described and a depositional model explaining the dominantly aggradational depositional pattern of the Karstryggen Formation is proposed.

Regional setting

Early Permian rifting between Greenland and Norway followed by Late Permian thermal contraction led to the formation of north-south trending depositional basin which was 400 km long and 80-100 km wide (Fig. 1) (Surlyk *et al.*, 1984, 1986). Towards the west the basin is separated from the stable Greenland craton by the post-Devonian main fault (Fig. 1). Maync (1961) further suggested that the basin was closed towards the south whereas it was open towards the north and northeast (Fig. 1).

The Foldvik Creek Group is considered to be of Late Permian age (Piasecki, 1984; Surlyk *et al.*, 1986). The biostratigraphical control of the lowermost part is, however, problematic and deposition may have been initiated in the latest Early Permian.

Widespread deposition of conglomerates initiated the Upper Permian deposition (Maync, 1961; Surlyk *et al.*, 1984, 1986) (Fig. 2). The Huledal Formation conglomerates were deposited initially in a system of coarse-grained braid-plains. The depositional environment changed as the Late Permian sea transgressed into the area. The upper part of the Huledal Formation is suggested to have formed in a protected marine bay dominated by fluviatile processes (Surlyk *et al.*, 1984, 1986).

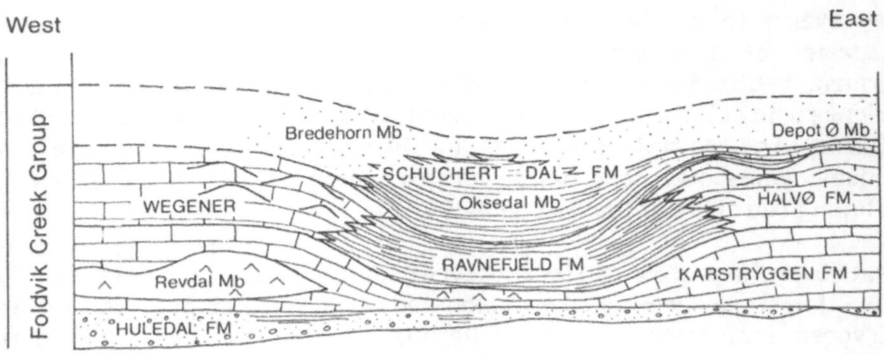

Fig. 2. Simplified E-W cross section of the Foldvik Creek Group in Jameson Land (modified from Surlyk *et al.*, 1986).

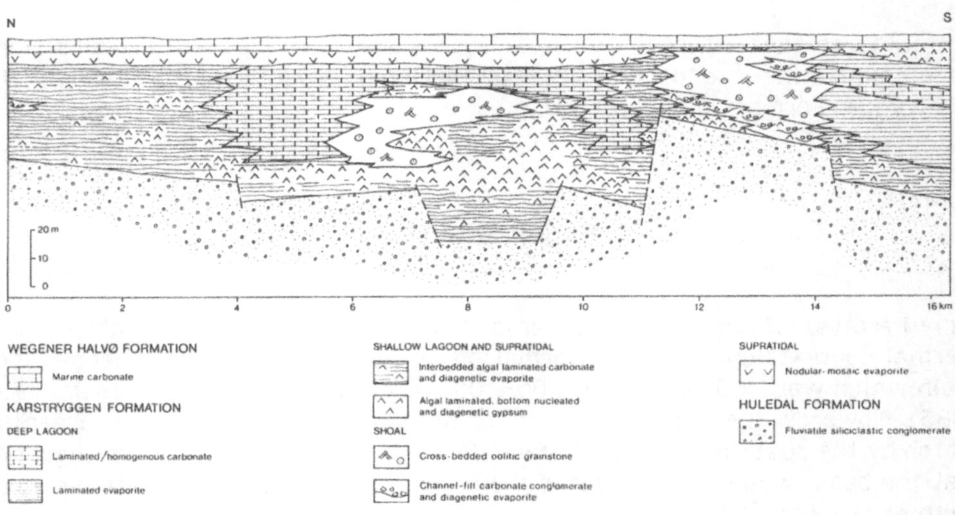

Fig. 3. North-south cross section of the Karstryggen Formation along the line A-A' on Fig. 1. The top of the formation is used as base line (modified from Surlyk *et al.*, 1986).

The overlying Karstryggen Formation (Figs. 2 & 3) records the beginning of a shallow hypersaline sea of mainly carbonate and sulphate deposition (Surlyk *et al.*, 1984, 1986). Sedimentation was controlled by the structural configuration of the basin (Fig. 1). In the northern part of the basin fully marine carbonates were deposited during the early stage of transgression (Maync, 1942). These are gradually replaced southward by non-fossiliferous, hypersaline limestone (Stemmerik, 1985; Surlyk *et al.*, 1984, 1986). The sulphates were deposited mainly along the western, down-tilted margin (Maync, 1942,1961; Stemmerik, 1980; Surlyk *et al.*, 1984, 1986). In Jameson Land (Fig. 1), Stemmerik (1985) described sulphate also to occur in small, structurally controlled basins away from the margin (Fig. 3).

The depositional pattern changed as the sea-level raised rapidly. Only carbonate production along margins of the basin was able to keep pace with the ongoing transgression. Here, marine fossiliferous limestone deposited and a carbonate shelf 10 km wide developed (Surlyk *et al.*, 1986). The carbonate shelf surrounded a more than 100 m deep oxygen-deficient mud basin (Surlyk *et al.*, 1986). Surlyk *et al.* (1986) erected the Wegener Halvø Formation to include the platform carbonates. The Ravnefjeld Formation comprises the black basinal shales (Fig. 2).

The Schuchert Dal Formation (Fig. 2) represents a final clastic infilling phase of the Late Permian basin. The sediments reflect a relative drop in sea-level or a marked progradation where subsidence was slower than the sediment influx (Surlyk *et al.*, 1986).

Sedimentary facies

Four distinct carbonate facies: intraclast conglomerate, oolitic grainstone, lime mudstone, and algal laminated limestone occur in the Karstryggen Formation. A number of types of diagenetic sulphate are also present. The limestone is partially dolomitized and the dolomite occurs as rhomboidal crystals less than 25 µm in size. The sulphate is now mainly alabastrine gypsum in outcrop, and samples from a shallow core still contain anhydrite. The complex late diagenetic history of the sulphate will not be discussed here.

Intraclast conglomerate

Description. This facies consists of lenticular 0-85 cm thick and 5-10 m wide bodies of intraclast conglomerate (Fig. 4A, B). The lower surface is erosive. The conglomerate is usually normally graded with 30-50% clasts near the base and 0-10% at the top.

The clasts range from 1-20 cm in size, but are usually less than 3 cm. The larger clasts are angular fragments of algal laminated limestone and early dolomitized lime mudstone (Fig. 5A). The smaller clasts are mainly rounded fragments of early dolomitized lime mudstone. Ooids are common in the upper clasts-poor part where planar cross-bedding is well developed (Fig. 4A,B). The lateral extension of this facies is only 200-300 m.

Fig. 4. Cyclically interbedded intraclast carbonate conglomerate and nodular-mosaic sulphate, cycle 3. Note the restricted lateral extension of the individual units **(A)**, the erosive lower surface and the planar cross-bedding in the upper part of the limestone **(B)**.

Interpretation. The erosional lower surface and the restricted lateral extension of this facies suggest deposition in shallow channels. The local origin and the angular shape of the clasts indicate short distance of transport and high energy conditions during deposition of the channel lag. The upper, ooid-rich part was deposited during lower flow regime conditions.

The early-dolomitized limestone clasts resemble intraclasts derived from recent supratidal carbonate flats. In recent tidal zones intraclasts form the lag in tidal channels comparable in size to those described here (Shinn, Lloyd & Ginsburg, 1969; Shinn, 1983). The intraclast conglomerate is therefore suggested to represent deposition in shallow tidal channels.

Lime mudstone

Description. This facies consists of 5-20 cm thick beds of homogenous lime mudstone and laminated, silty lime mudstone. Individual beds have a sharp, slightly undulating base, while the top is often more diffuse due to the diagenetic formation of sulphate in the overlying sediments. The lateral extension of individual beds is usually more than 200 m. The homogenous beds consist of micrite, now partly dolomite, with 10-15% silt-size siliciclastic material. The elongated grains are horizontally orientated with their long axes (Fig. 5B). Lamination is better developed with increasing content of terrigenous material. Horizons of 5-10 mm thick layers of normally graded plant debris occur regularly. The individual layers are separated by erosive surfaces.

Interpretation. The lime mudstone represents deposition in a protected low-energy environment. The absence of a benthic infauna is evident from the well-preserved lamination and the orientation of the terrigenous debris. These features result from increased rates of sedimentation, oxygen-deficient bottom conditions or extreme fluctuations in salinity.

Deposition occurred in a protected subtidal environment, occasionally punctuated by higher-energy pulses causing erosion and deposition of graded laminae. These pulses were probably related to storms.

Algal laminated limestone

Description. This facies is composed of 10-80 cm thick beds of laminated limestone. The lamination is undulating with 0.1-1 mm thick laminae of alternating brownish, organic-rich carbonate and pure carbonate. The lamination .(Fig. 5C) is usually horizontal, but occasionally the lamination forms up to 50 cm high domes (LLH). Internal disconformities, synsedimentary microfaulting, graded laminae, lenticular gypsum crystals and thin layers of intraclast conglomerate are seen within the laminated limestone.

Interpretation. The undulating lamination and the occurrence of domal structures suggest an algal origin for the lamination. This is supported by the appearance of algal threads within the organic-rich laminae. The algal laminated limestone is associated with a wide range of diagenetic sulphate described separately below (Fig. 6). In modern environments algal mats occur in the intertidal zone of protected lagoonal areas (Kinsman & Park, 1976). The apparent absence of a benthic fauna and the preservation of domal algal structures, suggest that this facies represents both the shallow subtidal and the intertidal parts of a lagoon (*cf.* Logan, Hoffman & Gebelein, 1974).

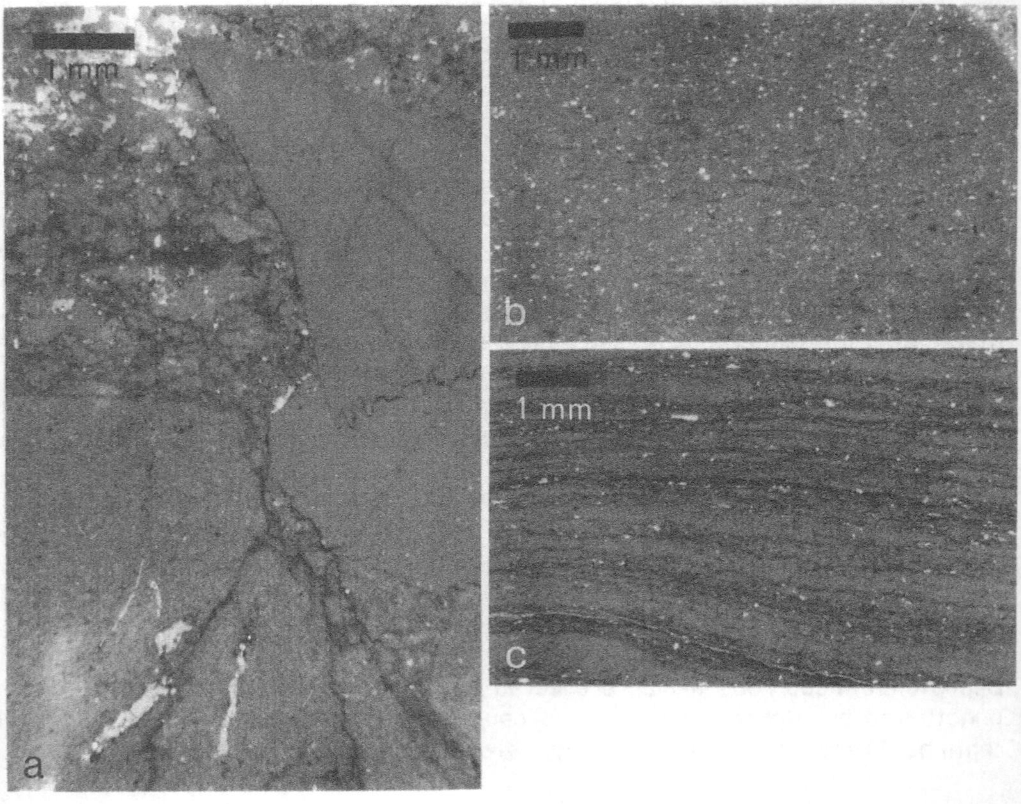

Fig. 5. Thin section microphotographs of **(a)** dolomitized intraclast conglomerate, **(b)** lime mudstone, **(c)** algal laminated limestone.

Oolitic grainstone

Description. This facies is composed of 10-90 cm thick beds of oolitic grainstone. The individual beds are sheet-like, and laterally traceable for several hundred metres. The grains are dominantly ooids with an average diameter of 0.25 mm. The ooids are preserved as fine-crystalline spar with only rare preservation of the concentric lamination. The intense diagenetic alteration combined with the uniform grain size obscures internal structures. Occasionally, planar cross-bedding occurs.

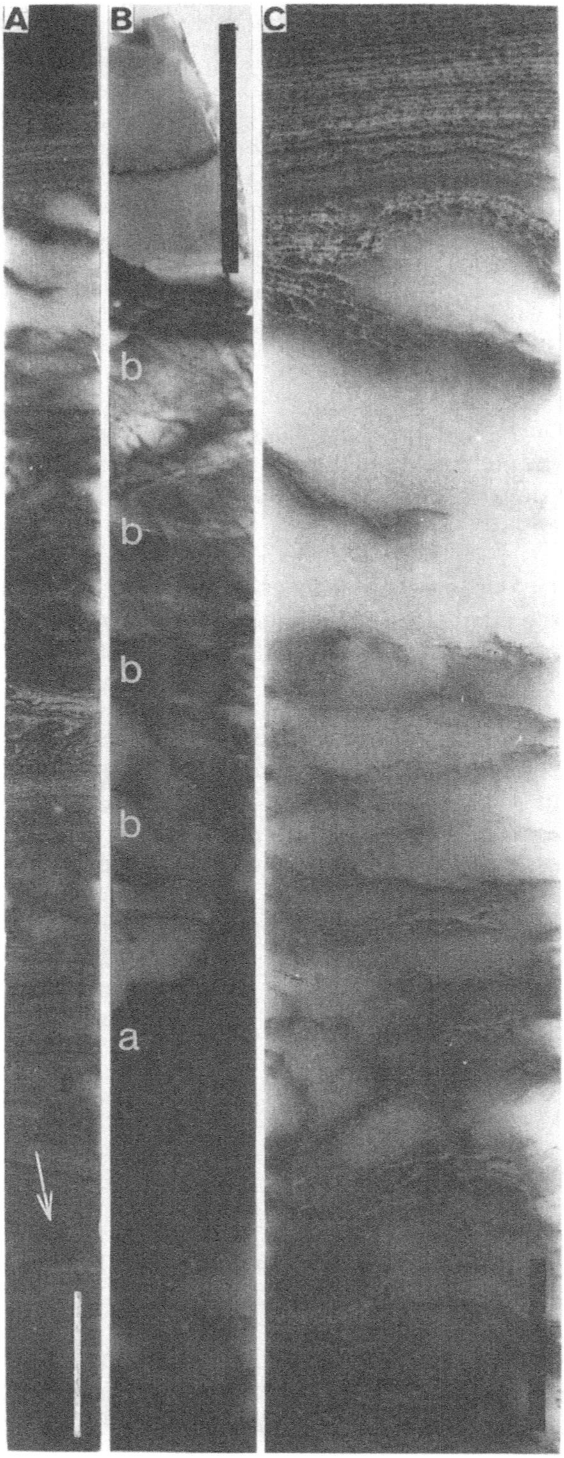

Fig. 6. (A) Cyclically interbedded algal laminated limestone and nodular-mosaic sulphate, cycle 2. Note the internal disconformities in the lower algal laminated unit (arrow). Scale bar = 5 cm. Core GGU 303105. **(B)** Algal laminated limestone with lenticular gypsum crystals (a) overlain by four horizons of enterolithic anhydrite. Scale bar = 5 cm. Core GGU 303105. **(C)** Nodular-mosaic sulphate formed in algal laminated limestone. Detail of Fig. 6A. Scale bar = 2 cm. Core GGU 303105.

Interpretation. Oolitic grainstone forms in shallow, high-energy environments (Wilson, 1975; Halley *et al.*, 1983). The cross-bedding, the sheet-like form and the bed thickness suggest deposition during migration of shallow oolitic sand belts (Halley *et al.*, 1983).

Diagenetic sulphate (nodular and nodular-mosaic sulphate)

Description. The nodular-mosaic sulphate is characterised by displacive growth of gypsum and anhydrite to form nodules and enterolithically folded layers (Fig. 6) (the classification follows Maiklem *et al.*, 1969). The nodules are horizontally elongated and vary in size from 0.5-5 cm. The thickness of the sulphate layers rarely exceeds 1 cm but due to folding, the enterolithical horizons can form layers up to 10 cm thick (Fig. 6B).

This type of diagenetic sulphate formed most commonly in algal laminated limestone, and the growth of the sulphate is often so extensive that the original fabric is destroyed and layers of nodular-mosaic (chicken-wire) sulphate are formed.

The thickness of this type of diagenetic sulphate ranges from 0-60 cm. The lower boundary is often diffuse as the number of nodules decreases downwards. The upper boundary is well defined and occasionally associated with minor erosion (Figs. 4B & 6B).

Interpretation. Gypsum and anhydrite nodules form in various sedimentary and diagenetic environments (e.g. Shearman, 1966; Dean *et al.*, 1975; West *et al.*, 1979; Butler *et al.*, 1982; Loucks & Longman, 1982). The erosional upper boundary suggests penecontemporaneous formation of this type of nodular evaporite. The displacive nodules and the enterolithically folded layers resemble recently formed diagenetic anhydrite in the supratidal sabkhas along the Persian Gulf. Here, the evaporite formation mainly occurs in the aeolian sediments above the ground water table (Butler *et al.*, 1982). The nodular-mosaic evaporite facies is suggested to represent evaporite formation in a subaerially exposed environment, which is diagenetically comparable to recent sabkhas.

Sedimentary cycles

The described facies occur in cyclic repetition forming four distinct types of cycles (Fig. 7).

Cycle 1 consists in ascending order of lime mudstone, algal laminated limestone, and nodular-mosaic evaporite (Figs. 7, 8). The individual facies in this 30-120 cm thick cycle form tabular bodies, laterally traceable for more than 200 m.

The depositional trend reflects a gradual shoaling upward. The protected, low-energy subtidal environment with deposition of non-fossiliferous, muddy limestone is followed by algal laminated limestone of the shallow subtidal and intertidal environment. The diagenetic sulphate formed in the uppermost part of the cycle indicates subaerial exposure of the algal laminated limestone (Butler *et al.*, 1982; Warren & Kendall, 1985).

This cycle in many aspects resembles the simple sabkha cycle (Shearman, 1966) and is therefore interpreted as a shallowing upward sequence formed during progradation of the intertidal algal laminated limestone into the deeper part of the lagoon. The non-fossiliferous appearance of the lagoonal lime mudstone indicates environmental conditions not condusive to animal colonisation and thus appears to be a more biologically stressful environment when compared to the present (Kendall & Skipwith, 1969). The sulphate was formed during subsequent exposure of the algal laminated limestone.

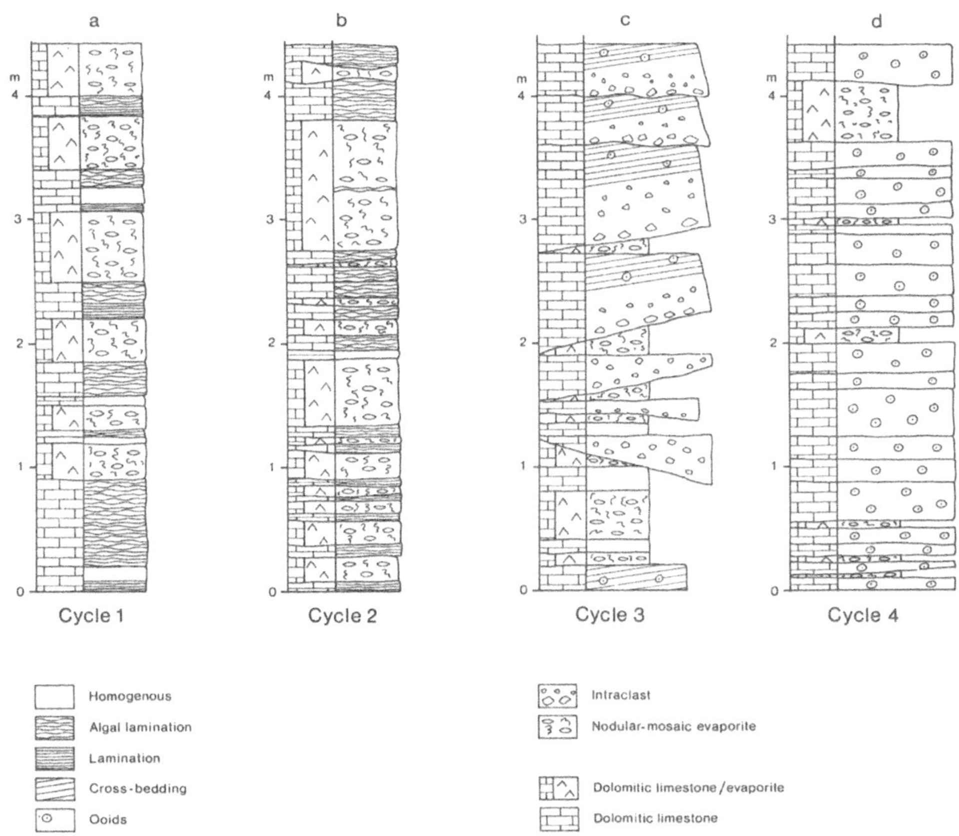

Fig. 7. Sedimentological logs of the different types of cyclically interbedded limestone and nodular-mosaic sulphate.

Cycle 2 is composed of cyclically interbedded algal laminated limestone and nodular-mosaic sulphate (Fig. 7). This cycle is widespread throughout the investigated area (Figs. 3 & 9). In some areas it interfingers with type 1 cycles producing 10-30 cm thick sequences, and it possibly represents deposition in the intertidal and shallow lagoonal environment succeeded by subaerial exposure and evaporite formation.

Cycle 3 consists of interbedded intraclast conglomerate and nodular-mosaic sulphate (Figs. 4A, B & 7). The individual beds in this less than 1 m thick cycle extend laterally for only 5-10 m. The cycles are stacked to produce less than 10 m thick carbonate-dominated sequences of very restricted lateral extension. Occasionally, these are seen to pass laterally into sequences composed solely of intraclast conglomerate.

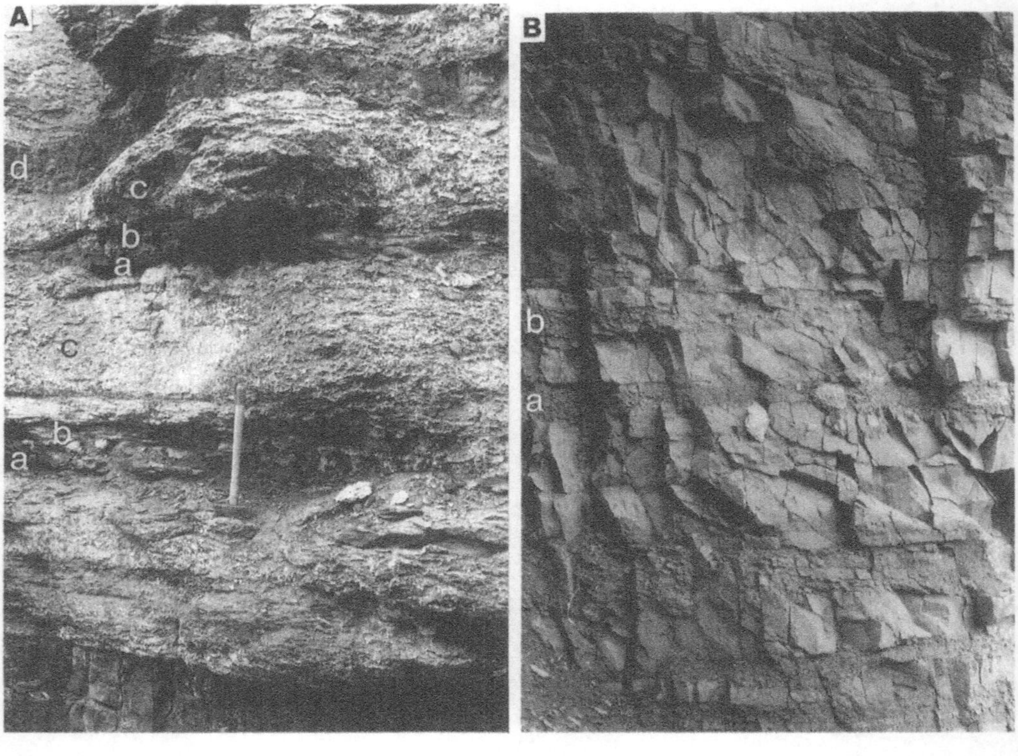

Fig. 8. (A) Cyclically interbedded lime mudstone (a), algal laminated limestone (b), and nodular-mosaic sulphate (c), cycle 1. The thicker darker beds (d) are Tertiary sills. **(B)** Cyclically interbedded oolitic grainstones (a) and nodular-mosaic sulphate (b), cycle 4.

The sedimentary cycle represents initial deposition of intraclast conglomerate in erosive channels. Subsequent exposure is indicated by evaporite formation. The erosional base of the succeeding channel rarely allows the preservation of the sulphate. Sequences of graded intraclast conglomerates separated by erosional surfaces dominate, therefore.

Cycle 4 is composed of interbedded oolitic grainstone and nodular-mosaic sulphate (Figs. 7 & 8B). The cyclic repetition of the two facies is less evident than in the former types of cycles. 10-200 cm thick units of bedded oolitic grainstone are interbedded with 10-20 cm thick beds of sulphate (Figs. 7, 8B). The lateral extension of this type of cycles is 1-2 km, but individual beds have not been traced over this distance.

This cycle, formed as oolitic shoals, occasionally became exposed.

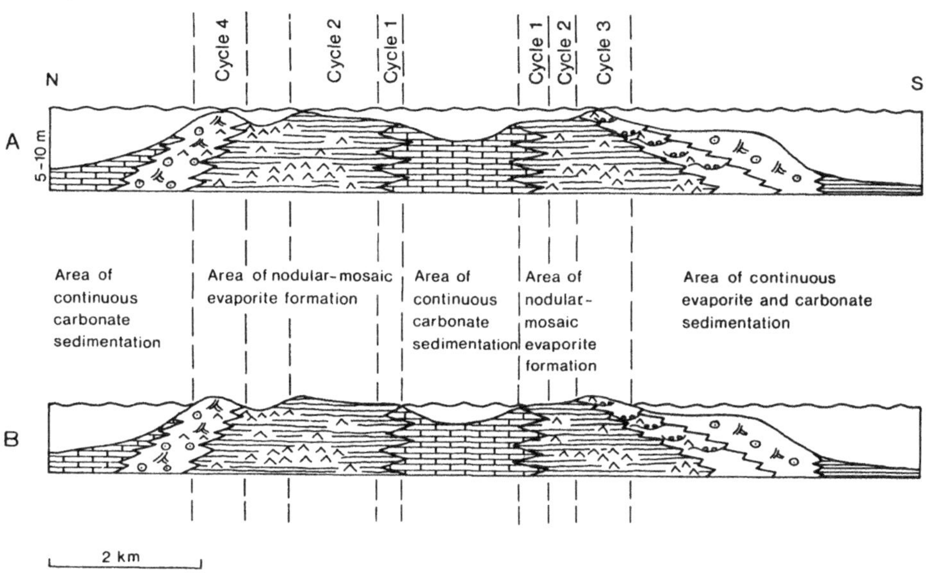

Fig. 9. Lateral facies variations during times of cyclic sedimentation. The section corresponds to the middle part of the Karstryggen Formation in section A-A' (interval 4-16 km, Fig. 3). For legend see Fig. 3.

Depositional model

The four different cycles occur in non-cyclic sequences of non-fossiliferous lime mudstone, algal laminated limestone, laminated gypsum, intraclast carbonate conglomerates, and oolitic grainstone (Fig. 3). The spatial distribution of the sediments indicates the occurrence of three well-defined depositional environments (Fig. 3).

The association of oolitic grainstone and intraclast conglomerate including type 3 and 4 cycles was deposited in a shallow lagoonal-supratidal high-energy shoal environment (Fig. 9). The protected intertidal and shallow subtidal environment is characterised by deposition of algal laminated limestone, shallow-water gypsum sediments and include type 2 cycles (Fig. 9). In the deep lagoonal environment, laminated gypsum and lime mudstone were deposited (Fig. 9). Type 1 cycles were developed along the fringe of protected lagoons (Fig. 9).

The progradational model proposed for recent sabkhas and the related sabkha cycle (Shearman, 1966) is determined by large-scale migration of various depositional environments. The spatial distribution of the sediments in the Karstryggen Formation (Fig. 3) suggests very restricted lateral migration of the depositional environments. The different types of cycles are dominated by aggradation, and facies progradation is only observed in type 1 cycles (Fig. 9). The sabkha model of Shearman (1966) is, therefore, not applicable in the present case.
In recent sabkha environments, evaporite precipitation mainly occurs in the aeolian deposits above the water table (Butler *et al.*, 1982; Patterson & Kinsman, 1982; Warren & Kendall, 1985). The various carbonate facies, therefore, have to be exposed during the formation of the evaporite. This indicates that the described types of cyclically interbedded carbonate and diagenetic evaporite partly formed as results of relative fluctuations of sea-level.

During periods of sea-level high stand, carbonate was deposited in various shallow-marine environments. The high-energy shoal environment was divided into an intertidal and supratidal area of erosion and conglomerate deposition, and a shallow subtidal area with deposition of oolitic grainstone (Fig. 9A). Algal laminated limestone was formed on the protected side of the shoals in the intertidal and shallow subtidal zone (Fig. 9A). Shallow-water gypsum sediments were deposited in small evaporite lagoons (Fig. 9A), and lime mudstone formed in the deep part of the protected lagoon (Fig. 9A).

The spatial distribution of these environments was governed by the underlying, partly structurally controlled topography (Fig. 3). When first established, the facies pattern apparently promotes itself. The biological and chemical production of algal lime and ooids enable the intertidal and shallow subtidal areas to keep pace with the drowning. In the deeper lagoonal areas, the scarcity of carbonate producing organisms makes the lime production too low to keep pace with the drowning. Progradation into these areas of algal laminated limestone and oolitic grainstone was prevented as the environmental conditions did not allow these facies to form. An aggradational depositional pattern developed, therefore (Figs. 3 & 9).

During relative lowering of sea-level, small scale progradation of algal laminated limestone towards the centre of the lagoon occurred (Fig. 9A). As the higher part of the shoals and the protected intertidal-shallow subtidal lagoonal area became exposed, diagenetic sulphate formed within the various carbonate sediments (Fig. 9B).

The exposed areas formed isolated islands during sea-level low stand (Fig. 9B). The 2-4 km wide islands were mainly composed of algal laminated limestone (Fig. 9B). The early lithification of the limestone and the restricted lateral extension of the exposed areas apparently prevented aeolian redeposition and the topography was, therefore, maintained.

During the succeeding sea-level high stand, the depositional pattern of the carbonate was governed by the preserved topography and a depositional pattern similar to that of the former cycle occurred (Fig. 9).

The simple pattern is superimposed on a larger pattern governed by different rates of subsidence and synsedimentary fault control of the sedimentation (Fig. 3).

The four different types of cycles are simply related to the repeated exposure of shallow subtidal-intertidal areas of carbonate sedimentation into an arid climate (Fig. 9). Cycles 1 and 2 represent the protected lagoonal and intertidal environments, whereas cycles 3 and 4 are formed in high-energy intertidal and shallow subtidal environments (Fig. 9).

Comparison to the Zechstein of NW Europe

The Upper Permian Foldvik Creek Group of East Greenland is not directly stratigraphically correlated to the Zechstein deposits of NW Europe. Therefore, the carbonates and evaporites of the Karstryggen Formation cannot be compared to a particular Zechstein carbonate unit.

The Karstryggen Formation is dominated by non-fossiliferous limestone and shallow subaqueous-supratidal sulphate deposits. Compared to the Zechstein of NW Europe, deposition apparently occurred in a less arid climate as no halite and potassium salt deposited in East Greenland (Stemmerik, 1985). Also, local tectonics significantly influenced sedimentation. Therefore, the present model is not directly applicable to any of the Zechstein carbonates in NW Europe even when the Zechstein carbonates developed in hypersaline facies.

Conclusions

Four types of cycles of shallow-marine carbonate and nodular-mosaic sulphate have been recognised within the Upper Permian Karstryggen Formation of East

Greenland. They differ in several ways from the simple sabkha cycle of Shearman (1966).

Cycle 1 formed in shallow to deep lagoonal setting (Fig. 9). The cycle consists, in ascending order, of lime mudstone, algal laminated limestone, and nodular-mosaic evaporite (Fig. 7). A cycle is 30-120 cm thick and laterally traceable for more than 200 m.

Cycle 2 formed in the protected shallow lagoonal and intertidal zone (Fig. 9). The cycle consists of interbedded algal laminated limestone and nodular-mosaic evaporite (Fig. 7). The thickness and lateral extension of this type of cycle is variable.

Cycle 3 formed in very shallow high energy settings (Fig. 9). The cycle is less than 1 m thick and consists of interbedded intraclast conglomerate and nodular-mosaic evaporite (Fig. 7). The lateral extension of this cycle is less than 20 m.

Cycle 4 formed in a shallow subtidal, high energy setting (Fig. 9). The cycle consists of interbedded oolitic grainstone and nodular-mosaic evaporite (Fig. 7). A cycle ranges from 20-200 cm in thickness. The lateral extension is 1-2 km.

The cycles formed as the result of repeated exposure of a shallow marine-intertidal area of carbonate production into an arid climate. The nodular-mosaic sulphate was formed by the growth of anhydrite and/or gypsum nodules during times of exposure.

The restricted lateral extension of the exposed island during sea-level low stand and the early lithification of the sediment prevented aeolian redeposition, and the primary topography was, therefore, maintained during times of exposure (Fig. 9).

Acknowledgements

I would like to thank Prof. Finn Surlyk for critically reading this manuscript which is part of a Ph.D. dissertation at the University of Copenhagen. Technical assistance from B. Sikker Hansen, J. Aagaard and I. M. Jensen is acknowledged.

This research is financially supported by the Danish Natural Research Council (SNF). This paper is published with the permission of the Director of the Geological Survey of Greenland.

References

Butler, G. P., Harris, P. M. & Kendall, C. G. St. C.,1982. Recent evaporite from the Abu Dhabi coastal flats. *SEPM Core Workshop* 3: 33-64.

Dean W. E., Davies, G. R. & Anderson, R. Y.,1975. Sedimentological significance of nodular and laminated anhydrite. *Geology* 3: 367-372.

Halley, R. B., Harris, P. M. & Hine, A. C.,1983. Bank margin environment. *Am. Assoc. Petroleum Geologists Memoir* 33: 463-506.

Kendall, C. G. St. C. & Skipwith, P. A. d'E., 1969. Holocene shallow-water carbonate and evaporite sediments of Khar al Bazam, Abu Dhabi, South-west Persian Gulf. *Am. Assoc. Petroleum Geologists Bull.*, 53: 841-869.

Kinsman, D. J. J. & Park, R. K., 1976. Algal belt and coastal sabkha evolution, Trucial Coast, Persian Gulf. In: Walter, M. R. (Ed.): *Stromatolites*, 421-433. Elsevier, Amsterdam.

Logan, B. W., Hoffman, P. & Gebelein, C. D.,1974. Algal mats, cryptalgal fabrics and structures, Hamelin Pool, western Australia. *Am. Assoc. Petroleum Geologists Memoir* 22: 140-194.

Loucks, R. G. & Longman, M. W.,1982. Lower Cretaceous Ferry Lake Anhydrite, Fairway Field, East Texas. *SEPM Core Workshop* 3: 130-173.

Maiklem, W. R., Bebout, D. G. & Glaister, R. P.,1969. Classification of anhydrite - a practical approach. *Bull. Can. Petrol. Geol.*, 17: 194-233.

Maync, W., 1942. Stratigraphie und Faziesverhältnisse der oberpermischen Ablagerungen Ostgrönlands (olim "Oberkarbon - Unterperm") zwischen Wollaston Forland und dem Kejser Franz Josephs Fjord. *Meddr. Grønland*, 115,128 pp.

Maync, W.,1961. The Permian of Greenland. In: Raasch, G. O. (Ed.): *Geology of the Arctic*, 1: 214-223, University of Toronto Press.

Mossop, G. D., 1979. The evaporites of the Ordovician Bauman Fiord Formation, Ellesmere Island, Arctic Canada. *Geol. Surv. Canada Bull.*, 298, 52 pp.

Patterson, R. J. & Kinsman, D. J. J., 1981. Hydrologic framework of sabkha along Arabian Gulf. *Am. Assoc. Petroleum Geologists Bull.*, 65: 1457-1475.

Piasecki, S., 1984. Preliminary palynostratigraphy of the Permian - Lower Triassic sediments in Jameson Land and Scoresby Land, East Greenland. *Bull. geol. Soc. Denmark* 32: 139-144.

Schreiber, B. C., 1986. Arid shorelines and evaporites. In: Reading, H. G. (Ed.): *Sedimentary Environments and Facies*, 189-228. Blackwell, Oxford.

Shinn, E. A., 1983. Tidal flat environments. *Am. Assoc. Petroleum Geologists Memoir* 33: 171-210.

Shinn, E. A., Lloyd, R. M. & Ginsburg, R. N., 1969. Anatomy of a modern carbonate tidal flat, Andros Island, Bahamas. *Jour. Sed. Petrology* 39: 1202-1228.

Shearman, D. J., 1966. Origin of marine evaporites by diagenesis. *Trans. Inst. Min. Metall.*, B 75: 729-760.

Stemmerik, L., 1980. Observations on Upper Permian sediments in southern Scoresby Land, East Greenland. *Rap. Grønlands geol. Unders.*, 100: 105-107.

Stemmerik, L., 1985. Sedimentære og diagenetiske processer i et karbonat/evaporit domineret subtidalt-supratidalt aflejringsmiljø, Øvre Perm, Østgrønland. (unpublished Ph. D. thesis, University of Copenhagen, in English).

Surlyk, F., Hurst, J. M., Piasecki, S., Rolle, F., Scholle, P. A., Stemmerik, L. & Thomsen, E., 1986. The Permian of the western margin of the Greenland Sea - a future exploration target. *Am. Assoc. Petroleum Geologists Memoir* 40: 629-659.

Surlyk, F., Piasecki, S., Rolle, F., Stemmerik, L., Thomsen, E. & Wrang, P.,1984. The Permian basin of East Greenland. In: Spencer, A. M. *et al.* (Eds.): *Petroleum Geology of the North European Margin,* 303-315. Graham & Trotman, London.

Warren, J. K. & Kendall, C. G. St. C.,1985. Comparison of sequences formed in marine sabkha (subaerial) and salina (subaqueous) settings; modern and ancient. *Am. Assoc. Petroleum Geologists Bull.*, 69: 1013-1023.

West,I. M., Ali, Y. A. & Hilmy, M. E., 1979. Primary gypsum nodules in a modern sabkha on the Mediterranean coast of Egypt. *Geology* 7: 354-358.

Wilson, J. L., 1975. *Carbonate Facies in Geologic History*, 471 pp. Springer Berlin.

THE UPPER PERMIAN (ZECHSTEIN) TUNSTALL REEF OF NORTH EAST ENGLAND: PALAEOECOLOGY AND EARLY DIAGENESIS

Neville T.J. Hollingworth and Maurice E. Tucker
Department of Geological Sciences
University of Durham
Durham DH1 3 LE, U.K.

Abstract: Detailed field studies and laboratory examination have identified the palaeocommunities of the Late Permian Tunstall Reef Complex of N.E. England, and have enabled accurate 3-dimensional diagrams to be constructed illustrating the palaeocommunity structure. The different palaeocommunities reflect the development of the reef through time, from reef-base coquina to lower reef core palaeocommunities, and then through space, in the laterally equivalent reef-flat, upper reef core, reef crest and fore-reef talus palaeocommunities, and the backreef patch reef palaeocommunity. An important factor in the formation of the Tunstall Reef was the precipitation of sea-floor cement. Aragonite particularly was an extensive and rapid precipitate in the reef-base coquina, as revealed in undolomitized limestones, and this type of cementation appears to have been widespread in other reef facies too, as judged by the rare occurrence of undolomitized limestones and by textures in the dolomites which can now be interpreted as relics of marine cements.

Introduction

The Permian Reef of North East England has been studied since the early nineteenth century and many fossils have been collected and figured in monographs and papers published over a hundred years ago. More recently the sedimentology and facies of this "Middle Magnesian Limestone Reef" have been described by Smith (1981). A major problem has been the obliterative dolomitization of the Zechstein carbonates, and the consequent loss of detail. The discovery of undolomitized reefal limestones by N.T.J.H. has thrown new light on both the fauna and early diagenesis of the reef. In this paper, the palaeocommunities of the reef complex are described and illustrated and then the early diagenesis is presented.

Lecture Notes in Earth Sciences, Vol. 10
T.M. Peryt (Ed.), The Zechstein Facies in Europe

Stratigraphy and location

The Upper Permian rocks of North East England comprise a complex sequence of carbonates and evaporites which were deposited along the western margin of the Zechstein Sea during five evaporite cycles (Smith, 1980). Rocks of the first two cycles are exposed near the city of Sunderland, with later cycles occurring farther south in the Seaham area and also beneath younger rocks in South Durham, Yorkshire and the North Sea. A new stratigraphic nomenclature for the English Zechstein proposed by Smith *et al.* (1986) is given in Fig. 1.

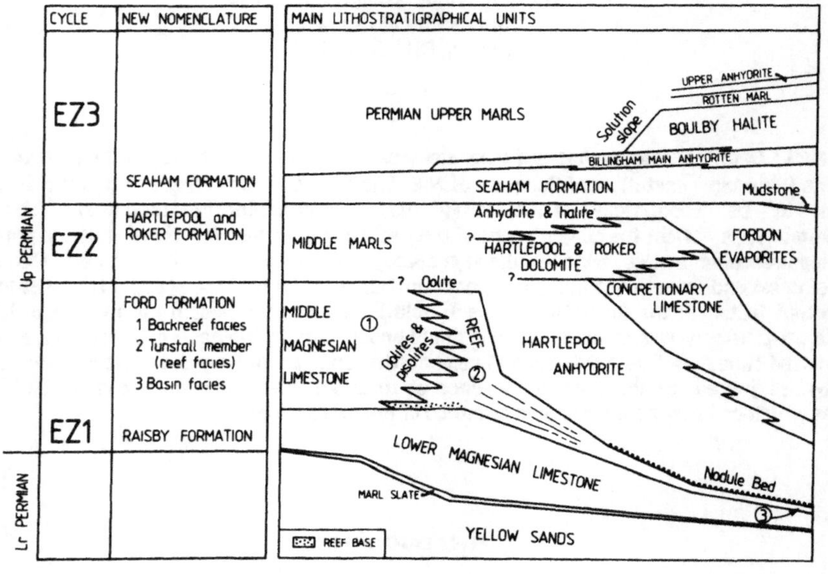

Fig. 1. Stratigraphic cross section and formation names of the Zechstein sequence in the Durham Province.

Deposition of Upper Permian sediments in North East England took place in two provinces separated by a palaeo-high (the Cleveland High): the Durham province to the north and the Yorkshire province to the south. The Permian magnesian limestones of the Durham province include a 100 m thick north-south linear reef complex, within the Ford Formation (Fig. 1). From the most northerly outcrop at Down Hill (see Fig. 2 for reef distribution), the reef can be traced southwards through the western suburbs of Sunderland into County Durham, where it is known to extend a short distance southwest of Hartlepool, a total distance of around 32 km (Trechmann, 1925; Smith, 1981). The reef is particularly well developed in, and forms, the Tunstall Hills; hence the Ford Formation reef is referred to as the Tunstall Reef. Tertiary uplift has given the

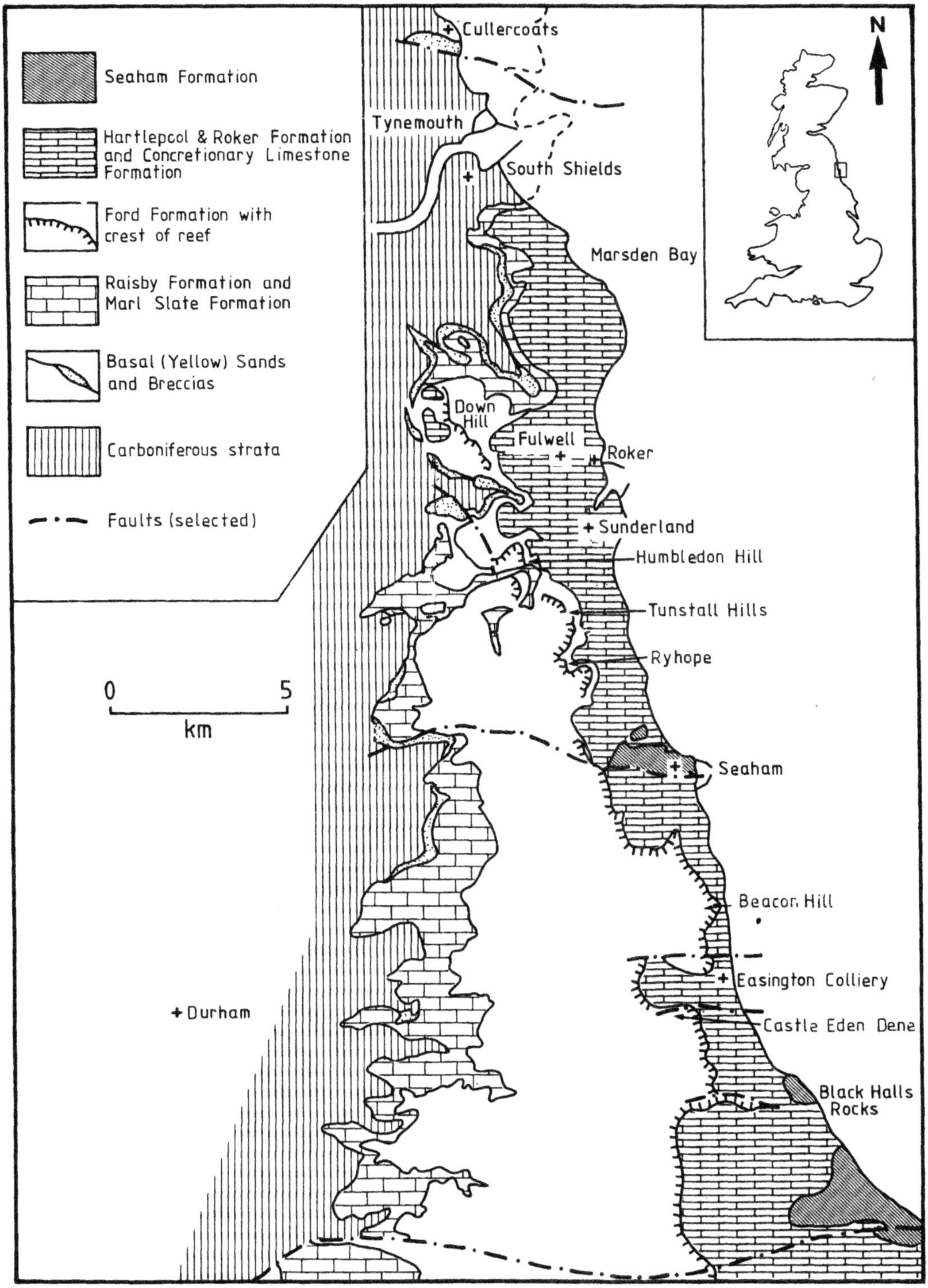

Fig. 2. Map showing geology of North East England, localities noted in text, and the position of the Tunstall Reef crest.

Stage 1. Foundation
Formation of coquina bank at edge of carbonate platform. Diverse and abundant cosmopolitan shelly biota. Water depths shallow in region of a few metres, with occasional and localised emergence of coquina. Abundant, voluminous inorganic marine cement precipitation between bioclasts. Majority of shelly fauna cemented in life position.

Stage 2. Establishment
Rapid subsidence of basin and submergence of coquina bank. Establishment and growth of bryozoans upon lithified substrates. Fauna diverse and abundant with some community differentiation. Bryozoans widely distributed. Abundant marine cement precipitation.

Stage 3. Growth
Continual basin subsidence matched by growth of reef vertically upwards and laterally basinwards. Initiation of talus wedge at edge of prograding reef. Evolution of niches within developing reef framework. Abundant marine cement precipitation.

Stage 4. Diversification
Reef dominated volumetrically by bryozoans. Distinct fore-reef slope. Evolution of communities within various reef sub-facies. Fauna diverse, but some genera, particularly spinose productids abundant in reef talus and lower reef slope. Some differentiation in bryozoan distribution across reef.

Stage 5. Specialisation
Development of reef flat. Upwards growth limited to subsidence rate. Rapid basinwards migration over reef talus. Distinct communities within reef sub-facies. Bryozoans specialised to various reef environments. Periodic subaerial exposure of reef flat in peritidal environment. Diverse fauna at reef crest. Development of surge channels.

Stage 6. Consolidation
Extensive development of reef flat. Steep outer reef slope with large talus wedge. Rapid lateral growth basinwards over talus contemporaneously eroded from reef crest. High stress environment over reef flat. Faunal diversity maintained at reef crest and slope areas only. Complete restriction of large spinose productids to lower reef slope and talus areas. Marine cement precipitation abundant in reef flat/crest areas.

Stage 7. Maturity
Maximum thickness (100 m) and width of reef (1-2 km). Highest diversity faunas found in reef crest and slope. Algal laminites abundant in reef flat. Extensive contemporaneous inorganic cementation of reef flat laminites.

Stage 8. Death
Rapid evaporative drawdown of Zechstein cycle 1 sea. Widespread catastrophic extinction of reef fauna. Subaerial exposure of reef flat, crest and slope. Deposition of anhydrite against steep outer reef slope. Early dolomitization of reef.

Fig. 3. Stages in the development and demise of the Tunstall Reef of the Ford Formation, Zechstein, Durham Province.

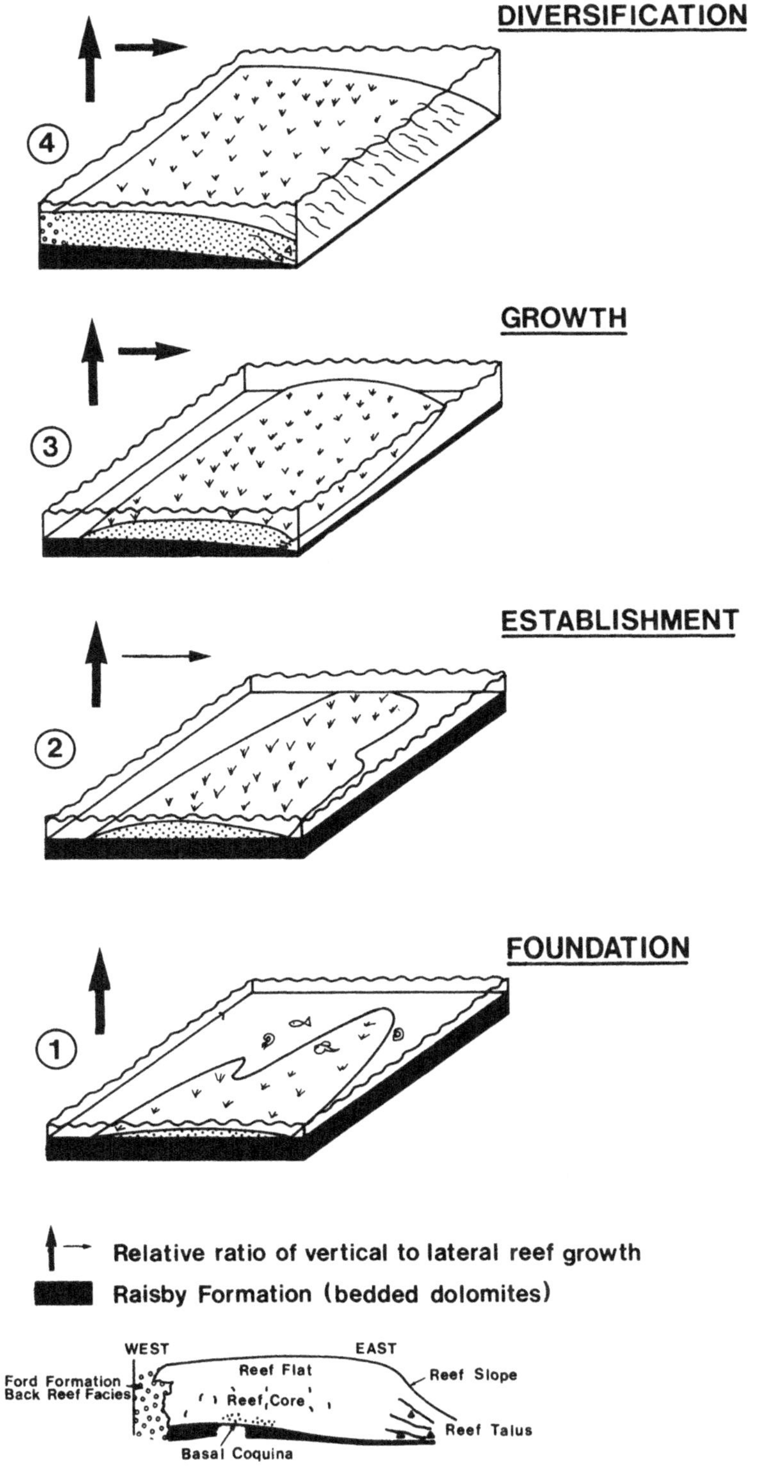

DIVERSIFICATION
4
GROWTH
3
ESTABLISHMENT
2
FOUNDATION
1
Relative ratio of vertical to lateral reef growth
Raisby Formation (bedded dolomites)
WEST
EAST
Ford Formation
Back Reef Facies
Reef Flat
Reef Core
Reef Slope
Reef Talus
Basal Coquina

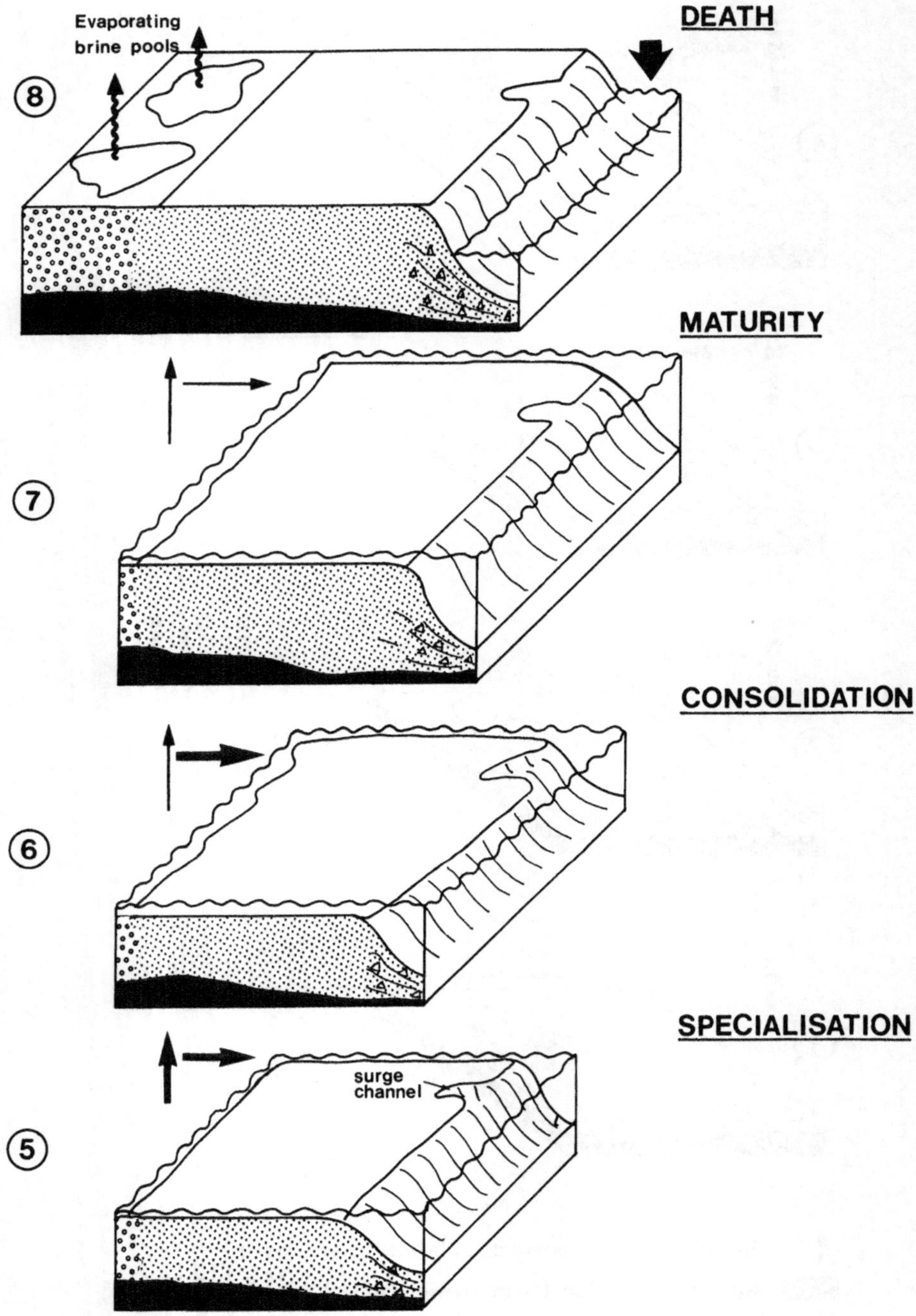
Evaporating
brine pools
DEATH
8
MATURITY
7
CONSOLIDATION
6
SPECIALISATION
surge
channel
5

Tunstall Reef a gentle southerly tilt and consequently most of the main exposures of the reef base and reef core facies occur in the Sunderland area, while farther south successively higher stratigraphic levels are exposed.

Most of the Permian carbonates in North East England have undergone extensive dolomitization and this has obliterated many primary structures, particularly within the Tunstall Reef. As a consequence of the often poor preservation of fossils there has been much discussion of the roles played by the organisms in reef construction. Evidence from recently discovered undolomitized parts of the reef suggests that seafloor cement precipitation was a major factor in the construction of a wave resistant structure (later section and Tucker & Hollingworth, 1986).

Evolution of the Tunstall Reef

The geological history of the Tunstall Reef of the Ford Formation can be divided into eight stages (Fig. 3). After the initial accumulation of a coquina rich in bryozoans, brachiopods and molluscs (Stage 1), the reef was established through the baffling action of bryozoans and other carbonate skeletons (Stage 2), leading to the development of a distinct reef core and fore-reef slope and talus facies (Stages 3 and 4). A reef flat (Stage 5) then developed in the back-reef area where there was periodic exposure. Rapid basinward migration of the reef crest over the talus apron (Stages 5, 6 and 7) was accompanied by extensive planar stromatolite formation on the reef flat. Reef growth was terminated (Stage 8) by evaporative drawdown of the Zechstein Sea, and exposure of the carbonate shelf.

In terms of lithofacies, four broad categories can be distinguished in the reef complex: reef-base coquina, reef core, reef flat, and fore-reef talus facies (see Smith, 1981), In detail, these can be divided into various subfacies. There is also a distinct backreef facies, with patch reefs.

The reef-base coquina, which is the foundation of the Tunstall Reef, accumulated upon an irregular topography with slopes reaching 45° in places. This irregular surface was produced by a major submarine slide which occurred after the deposition of the underlying Raisby Formation (Smith, 1981). In the Tunstall Hills, where this lithofacies has locally escaped dolomitization, the reef-base coquina appears to have accumulated upon a topographic high. The coquina consists entirely of bioclasts, and accumulated in a high-energy shallow-water location. Seafloor cementation was pervasive in the coquina. The reef core facies consists of massive dolomite with bryozoans in life position and many other faunal elements (next section). In spite of the dolomitization there is evidence for extensive seafloor cementation. The reef flat facies is bedded dolomite with bioclastic debris and stromatolites, especially towards the top. The fore-reef talus facies contains large blocks of reef material and has an original sedimentary dip reaching 30° and more. The backreef facies consists of cross-bedded oolites, thin-bedded micrites, horizons of evaporite pseudomorphs, and patch reefs. These were deposited in near-reef sand shoals (the oolites), in the quieter-water shelf lagoon, and in tidal flats along the inner shelf margin and behind the reef.

Fig. 4. Reef base palaeocommunity. **A.** *Schizodus obscurus.* **B.** *Permophorus costatus.* **C.** *Horridonia horrida.* **D.** *Naticopsis minima.* **E.** *Edmondia elongata.* **F.** *Mourlonia antrina.* **G.** *Synocladia virgulacea.* **H.** *Coelostyalina permiana.* **I.** *Liebea squamosa.* **J.** *Kingopora ehrengerbi.* **K.** *Cyathocrinites ramosus.* **L.** *Stenoscisma* sp. **M.** *Bakevellia* sp. **N.** *Pterospirifer alatus.* **O.** *Parallelodon striatus.* **P.** *Meekospira symmetrica.* **Q.** *Pseudomonotis speluncaria.* **R.** *Fenestella* sp. **S.** *Streptorhynchus pelargonatus.* **T.** *Dielasma elongatum.* **U.** *Dyscritella columnaris.* **V.** *Yunnania tunstallensis.* **W.** *Acanthocladia* sp.

Reef palaeocommunities

Field observations and quantitative bulk sampling have identified seven major palaeocommunities within the Tunstall Reef: reef-base, lower reef core, higher reef core, reef crest, reef flat, reef talus and patch reef palaeocommunities. The last occurs in the back reef lagoon, several km from the reef complex itself, where all the others occur. The distribution of the reef palaeocommunities is related to physical environmental parameters, such as substrate type and water depth, and to microenvironmental regimes created by growth of the reef itself. The Tunstall Reef contains a diverse and abundant fauna, that varies both laterally (that is in space) and vertically (through time), within the different reef lithofacies. Similar faunal variations have been described from the Silurian Niagaran Reefs of North America by Walker and Alberstadt (1975). Although some of the taxa in the Ford Formation are widely distributed, the brachiopod *Dielasma* for example, others such as *Cyathocrinites*, *Stenoscisma* and *Horridonia* are characteristic of certain horizons within the reef.

The **reef-base palaeocommunity** (Fig. 4) contains a high diversity fauna, though numbers of each individual species are relatively low. The majority of the fauna is dominated be epifaunal genera , particularly the pedunculate brachiopod *Dielasma* (which may reach a large size of 30 mm) and epifaunal epibyssate bivalves such as *Bakevellia*, *Parallelodon*, *Liebea* and *Pseudomonotis*. Infaunal bivalves are usually found in monospecific masses and were probably confined to fine grained skeletal sand pockets within the carbonate gravel bank.

Bryozoanas are abundant in the reef-base community, and include *Acanthocladia*, *Fenestella* and the trepostome *Dyscritella*. Many bryozoan fronds are broken. Gastropods, mainly browsing herbivores, form an important constituent of the reef-base fauna and include *Mourlonia*, *Meekospira*, *Yunnania* and *Naticopsis*. Isolated ossicles of the crinoid *Cyathocrinites* also occur, but with the echinoid *Miocidaris*, form a minor constituent of the fauna. This community lived in a high energy, shallow-water environment where the precipitation of sea-floor cements was extensive (later section).

The **lower reef core palaeocommunity** (Fig. 5) is numerically dominated by the fenestrate and trepostome bryozoans, *Fenestella*, *Acanthocladia* and *Dyscritella*, and displays a wide degree of palaeocommunity overlap. *Cyathocrinites* occurs abundantly in lower reef core lithologies where ossicles are commonly present in sufficient numbers to form crinoidal packstone lenses.

Distinct faunal communities evolved when the reef was fully established and was migrating rapidly basinwards over its own talus. In stratigraphically higher reef core lithologies certain brachiopods such as *Horridonia* and *Pterospirifer* and the crinoid *Cyathocrinites* became progressively rarer and were eventually confined to the reef slope and talus areas.

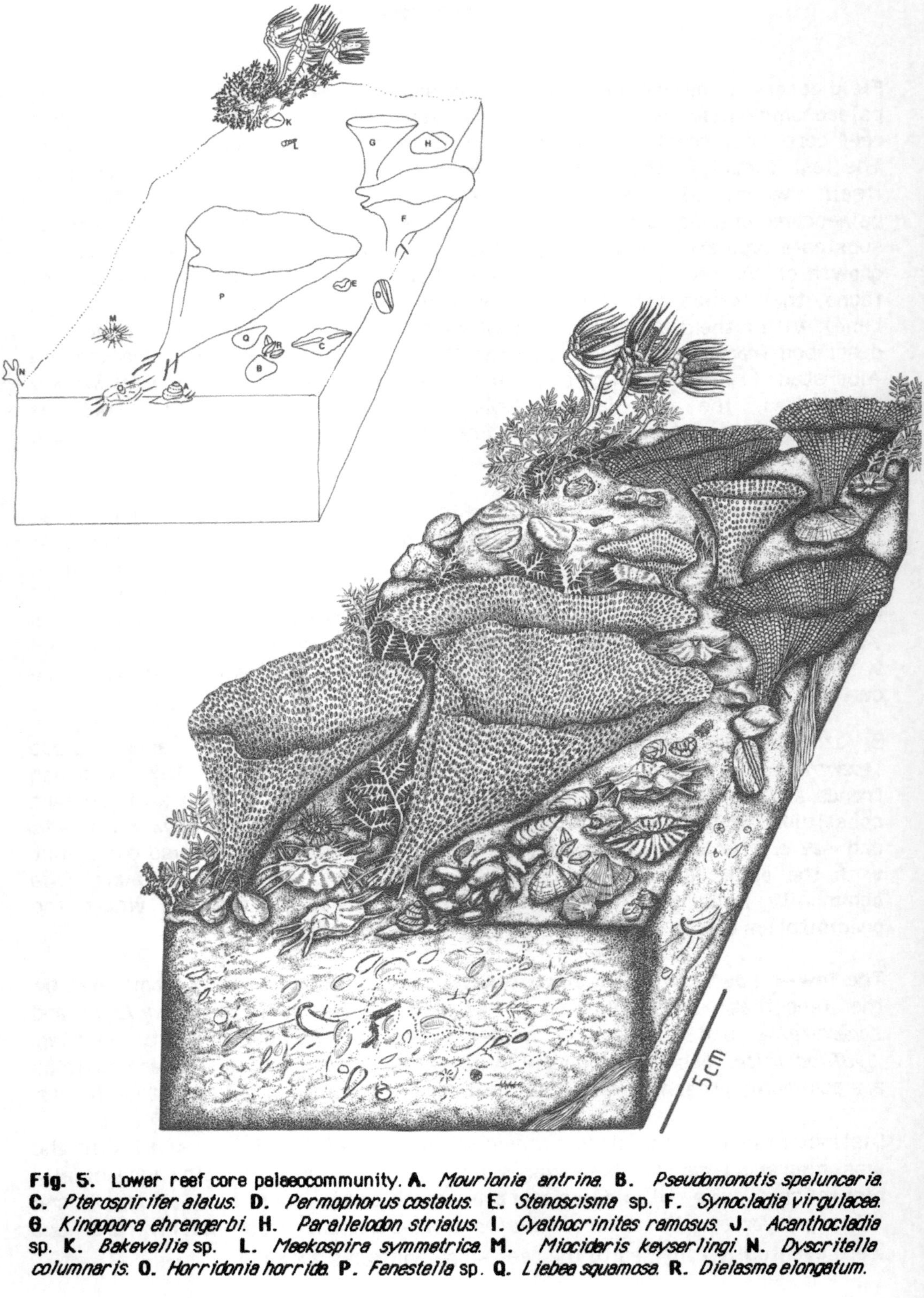

Fig. 5. Lower reef core palaeocommunity. **A.** *Mourlonia antrina.* **B.** *Pseudomonotis speluncaria.* **C.** *Pterospirifer alatus.* **D.** *Permophorus costatus.* **E.** *Stenoscisma* sp. **F.** *Synocladia virgulacea.* **G.** *Kingopora ehrengerbi.* **H.** *Parallelodon striatus.* **I.** *Cyathocrinites ramosus.* **J.** *Acanthocladia* sp. **K.** *Bakevellia* sp. **L.** *Meekospira symmetrica.* **M.** *Miocidaris keyserlingi.* **N.** *Dyscritella columnaris.* **O.** *Horridonia horrida.* **P.** *Fenestella* sp. **Q.** *Liebea squamosa.* **R.** *Dielasma elongatum.*

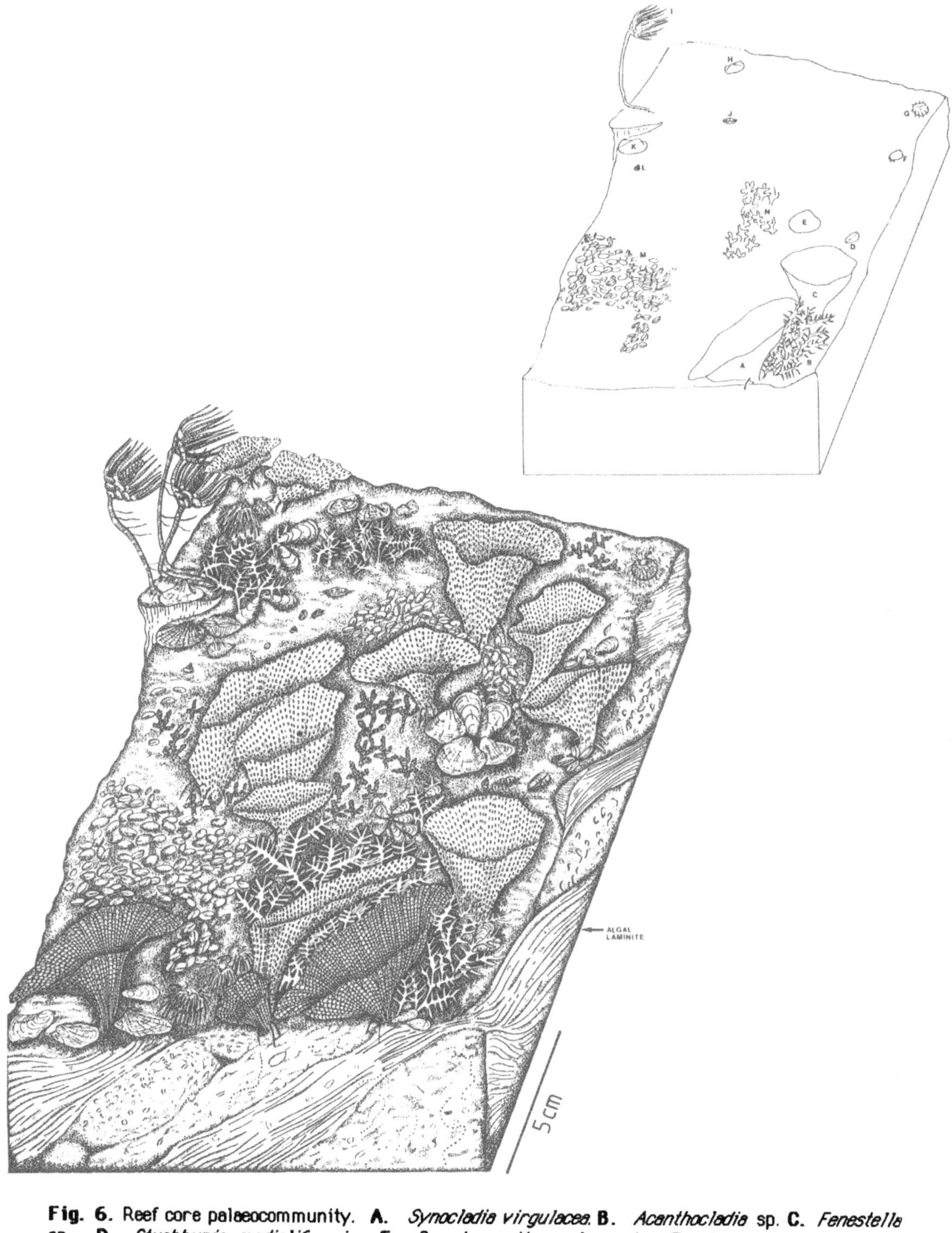

Fig. 6. Reef core palaeocommunity. **A.** *Synocladia virgulacea.* **B.** *Acanthocladia* sp. **C.** *Fenestella* sp. **D.** *Stuchburia modioliformis.* **E.** *Pseudomonotis speluncaria.* **F.** *Liebea squamosa.* **G.** *Strophalosia* sp. **H.** *Bakevellia* sp. **I.** *Cyathocrinites ramosus.* **J.** *Spiriferellina cristatus.* **K.** *Parallelodon striatus.* **L.** *Yunnania tunstallensis.* **M.** *Dielasma elongatum.* **N.** *Dyscritella columnaris.*

Fig. 7. Reef crest palaeocommunity. **A.** *Bakevellia* sp. **B.** *Dielasma* sp. **C.** *Pseudomonotis speluncaria.* **D.** *Acanthocladia* sp. **E.** *Liebea squamosa.*

Fig. 8. Reef talus palaeocommunity. **A.** *Mourlonia antrina*. **B.** *Cleiothyridina pectinifera*. **C.** *Horridonia horrida*. **D.** *Liebea squamosa*. **E.** *Stenoscisma* sp. **F.** *Synocladia virgulacea*. **G.** *Pterospirifer alatus*. **H.** *Acanthocladia* sp. **I.** *Streblochondria pusilla*. **J.** *Pseudomonotis speluncaria*. **K.** *Parallelodon striatus*. **L.** *Spiriferellina cristatus*. **M.** *Fenestella* sp. **N.** *Dyscritella columnaris*.

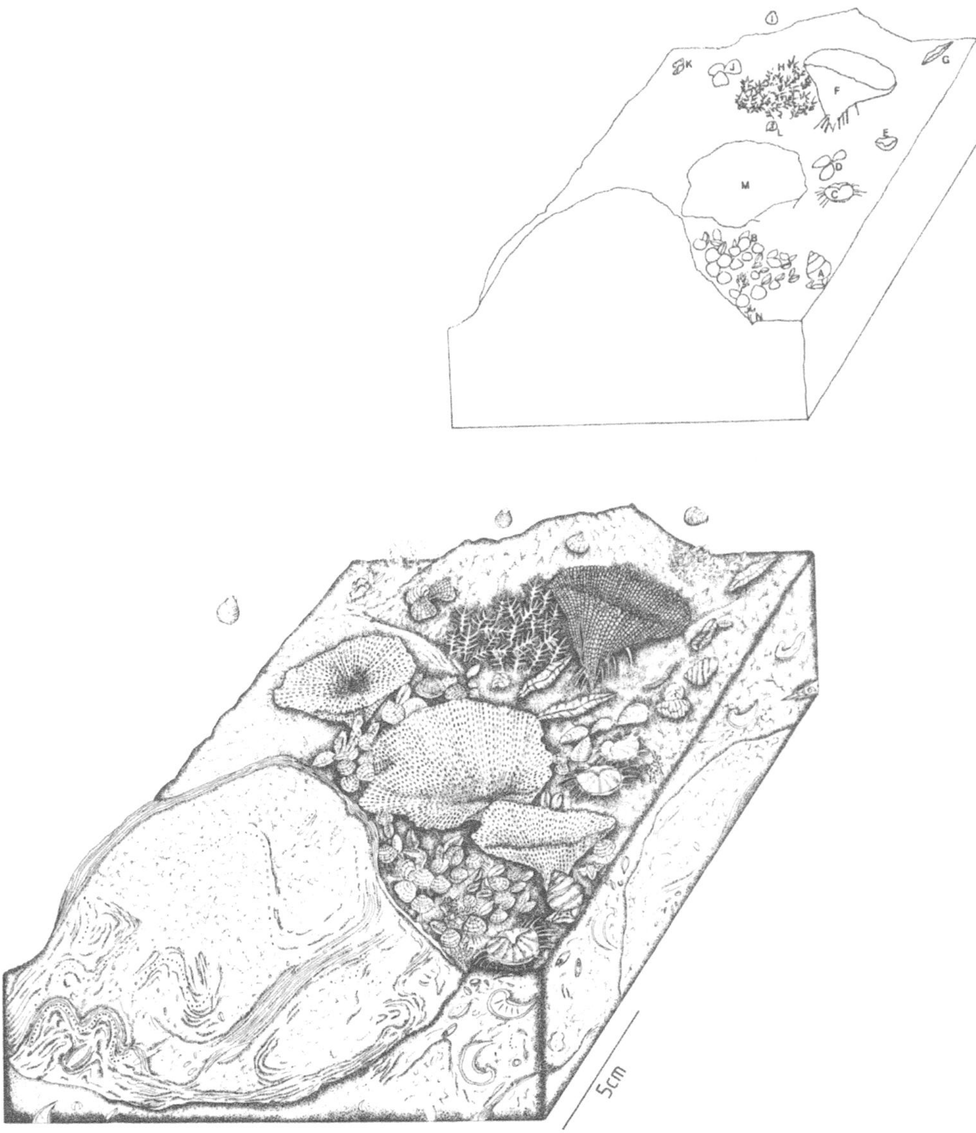

Fig. 9. Patch reef palaeocommunity. **A.** *Stenoscisma* sp. **B.** *Strophalosia* sp. **C.** *Bakevellia* sp. **D.** *Parallelodon striatus.* **E.** *Fenestella* sp. **F.** *Pseudomonotis speluncaria.* **G.** *Kingopora ehrengerbi.* **H.** *Acanthocladia* sp. **I.** *Liebea squamosa.* **J.** *Edmondia elongata.* **K.** *Janeia biarmica.*

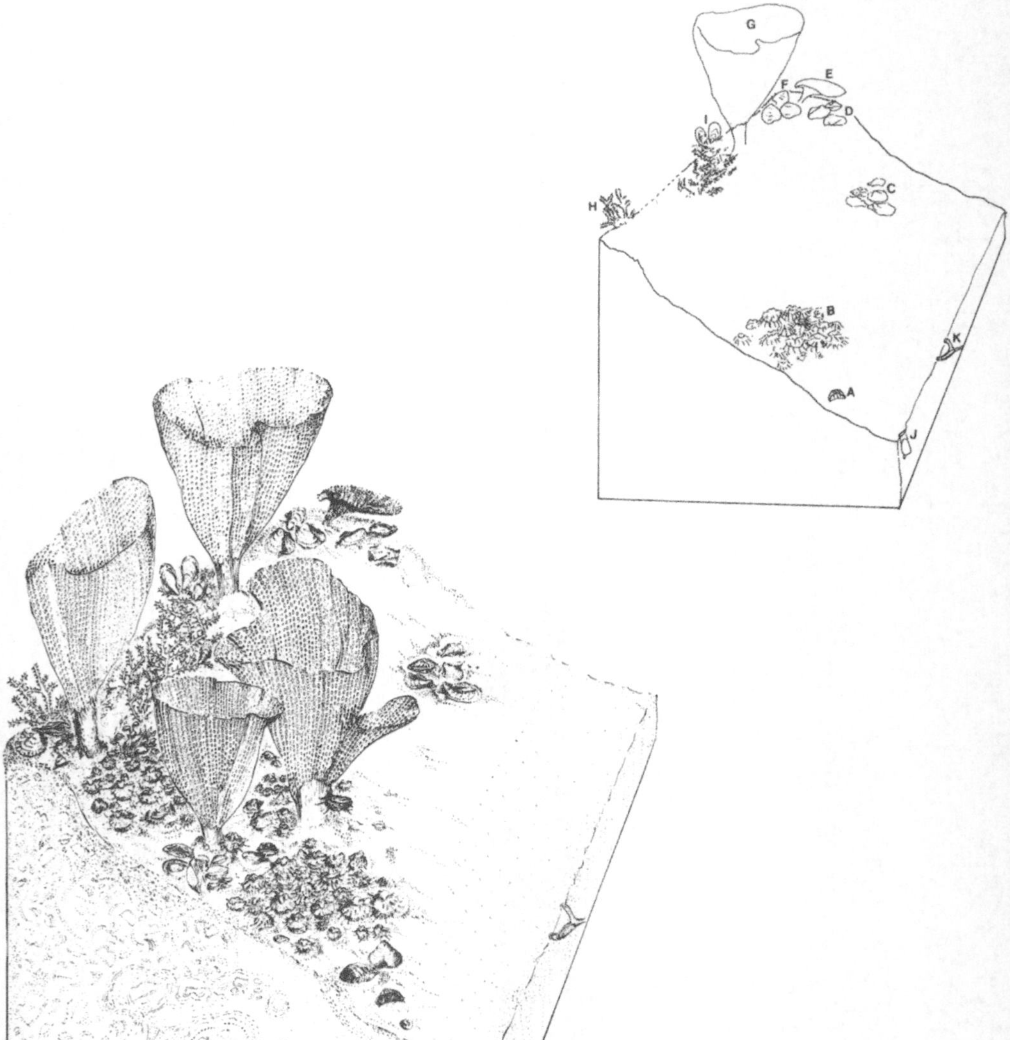

Reef core palaeocommunities (Fig. 6) are dominated by small fenestrates and encrusting forms of the trepostome *Dyscritella*. The bivalve fauna tends to be dominated by *Bakevellia* and *Liebea*, which are epibyssate. Quasi-infaunal brachiopods such as *Horridonia* are absent. *Dielasma*, though abundant, occurs as nests composed entirely of juveniles, which indicates some degree of environmental stress.

Reef flat palaeocommunities may contain large numbers of individuals, such as the bivalve *Bakevellia* and the gastropod *Anomphalus*. Diversity, however, is lower and may indicate that only certain genera could tolerate the high stress conditions and possible subaerial exposure in a peritidal environment.

The **reef crest palaeocommunity** (Fig. 7) is dominated by *Acanthocladia* which forms large bush-like colonies, and to a lesser extent by the ramose growth form of *Dyscritella*. Numerous epibyssate genera such as *Bakevellia*, *Parallelodon* and *Pseudomonotis* probably inhabited bryozoan frameworks which provided shelter and possible protection from predation.

The reef slope is dominated by crinoids where they are encountered in large numbers. The **reef talus palaeocommunity** (Fig. 8) is dominated by the athyrid *Cleiothyridina* which may have been attached in nest-like masses to allochthonous blocks. Quasi-infaunal brachiopods such as *Horridonia*, *Strophalosia* and the free-lying spiriferid *Pterospirifer* are also abundant in the talus apron, probably living in shelly sand pockets between derived blocks.

Bryozoans, particularly the delicate fenestrate *Fenestella*, often reaches a large size in the reef talus, forming large steeply erect colonies. The reef talus palaeocommunity is indicative of a relatively low energy deepwater environment where water depths were probably in excess of 90 metres during the latest stages of reef development. This would be ideal for delicate fenestrate bryozoans and free lying brachiopods that could not tolerate high energy environments.

Within the bedded dolomites of the back reef facies, there occur fossils which are not present in the main reef complex. These include infaunal deep-burrowing suspension feeding bivalves such as *Janeia* and *Wilkingia*, and endobyssate genera such as *Aviculopinna*.

The **patch reefs** (Fig. 9) contain a fauna which is distinct from surrounding bedded dolomicrites and the main reef complex. The bryozoan *Kingopora* is entirely confined to the patch reef community and forms steeply erect cone-shaped colonies with a thick calcified holdfast (Fig. 9). The quasi-infaunal brachiopod *Strophalosia* is also abundant and *Kingopora* is commonly in life position, attached to strophalosiid shells. The cores of patch reefs are dominated by *Acanthocladia* and the trepostome *Dyscritella*. An interstitial fauna of epibyssate bivalves such as *Pseudomonotis* and *Parallelodon* also occurs. The highest diversity faunas usually occur on the flanks of patch reefs and include shallow infaunal bivalves such as *Schizodus*, which is found in large numbers.

The vertical decrease of diversity accompanied by "dwarfing" in the reef core and reef flat faunas has been used in the past to indicate increasing salinity during reef growth (see Trechmann, 1925). However, data collected during this study suggest that salinity only became an important factor in causing faunal extinction in reef core and reef flat lithofacies during the latest stages of growth. Earlier ideas of decreasing faunal diversity and size were in part due to a lack of continuous lateral exposure within time related reef core, crest and slope lithologies. It is more likely that increased salinity was more localised than widespread, probably confined to reef flat and reef core areas where increased environmental stress was also an important factor.

The termination of reef growth was effected by a rapid increase in salinity at levels well above the tolerance limits of reef dwelling and building genera. This was brought about by evaporative drawdown of the Zechstein Sea, which terminated carbonate sedimentation and resulted in gypsum-anhydrite deposition in the adjoining North Sea Basin.

Carbonates are well developed in the succeeding second and third Zechstein cycles, but the faunas do not reach the same level of diversity as seen in the Tunstall Reef.

Diagenesis of the Tunstall Reef

Much of the early diagenesis of the Tunstall Reef has been oblitarated by pervasive dolomitization, which is a general feature of Zechstein carbonate facies (e.g., Clark, 1980). Evaporite replacement and pore filling, and late stage dedolomitization are also important diagenetic events in reef facies. Where sediments have escaped major alteration of these types and are still primary limestones, there is abundant evidence for extensive marine cementation by aragonite and, to a lesser extent, high-Mg calcite (Tucker & Hollingworth, 1986).

Marine cementation

Undolomitized limestones occur in the reef base coquina of the Ford Formation and in small lenses within the reef facies itself. The limestones are compact, with a crystalline appearance and a grey to brown colour, contrasting with the dolomites, which are mostly vuggy and porous yellow rocks with few textural features remaining. Three types of marine cement are distinguished in the limestones: (a) aragonite, now replaced by calcite and dolomite, occurring as botryoids and isopachous crusts, (b) acicular high-Mg calcite (now calcite) as isopachous crusts and peloids, and (c) calcite crystal fans.

Calcitized aragonite botryoids and crusts

Much inter- and intra-granular porosity of the Zechstein primary limestones is occluded by honey-coloured calcite crystal mosaics which are replacements of aragonite. Relics of aragonite are present and can be seen in thin section and on lightly-etched, polished surfaces under the SEM (Fig. 10). A small Sr peak is obtained from the inclusions with EDAX and the principal aragonite reflection is recorded with XRD.

The original aragonite was precipitated in the form of discrete botryoids and as crusts and layers upon bioclasts. The botryoids are most prominent within articulated brachiopod and bivalve shells (Figs. 11 & 12) and they reach 20 mm in diameter. The botryoids consisted of closely-packed, fanning aragonite needles and blades. In some instances square-ended terminations are present at the botryoid margin; in other cases steep-sided faces were developed (Fig. 11). The botryoids are now chiefly composed of coarse calcite crystals (neomorphic spar or pseudospar), which cross-cut the original fanning pattern of the precursor aragonite, but preserve the latter through inclusion patterns. Some botryoids have no relic structure, but are composed of equant drusy calcite spar which filled the void after wholesale aragonite dissolution.

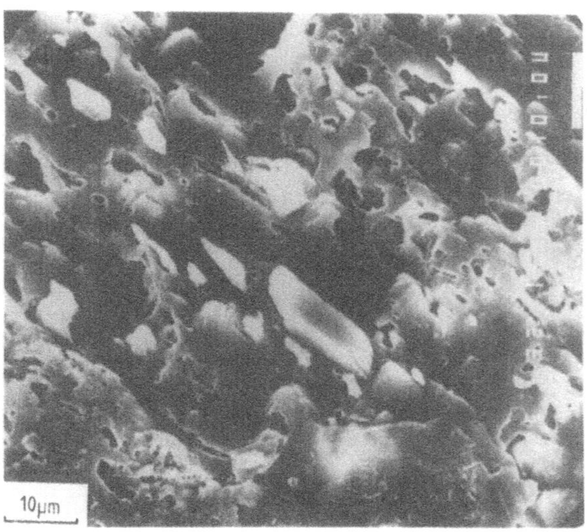

Fig. 10. Relics of aragonite in calcite from a calcitized aragonite botryoid. SEM view.

Aragonite crusts are up to 30 mm thick and occur around bioclasts, especially fenestellid bryozoans (Fig. 13). Some crusts show a millimetre growth banding. The outer surface of the thicker crusts is mammillated. The crusts also show a relic texture of parallel and fanning acicular and columnar crystallites, preserved by inclusion of aragonite and organic matter. These former aragonite-crusts are the most important marine cement; indeed, in some thin sections they comprise more than 80% of the rock, in a bryozoan framework-cement microfacies (Fig. 13).

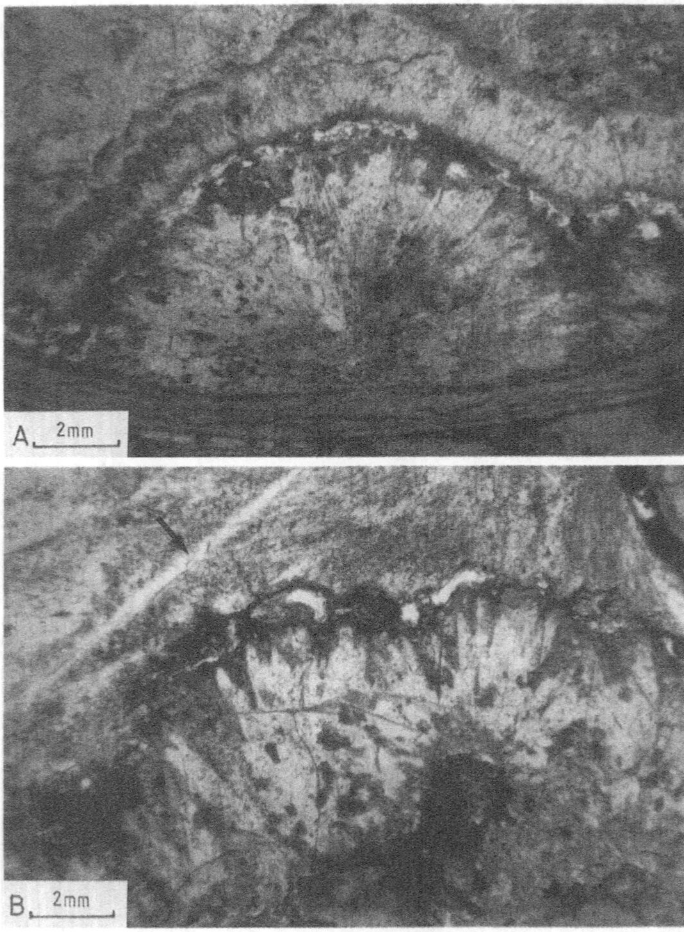

Fig. 11. (A) Botryoid, formerly of aragonite, within a brachiopod shell, overlain by an isopachous layer of acicular calcite. **(B)** Fans of former aragonite, replaced by cross-cutting calcite. The later void-filling cloudy calcite at top is also after aragonite. The clear, needle-like structure (arrowed) crossing the later aragonite is a calcite pseudomorph after gypsum.

There are few cases of the originally aragonite cements being bored. Borings in the bioclasts are also relatively rare. However, in the aragonite crusts of the reef base coquina, there is evidence for some early dissolution of the outer part of the aragonite crusts and some vadose-type cement geometries (Tucker & Hollingworth, 1986). These features are interpreted as indicating short periods of subaerial exposure, when reef debris was piled up to form small and probably temporary islands.

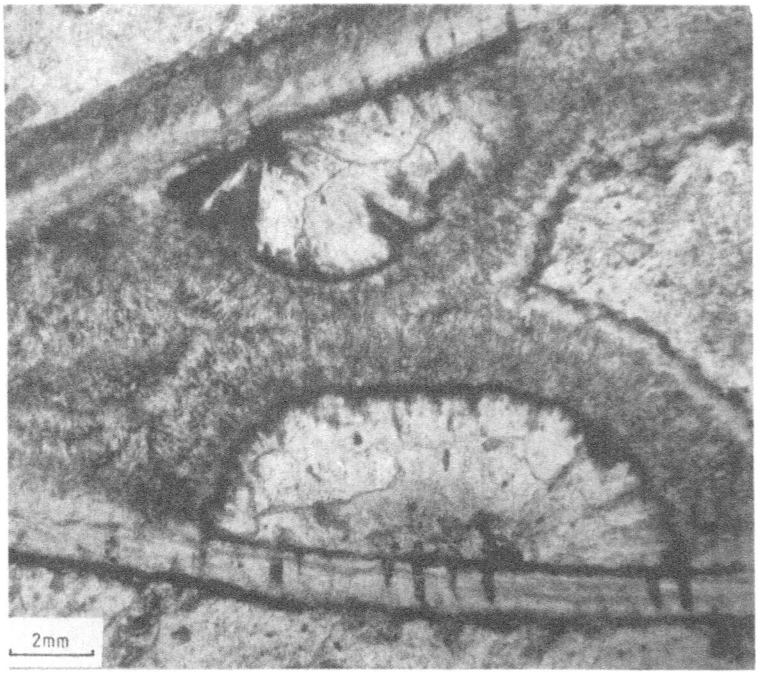

Fig. 12. Calcitized aragonite botryoids with feathery edges and a micrite coat (possibly some evidence of dissolution in the upper one), occurring within a brachiopod shell showing good fabric preservation with puncti. The botryoids are succeeded by acicular calcite and then calcite spar occludes the final porosity.

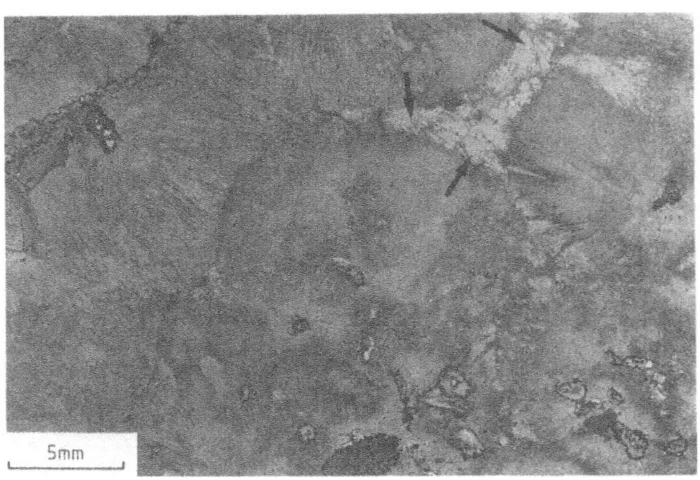

Fig. 13. Masses of aragonite, now calcitized, occurring around fenestellid bryozoans. As can be seen, the original aragonite filled most of the space between the bryozoan fronds. There was little or no fine skeletal debris (matrix). Calcite crystal fans, the last marine precipitate, are arrowed.

The crusts have mostly been replaced by neomorphic calcite spar, and this shows a range of fabrics. Irregular to equant cloudy pseudospar with straight extinction replaces many crusts (Fig. 14), but some consist of columnar calcite crystals with extinction ranging from fascicular-optic (divergent optic axes) to unit to radiaxial (convergent optic axes) (Fig. 14). Complex subcrystals occur in large columnar crystals and curved twin planes may be present. Within some crusts there are clear areas of calcite spar (e.g., Fig. 15) with crystal growth zones revealed under cathodoluminescence. These are the fills of larger voids which developed during the calcite replacement of the aragonite, which for the most part was a thin-film, calcitization process.

It was noted earlier that dolomitization has obliterated much of the details in the Ford Formation but there are locations where dolomitized aragonite botryoids can be identified (Fig. 16). They are seen in the reef flat facies at Castle Eden Dene, for example. It is considered likely that aragonite cementation was pervasive throughout the fore-reef and reef-core facies, as it clearly was in the basal reef coquina, and that this very early sea-floor cementation contributed towards reef stabilization.

The original cement of the Tunstall Reef has many parallels with modern aragonite botryoids and crusts described by James and Ginsburg (1979) from Belize and Aissaoui (1985) and Aissaoui *et al.* (1986) from French Polynesia. In the geological record, similar formerly aragonite cements are well documented from the Permian Capitan Reef of Texas (Mazzullo & Cys, 1979) and from the Triassic reefs of the Alps (Scherer, 1977). The pseudospar replacement mosaics of the Ford Formation Reef aragonites are comparable with those of the Capitan (Mazzullo & Cys, 1979) and of replaced aragonite bioclasts (Sandberg & Hudson, 1983). The columnar calcite with undulose extinction, however, is reminiscent of radiaxial and fascicular fibrous calcite which is a common cavity fill in Palaeozoic and Mesozoic limestones. The fibrous cements were long considered replacements of acicular precursors (e.g., Kendall & Tucker, 1973) but they have recently been reinterpreted as primary precipitates, with the peculiar fabrics the result of split growth (Kendall, 1985). Heinrich and Zankl (1986) have adhered to a replacement origin for radiaxial fibrous calcite in the Upper Triassic Wetterstein reefs of the Alps.

Acicular high-Mg calcite fringes, fans and peloids

Acicular calcite cements occur as fringes around and within bioclasts and upon botryoids originally of aragonite (Figs. 12 & 17). They are mostly less than a millimetre thick, but several generations may lead to thicker fringes. Small fans of acicular calcite have a radial-fibrous extinction pattern. Although isopachous fringes dominate, some show asymmetries or a preferred location where bioclasts are in contact, suggesting precipitation in a vadose zone. Crystallites are chiefly many 10's to several 100 microns long and a few microns wide. Under SEM microdolomite crystals can be located, less than 5 µm across. Within bioclasts and locally between shells, there are commonly peloids of radially-arranged acicular calcite crystals (Fig. 18).

Fig. 14. (A & B) Aragonite replaced by an inclusion-rich irregular neomorphic calcite spar. The original near-parallel arrangement of aragonite crystallites is clearly visible. A: ppl, B: xp. **(C)** Aragonite replaced by columnar calcite with undulose extinction (radiaxial) and curved twin planes (concave away from substrate). The original near-parallel arrangement of aragonite crystallites is still visible. xp.

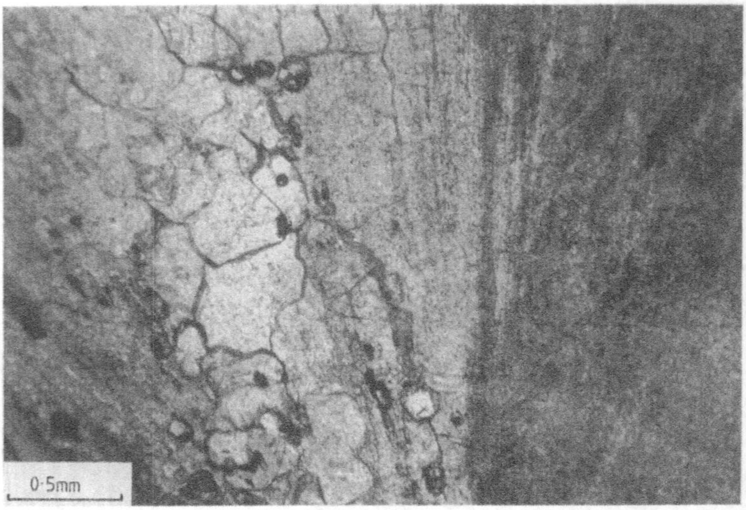

Fig. 15. Aragonite botryoid from within a brachiopod shell replaced by neomorphic spar, but in central part, complete dissolution resulted in a void filled by drusy equant spar.

These cements of acicular calcite are interpreted as originally high-Mg calcite marine precipitates, and are reminiscent of the bladed HMC cements, well documented from modern reefs (e.g., Schroeder, 1972; James & Ginsburg, 1979). The peloidal form is also a common precipitate in Recent reefs (Macintyre, 1985).

Calcite crystal fans

These cements are the last early precipitate and occur upon aragonite crusts as fans of calcite with prominent radiating columnar subcrystals (Fig. 19). In some instances the outer part of these crystals has replaced aragonite. This cement appears to be a "new type". It is interpreted as a marine precipitate, although it probably formed during very shallow burial. The characteristic fanning subcrystal fabric is probably the result of split growth (*cf.* Kendall, 1985).

The three types of cement described above were all synsedimentary precipitates and contributed towards the early stabilization of the Ford Formation Reef. Although precipitated from marine pore-fluids there is evidence in the form of cement geometries for the vadose environment. In addition, minor dissolution textures indicate occasional emergence of reef debris and contact with meteoric water.

Fig. 16. (A) Dolomitized reef-rock with relic aragonite botryoids and much dissolutional porosity. **(B)** Dolomitized aragonite crystallites of a botryoid in Fig. 16A, showing fine dolomite crystals and vugs.

Dolomitization

The main phase of dolomitization of the Tunstall Reef was early, before ány significant compaction. The dolomite is mostly a coarse xenotopic mosaic of anhedral-subhedral crystals, with the faunal remains scattered throughout having a moderate to good fabric preservation (e.g., Fig. 20). The most common skeletons, fenestellid bryozoans, mostly consist of very fine dolomite crystals, resulting in a high degree of fabric retention. Brachiopods and foraminifera are also well preserved, contrasting with the originally aragonitic bivalves and gastropods which rarely show any original internal structure, and are mostly seen as moulds. Much of the matrix of the dolomitized rock is presumably a replacement of fine sediment and/or marine cement. Commonly the dolomite of the matrix has undulose extinction, although terminations occur where these crystals line voids (Fig. 20). In some localities, botryoids of fanning crystals, presumably aragonite originally, have been replaced by dolomite (Fig. 16). One important consequence of the dolomitization is the generation of a very porous rock with mm-scale vugs.

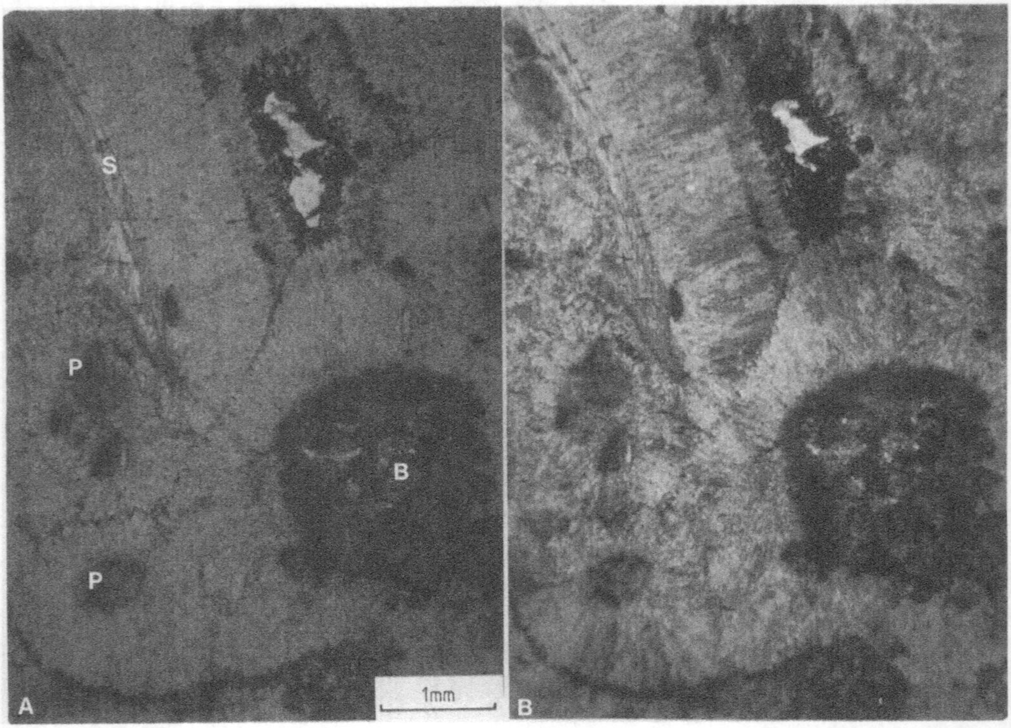

Fig. 17. Acicular calcite fringes occurring around a bryozoan (**B**), a brachiopod shells (**S**) and peloids (**P**). Note junctions between cement layers. A: ppl, B: xp.

Zechstein dolomitization is generally attributed to seepage-reflux with brines generated during an evaporative phase after carbonate deposition, descending into the carbonate pile (Clark, 1980; Smith, 1981). In the case of the Ford Formation Reef, a lagoon and sabkha developed landward of the reef and this may have been the source of dolomitizing fluids. The latter could have been drawn through the reefal carbonates during the sea-level fall which finally terminated reef growth and resulted in precipitation of the Hartlepool Anhydrite in prograding sabkhas and lagoons on the basinward side of the reef.

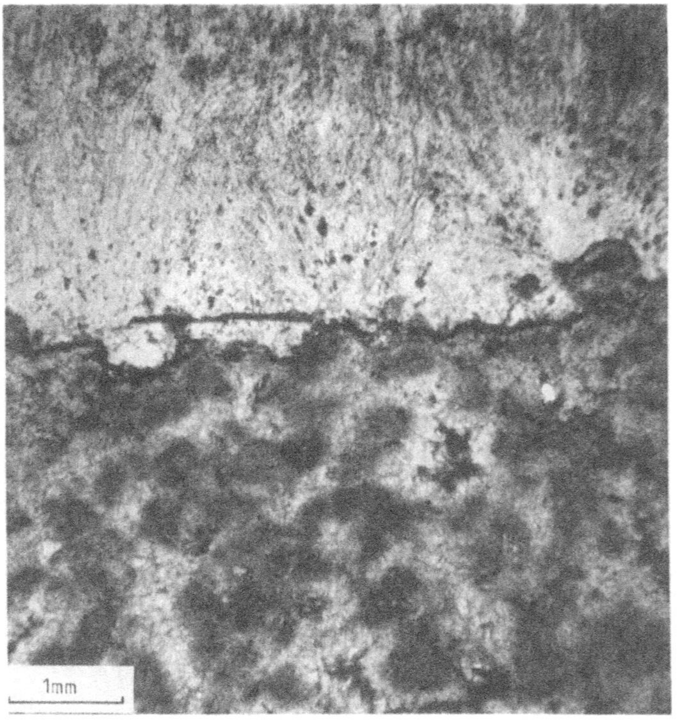

Fig. 18. Peloids of acicular calcite overlain by aragonite fans, now replaced by calcite. Note irregular junction between the two, possibly the result of some dissolution through subaerial exposure.

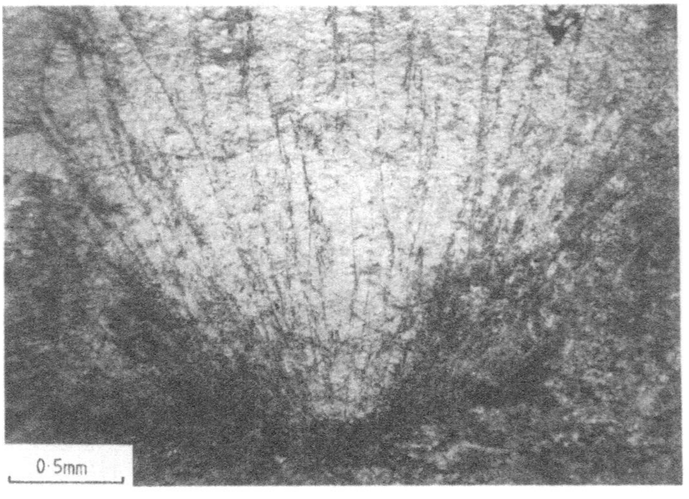

Fig. 19. Calcite crystal fan, occurring upon calcitized aragonite.

During ther second Zechstein cycle (EZ2), evaporites (the Middle Marls) were precipitated above the first cycle carbonates while the Hartlepool and Roker Dolomite was deposited at the shelf margin. These evaporites could also have been a source of dolomitizing fluids. It is possible that dolomitization took place in a meteoric-marine mixing zone, which traversed the carbonate wedge during the evaporative drawdown which terminated each Zechstein cycle.

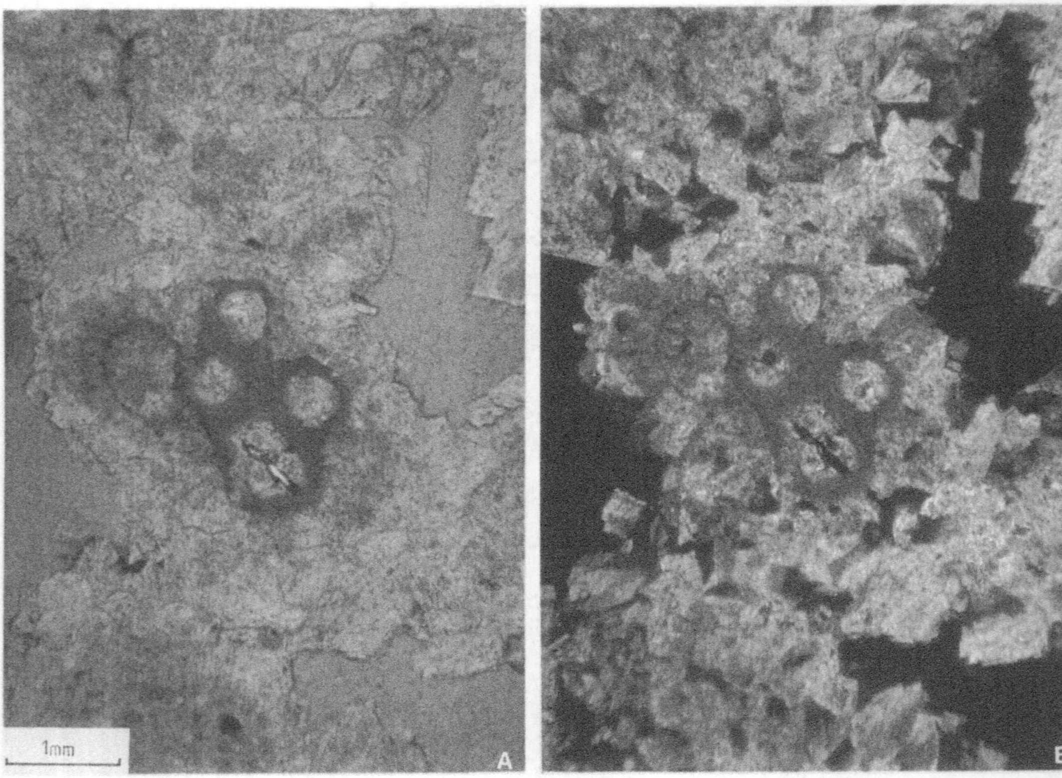

Fig. 20. Dolomitized reef-rock with bryozoan preserved in micritic dolomite and matrix of coarse dolomite crystals showing undulose extinction, found by replacement of aragonite cement. A: ppl, B: xp.

Later diagenesis

Later diagenesis of the Tunstall Reef involved calcitization of the aragonite, and locally of the dolomite. too. The latter has given rise to a range of dedolomite rocks, especially cannonball and pseudocoralline limestones with coarse, commonly fibrous calcite mosaics. Evaporite replacement and filling of vugs is also a later diagenetic phase. This could be contemporaneous with precipitation of the Hartlepool Anhydrite, or a later event still. Within the calcitized aragonite mosaics, there are clear crystal shapes, which are interpreted as gypsum pseudomorphs (Fig. 21). The gypsum was replaced by calcite during the calcitization of the aragonite. These later burial processes have been discussed by Clark (1980) and Smith (1981).

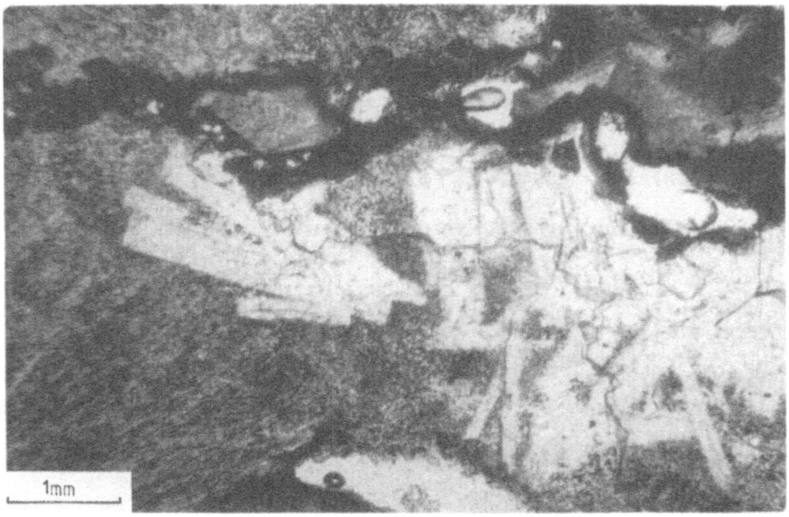

Fig. 21. Pseudomorphs after gypsum occurring within former aragonite botryoids. The aragonite was replaced by gypsum and then both were calcitized during later burial.

Conclusions

Detailed field studies and laboratory examination have identified the palaeocommunities of the Late Permian Tunstall Reef Complex of N.E. England, and have enabled accurate 3-dimensional diagrams to be constructed illustrating the palaeocommunity structure. The different palaeocommunities reflect the development of the reef through time, from reef-base coquina to lower reef core palaeocommunities, and then through space, in the laterally equivalent reef-flat, upper reef core, reef crest, and fore-reef talus palaeocommunities, anf the backreef patch reef palaeocommunity. An important factor in the formation of the Tunstall Reef was the precipitation of sea-floor cement. Aragonite particularly was extensive and rapid precipitate in the reef-base coquina, as revealed in undolomitized limestones, and this type of cementation appears to have been widespread in other reef facies too, as judged by the rare occurrence of undolomitized limestones and by textures in the dolomites which can now be interpreted as relics of marine cements.

Acknowledgements

We are most grateful to Carole for her patience and understanding in producing the typescript of this paper.

References

Aissaoui, D.M., 1985. Botryoidal aragonite and its diagenesis. *Sedimentology* 32: 345-361.

Aissaoui, D.M., Buigues, D. & Purser, B.H., 1986. Model of reef diagenesis. In: Schroeder, J.H. & Purser, B.H. (Eds.), *Reef Diagenesis*, 27-52. Springer, Berlin.

Clark, D.N., 1980. The diagenesis of Zechstein carbonate sediments. *Contr. Sedimentology* 9: 167-203.

Heinrich, R. & Zankl, H., 1986. Diagenesis of Upper Triassic Wetterstein Refs of the Bavarian Alps. In: Schroeder, J.H. & Purser, B.H. (Eds.), *Reef Diagenesis*, 245-268. Springer, Berlin.

James, N.P., & Ginburg, R.N., 1979. The seaward margin of Belize Barrier and Atoll Reefs. *IAS Spec. Publ.*, 3:191 pp.

Kendall, A.G., 1985. Radiaxial fibrous calcite: a reconsideration. *SEPM Spec. Publ.*, 36: 59-77.

Kendall, A.C. & Tucker, M.E., 1973. Radiaxial fibrous calcite: a replacement after acicular carbonate. *Sedimentology* 20: 365-389.

MacIntyre, I.G., 1985. Submarine cements - the peloidal question. *SEPM Spec. Publ.*, 36: 109-116.

Mazzullo, S.J. & Cys, J.M., 1979. Marine aragonite sea-floor growths and cements in Permian phylloid algal mounds, Sacramento Mountains, New Mexico. *Jour. Sed. Petrology* 49: 917-936.

Sandberg, P.A. & Hudson, J.D., 1983. Aragonite relic preservation in Jurassic calcite replaced bivalves. *Sedimentology* 30: 879-892.

Scherer, M., 1977. Preservation, alteratiom and multiple cementation of aragonitic skeletons from the Cassian beds (U. Triassic, southern Alps): petrographic and geochemical evidence. *N. Jb. Geol. Paläont. Abh.*, 154: 213-262.

Schroeder, J.H., 1972. Fabrics and sequences of submarine carbonate cements in Holocene Bermuda cup reefs. *Geol. Rundschau* 61: 708-730.

Smith, D.B., 1970. Submarine slumping and sliding in the Lower Magnesian Limestone of Northumberland and Durham. *Proc. Yorks. Geol. Soc.*, 38: 1-36.

Smith, D.B., 1980. The evolution of the English Zechstein basin. *Contr. Sedimentology* 9: 7-34.

Smith, D.B., 1981. The Magnesian Limestone (Upper Permian) Reef Complex of Northeastern England. *SEPM Spec. Publ.*, 30: 187-202.

Smith, D.B., Harwood, G.M., Pattison, J. & Pettigrew, T., 1986. A revised nomenclature for Upper Permian strata in eastern England. *Geol. Soc. Lond. Spec. Publ.*, 20: 9-17.

Trechmann, C.T., 1925. The Permian formation in Durham. *Proc. Geol. Assoc. Lond.*, 36: 135-145.

Tucker, M.E. & Hollingworth, N.T.J., 1986. The Upper Permian Reef Complex (EZ1) of Northeast England: Diagenesis in a marine to evaporitic setting. In: Schroeder, J.H. & Purser, B.H. (Eds.), *Reef Diagenesis*, 270-290. Springer, Berlin.

Walker, K.R. & Alberstadt, L.P., 1975. Ecological succession as an aspect of structure in fossil communities. *Palaeobiology* 1: 238-257.

DISSOLUTION EFFECTS AND REEF-LIKE FEATURES IN THE ZECHSTEIN ACROSS THE MID NORTH SEA HIGH

Malcolm K. Jenyon
Seismograph Service (England) Limited,
Holwood, Keston, Kent, U.K.
and
John C. M. Taylor
V. C. Illing & Partners
Cuddingron Croft, Ewell Road, Cheam, Sutton,
Surrey, U.K.

Abstract: Observations made in recent seismic data, combined with information from released wells, suggest that many structural features in the Zechstein interval across the Mid North Sea High are due to dissolution and removal of salt rock where it thinned against the upper flanks of the structure produced by the underlying High. It appears that radial run-off of intraformational, undersaturated water from the culmination of the structure at the level of Post-Permian (Mesozoic) sediments has occurred.

Salt-edge, and *Salzspiegel* -type areal dissolution both probably took place in some parts of the MNSH area. The complexity of the process was increased by the presence of Pre-Permian palaeogeomorphic features - faults and horsts - which interrupted, or modified, the intraformational drainage pattern. Subsequent settlement of the overlying Mesozoic sediments into the space vacated by the salt has revealed, through subsidence-drape, the presence of positive features resting on basal Zechstein units; certain of these features consist of material of much higher velocity than that of salt rock, and although they may be bryozoan-algal reefs from their form and location, it is equally possible that they may be primary or relict pods of anhydrite. Others are certainly Upper Carboniferous or Devonian horsts projecting up through the Zechstein interval. These horsts are of lower seismic velocity than the Zechstein interval material, which can be diagnostic.

Other striking, reef-like features in off-shelf locations have been observed which, however, may be local thickenings of non-reefal material in the basal Zechstein units, although buildups of porous carbonates, together with several other alternatives, cannot positively be ruled out.

Lecture Notes in Earth Sciences, Vol. 10
T. M. Peryt (Ed.), The Zechstein Facies in Europe

Introduction

The Mid North Sea/Ringkøbing-Fyn High (MNSH) has in the past been considered as a substantial barrier separating the Northern and Southern Zechstein sub-basins. More recently, it has been recognised that this barrier was breached, perhaps in several places, at the time of the Zechstein inundation. Although some questions may still remain as regards the exact ages of the Central and Horn Graben with respect to the latter event, it has been demonstrated (Jenyon *et al.*,1984) that a major channel through the barrier, filled with Z2 salt and located in U.K. Quadrant 37, certainly existed in pre- or very early-Zechstein times, and there may have been other breaches of the MNSH apart from those mentioned.

Even if not an impenetrable barrier, the MNSH and its extension eastward clearly represented a major modification to the depositional environment of the Zechstein carbonates and evaporites; both seismic and well data in the area have disclosed a complex and highly-variable situation. One of the greatest difficulties facing the geologist in attempting to come to grips with Zechstein geology in this area is the comparative sparsity of available borehole information. Since early work revealed the nature of the MNSH, practically no further drilling for hydrocarbon prospects was carried out until recently, when interest was awakened in possible Pre-Permian hydrocarbon plays. This has resulted in a few more wells being drilled, but since the resulting information will not come into the public domain for several years, data from the few old wells available must still be relied on. In such a situation, and in a study involving a high degree of lateral variation in a stratigraphic interval, much dependence must be placed on the available seismic data. In this work, the seismic information has been combined with results from the limited number of released wells in the area to draw certain conclusions, and to suggest a rather simple model which might explain the structures present in the Zechstein over a part of the Mid North Sea High.

The study area

The sketch map in Fig. 1 shows some of the principal features of the western end of the Zechstein Basin, with the study area bounded by heavy lines. In Fig. 2, a more detailed sketch-map is seen for this part of the MNSH area. The main features of this map are indications of the Zechstein salt-edge dissolution slope to be discussed, which delimitates areas in which Zechstein salt is present from others where it is absent due to dissolution. The heavy dashed lines indicate the edge-dissolution slope where it forms the western and eastern boundaries of the salt channel through the MNSH referred to previously

(Jenyon *et al.*, 1984), while the dot-dashed lines indicate the continuations to east and west of the boundary slopes of the channel. The approximate position of the crest of the MNSH structure at Upper Permian/Lower Mesozoic level is also indicated, and its location with respect to the channel configuration suggests that the MNSH is an asymmetrical structure, with steeper dip on the southern flank. The letters mark locations which are illustrated and/or referred to in the text.

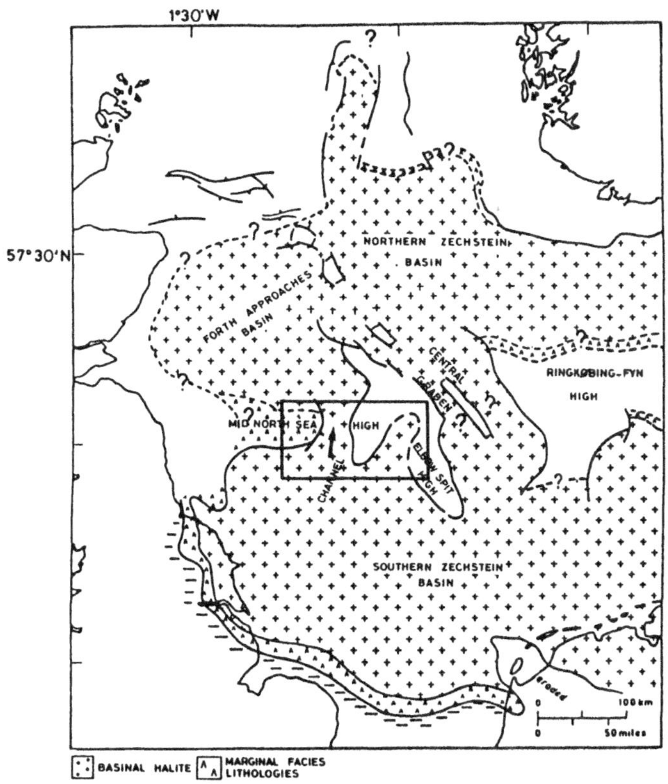

Fig. 1. Sketch-map showing principal features at the western end of the Zechstein Basin. Study area bounded by heavy lines. Modified from Jenyon *et al.*, 1984 , after Taylor, 1984.

The model

Study of the seismic data in the general area has led to the visualisation of a model of dissolution of Zechstein salt, and Mesozoic overburden collapse, which is closely related to the structure imposed at least locally by the underlying, Lower Palaeozoic-cored MNSH.

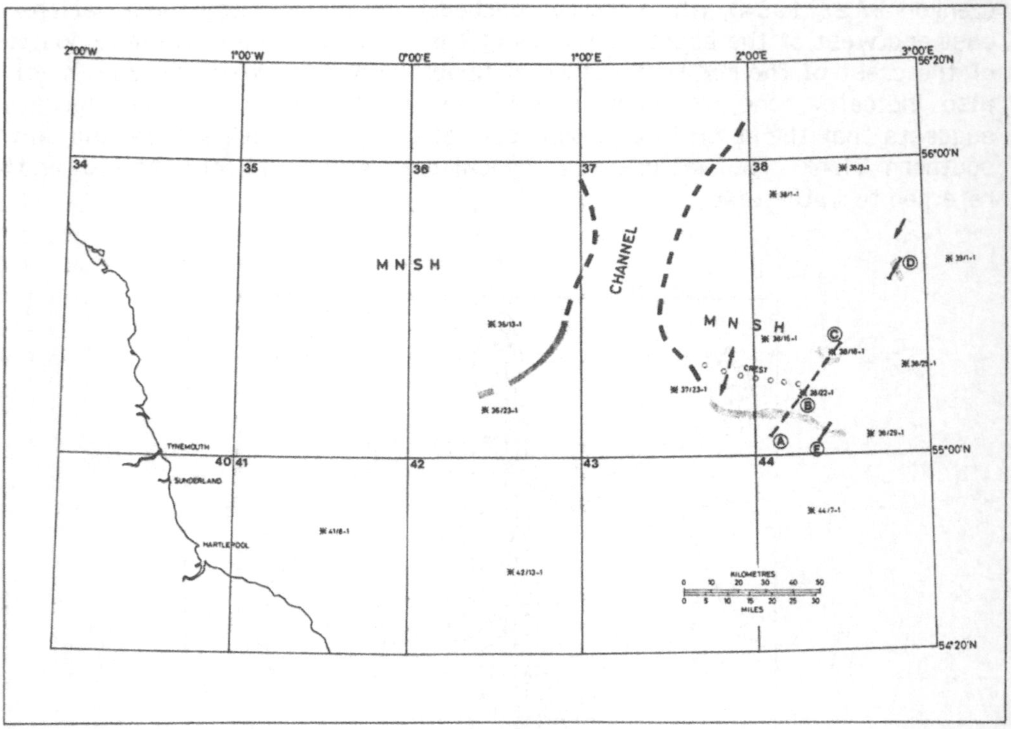

Fig. 2. Sketch-map of part of the Mid North Sea High. Segments of seismic lines lettered are referred to in text. Local crest of MNSH indicated by open circles and dip arrows. Dip arrow near to 'D' indicates where regional dip reverses to a southerly direction. Some well locations marked.

The sketches in Fig. 3 show a simplified illustration of the model in the form of a schematic N - S section across the MNSH. Sketch 3(a) shows the situation before any dissolution has affected the salt, with the Pre-Permian surface of the MNSH overlain by the Zechstein salt interval thickening into both northern and southern sub-basins from the crest of the High. The unit marked 'A' is an indicator band within the Mesozoic section. It is supposed that at some stage, undersaturated brine entered Mesozoic units resting on the Zechstein salt in the crestal area of the structure, and flowed intraformationally (or between salt and clastics) down both flanks, dissolving and removing the upper part of the salt interval, as suggested by the arrows at 'W'. There are several possible reasons why this may have occurred:

(i) Compaction of Mesozoic across the MNSH axis may have induced fracturing along the axial zone of the structure.

(ii) Pre-Permian faults trending along the axial zone of the MNSH may have resulted in zones of weakness and fracturing in the compacting Mesozoic sediments.

(iii) The cut-down of the Late Cimmerian (CU) unconformity may, perhaps only in the later stages, have allowed water to percolate down through the unconformity surface in the axial zone of the MNSH.(Some relatively minor subsidence effects are seen at CU level, but these are on a much smaller scale than the main collapse features).

(iv) Salt may have been dissolved from below, aided by dislocations in the Pre-Zechstein strata, or by the Zechstein carbonates acting as aquifers - an inverted *Salzspiegel* situation.

As the salt was removed, the Mesozoic overburden would have subsided, leading to the situation in sketch (b) of Fig. 3, where the retreating salt terminates on both flanks in a salt-edge dissolution slope 'D', and the Late Cimmerian erosion has cut down into the Mesozoic (unconformity CU). This is the simplest situation which can be envisaged, and assumes a completely uniform Zechstein salt interval.

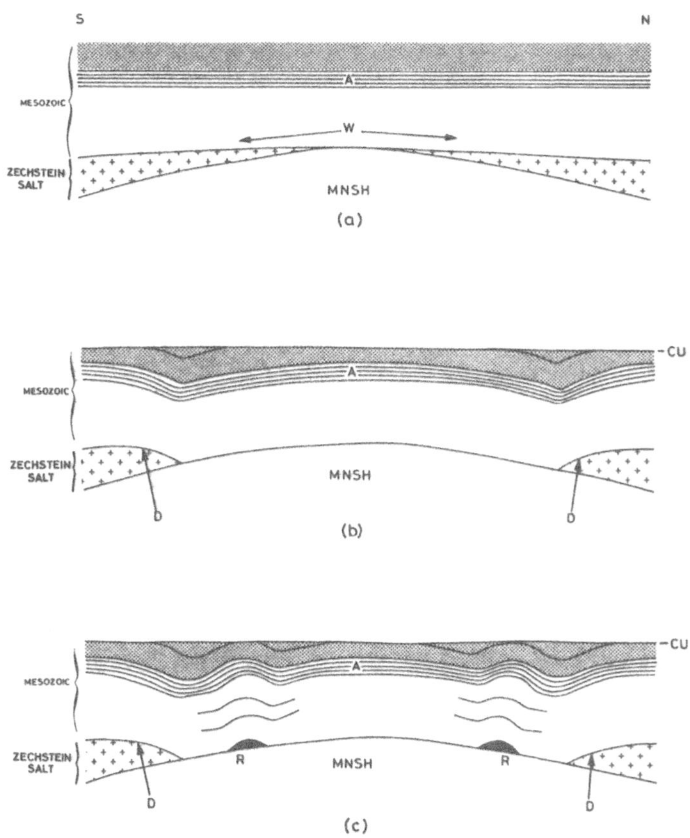

Fig. 3. Schematic model across the Mid North Sea High, discussed in the text.

Sketch (c) in Fig. 3 shows schematically what the *actual* situation appears to be, from interpretation of the seismic data. The Zechstein salt interval was *not* uniform; positive irregularities ('R') are present at basal Zechstein level which, after removal of the salt, produce strong effects on the seismic section, having imparted subsidence-drape to the collapsed Mesozoic units, which persists up

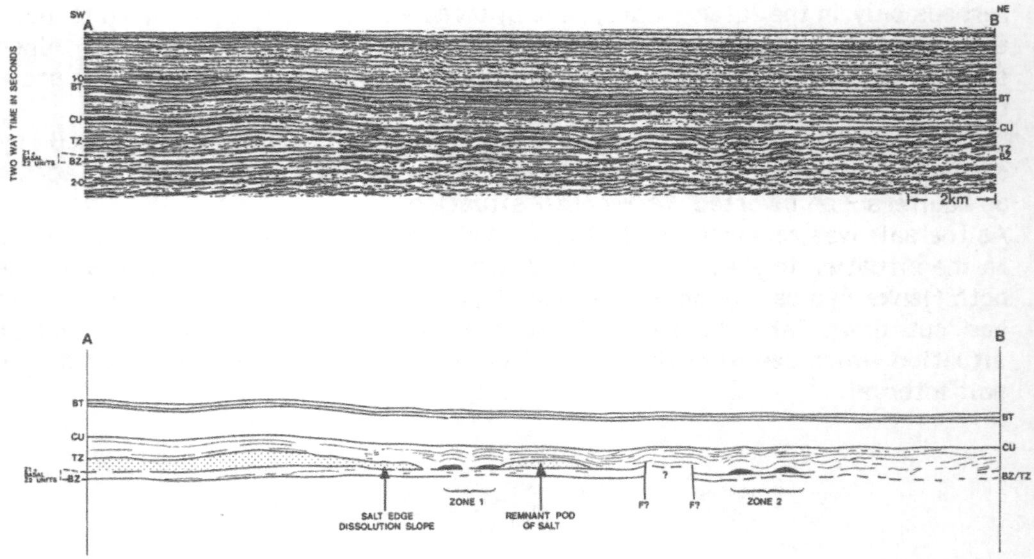

Fig. 4. Seismic section with sketch interpretation below, of the line segment A - B in Fig. 2. Seismic example by courtesy of Seismograph Service (England) Limited - SS(E)L.

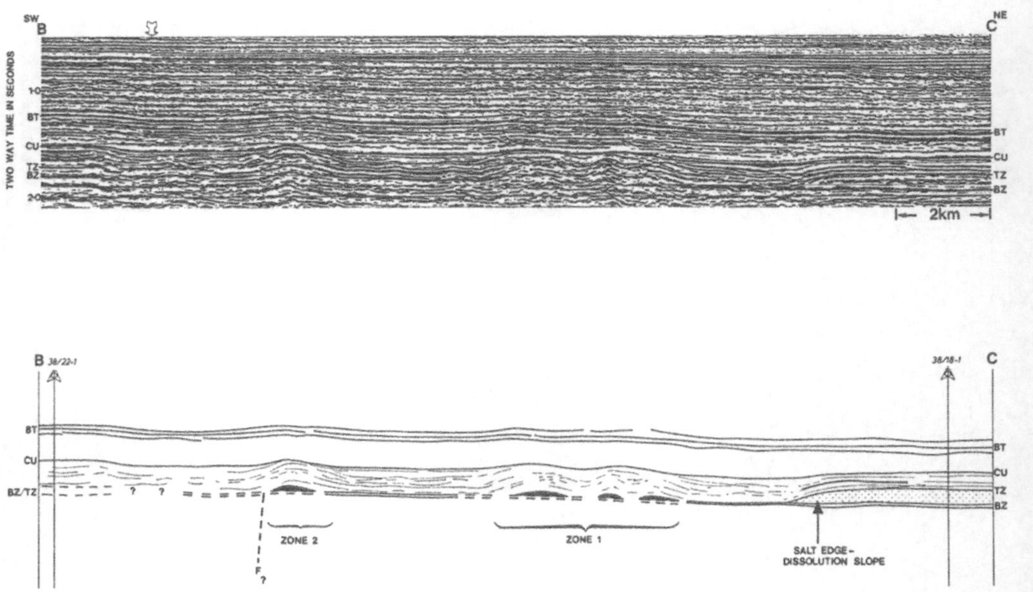

Fig. 5. Seismic section with sketch interpretation below, of the line segment B - C in Fig. 2. Open arrow indicates approximate position of crest of MNSH at Permian level. Locations of two wells referred to in the text are shown on the interpretation. Seismic example by courtesy of SS(E)L.

to the CU unconformity (and above, in places). The possible nature of these irregularities will be discussed at a later stage.

Comparison of seismic data with model

Referring to the map in Fig. 2, locations A, B, and C are three points on a continuous portion of migrated seismic profile. Segment A - B is illustrated in Fig. 4, and B - C in Fig. 5. The approximate position of the crest of the MNSH as it was in Upper Permian/Lower Mesozoic times (the area was subjected to tilting in the early/mid Tertiary, as can be seen from the attitude of the Late Cimmerian unconformity, CU, and shallower horizons in these examples) is indicated in Fig. 5 by the open arrow, and the locations of two wells are marked. The upper part of each figure shows the seismic data, and the lower part is a sketch interpretation showing the Base Zechstein (BZ), Top Zechstein (TZ), Late Cimmerian unconformity (CU), and Base Tertiary (BT).

Examining the section in Fig. 5 first, as being the less complex of the two, Well 38/18-1 logs a thickness of approximately 300 m (1000 ft.) of Zechstein salt, which at the normal seismic interval velocity for salt in this area (4500 m/s) gives a two-way-time thickness of 130 ms, which fits well with the salt interval within BZ-TZ in the example. It can be seen that the salt terminates in an edge-dissolution slope about 3 km to the southwest of the well, in a location some 14 km down the north flank of the MNSH. It is belived that complete dissolution has resulted from intraformational run-off of undersaturated water from the MNSH crestal area (open arrow) down this flank, leading to collapse of the Mesozoic sediments seen in the TZ-CU interval. These are mainly Triassic clastics in this zone, with a thin Jurassic sequence being present in places.

From inspection of this cross section, it is possible to deduce certain points of importance:

(i) South-westward from Well 38/18-1, the salt thickness decreases from about 300 m to zero over a distance of little more than 1 km. (This interpretation is supported by evidence from other wells in the area - e.g., 38/22-1 - in which no Zechstein salt was found). In spite of this, there is no evidence of any velocity anomaly, either positive or negative, at and below the level of BZ across this zone.This is strong evidence that the seismic velocity in the collapsed Mesozoic sediments adjacent to the salt is approximately the same as the salt velocity - i.e., 4500 m/s.

(ii) The strong BZ evént, produced by the Z1 and basal Z2 anhydrite/carbonate units, is continuous everywhere below the salt, and also below the collapsed Mesozoic clastics, but beneath the section segments marked Zones 1 and 2 there are discontinuities.

(iii) In zones 1 and 2, the BZ event disappears due to strong "pull-up" velocity anomalies. This is clear evidence that the lower part of the

BZ-CU interval includes material with markedly higher velocity than either the Zechstein salt or the Mesozoic clastics. The velocity of this material is probably within the range of 5000 to 6000 m/s.

(iv) The cross sectional shapes of this unknown material in Zones 1 and 2 can be deduced from the shapes taken up by the collapsed overlying Mesozoic events. The upward persistence of drape (see Jenyon, 1986) in these zones has produced positive features at the surface of unconformity CU. This is particularly well seen in Zone 2. These drape effects persist upward in the section through the Cretaceous interval CU-BT, and into the early Tertiary.

The position, shape and size of the features consisting of the unknown material have been suggested by the solid black areas in the lower part of Fig. 5; they have been marked as though the velocity anomaly were removed and the estimated position of the basal Zechstein anhydrite/carbonate units is shown as dashed lines. It is not clear whether these features are resting on Z1 or on Z2 basal units. From the order of probable seismic velocity of the material, however, it seems likely that they consist of either dense anhydrite, dolomite, or both.

Referring now to Fig. 4, on the south flank of the MNSH, a similar situation is apparent. In this example, basinal salt at the southwest end of the profile terminates northeastward in a salt-edge dissolution slope which is situated approximately as far south of the MNSH crest (Fig. 5) as the slope in Fig. 5 is north of the crest. Just to the northeast of the slope in Fig. 4 there is a remnant pod of salt which has been left behind as the dissolution "front" advanced down the flank from the crestal area. The basal Zechstein units are of substantial thickness - 300 to 350 m (1000 to 1200 ft.) which, in Well 38/22-1 on segment B - C (Fig. 5) consists mainly of anhydrite. These basal units produce a strong seismic group event about 100 ms long (down the seismic trace), as indicated at the southwest end of the section; this maintains excellent seismic character beneath the salt at the southwest end of the section, and under the remnant salt pod, but as in Fig. 5, is distorted by pull-up anomalies in Zones 1 and 2, and also in the zone indicated in the lower part of the figure as a (?) horst block.

The local pattern of salt dissolution

From the observations made, it seems that for a distance of over 25 km across the crest of the MNSH in the direction of the profile A-B-C in Figs. 3 and 5, the Zechstein interval has been completely denuded of salt by dissolution. The resulting concentrated brine may have flowed under gravity (see Anderson & Kirkland, 1980) down the flanks within the Lower Mesozoic units present above the Zechstein, or at the salt/clastic interface, to escape under artesian pressure through fractures further down in the two sub-basins. If the salt was attacked via aquifers below or within the Zechstein, it may have been removed

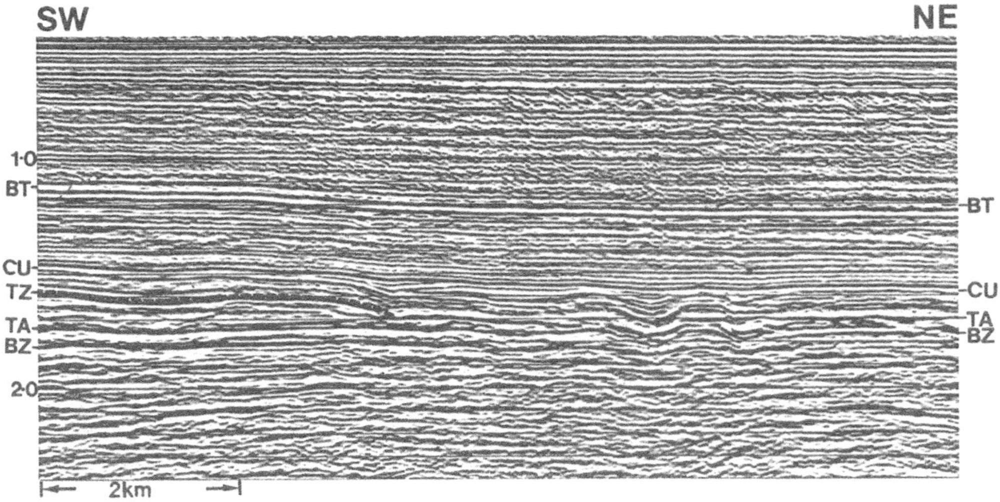

Fig. 6. Seismic line segment at 'E' in Fig. 2. Courtesy of SS(E)L.

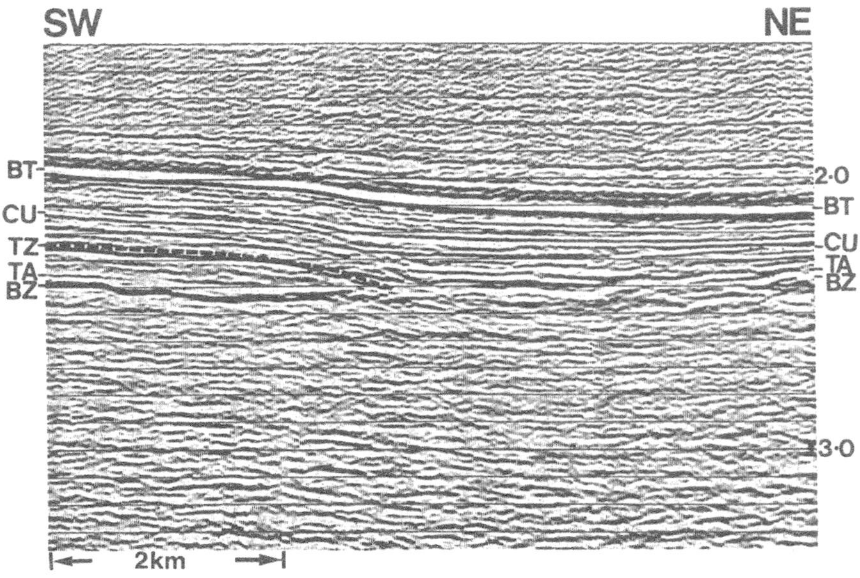

Fig. 7. Seismic line segment at 'D' in Fig. 2. Courtesy of SS(E)L.

or replenished by density-driven convection cells either to be expelled upward elsewhere after dilution, or lost by compression in the vast volume of North Sea Basin pore fluids.

More detailed observation of the salt slope seen in Fig. 4 can be achieved in the enlarged portion of an adjacent line seen in Fig. 6, and located at 'E' on the Fig.

2 map. The strong similarities between this slope and the one seen on the north flank of the High in Fig. 5 should be noted, as should the details and location of the irregularities at BZ level between the slope and the northeast end of the section (*cf.* with Fig. 4). In conjunction with the salt slope on the north flank, a mirror-image situation of dissolution is suggested, symmetrical about the axial plane of this part of the MNSH. In Fig. 6, the lettering indicates the Base Zechstein (BZ), the top of the Z1/basal Z2 anhydrite-dolomite-anhydrite "sandwich" (TA), and at the southwest end of the section, the top of the Zechstein salt (TZ). The Late Cimmerian unconformity is at CU, and the Base Tertiary at BT.

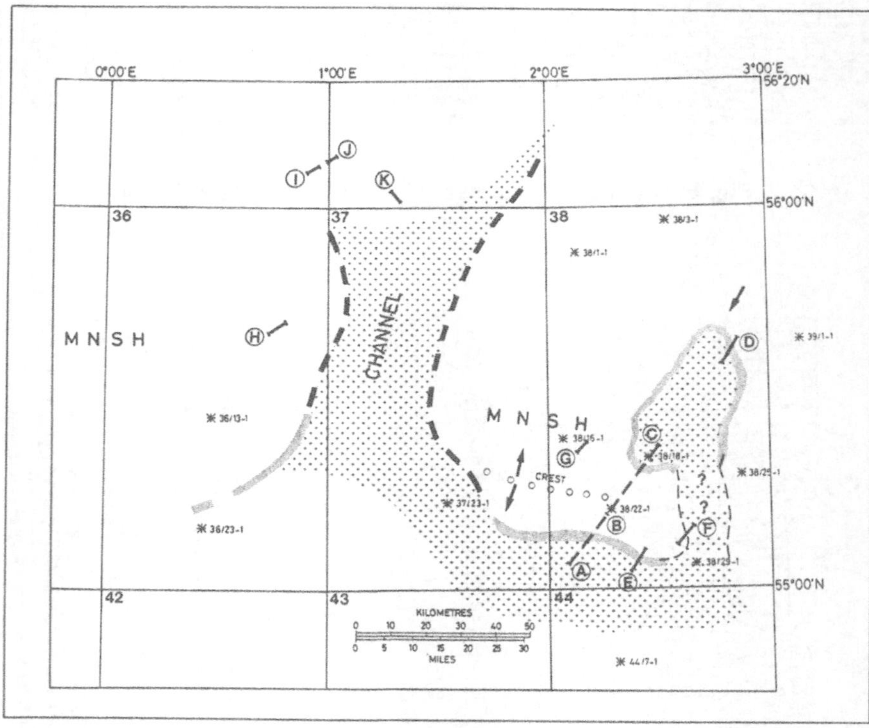

Fig. 8. Sketch-map based on that in Fig. 2, showing in more detail the distribution of Zechstein salt to the east of the channel, and indicating the possible connection of the apparently isolated patch of salt to the east (between 'C' and 'D') with a salt embayment in the south flank of the High (at 'F'). Additional seismic line locations lettered 'F' to 'K'.

Details of the salt on the north flank are of considerable interest also. The salt continues down flank as a uniform layer from the northeast end of Fig. 5, and terminates again at a (in this location) north-facing salt-edge dissolution slope which is illustrated in Fig. 7 (location 'D' in Fig. 2). It is at this location that the regional dip at BZ level reverses (to south dip), and, presumably, run-off of water from the north down this south-dipping slope is responsible for the presence of salt slope's position. The key lettering in Fig. 7 is as for Fig. 6.

Thus there appears at first sight to be an isolated patch of salt on the north flank of the MNSH; however, this isolation may be more apparent than real. Other data in the area suggest that this patch of salt may be connected in some way to an embayment of salt in the south flank of the High (the feature shown immediately to the west of the Elbow Spit High in Fig. 1), and this is indicated as a possibility in the sketch-map of Fig. 8.

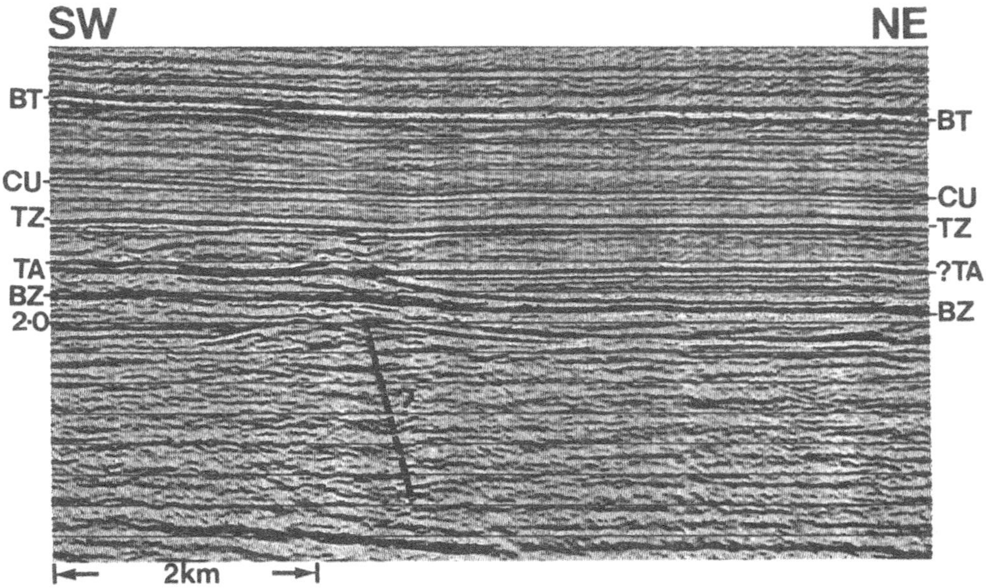

Fig. 9. Seismic section from location 'F' in Fig. 8, showing ramp-like thickening of the basal Zechstein from northeast to southwest. BZ=Base Zechstein; TA=top of the Z1-basal Z2 "sandwich" of anhydrite-dolomite-anhydrite; TZ=top Zechstein (salt); CU=Late Cimmerian unconformity; BT=Base Tertiary. The TA correlation at the northeast end of the section is not secure in the absence of any nearby well data. Courtesy of SS(E)L.

The exact nature of the embayment just mentioned is not determinable from the available data. Well 38/29-1 is located just in the mouth of the embayment, and records a thickenss of over 300 m (1000 ft.) of the basal "sandwich"mentioned previously (though here in "shelf" facies), overlain by 230 m (760 ft.) of salt with a thin shale and anhydrite capping. Lack of available well data further into the embayment precludes detailed interpretation, but its western boundary shows a remarkable thickening of the lowermost Zechstein interval (Z1 anhydrite) from east to west, as seen in Fig. 9 (location 'F' on the map of Fig. 8). There seems to be no faulting at Zechstein base level which would relate to this sudden, ramp-like thickening, but this does not rule out the possibility that the embayment is controlled by older faulting. However, the feature seems to indicate that the "shoreline" of the embayment acted as a true basin-edge during deposition of the basal Zechstein units. The position of 'TA' at the northeast end of the setion in Fig. 9 is conjectural, and the nature of the units represented by the strong, continuous series between BZ and TA is problematic, and cannot be resolved at present without well information.

The connection between the salt in the embayment and the patch of salt on the north slope of the MNSH is not well defined in the data, and has been left as a possibility only in the Fig. 8 map. The eastern boundary of the embayment is also ill-defined, and appears to be complicated by faulting. Apart from the complexity of the geology and the absence of available well information, the seismic grid here is too widely spaced to allow more detailed examination of the embayment. If the connection does exist, it suggests yet another breach which was present across the MNSH during Zechstein deposition, possibly leading towards the area now occupied by the Central Graben.

A further problem which must be pointed out as regards the possible connection between the salt embayment and the salt patch on the north slope of the MNSH relates to the interpretation of the information in Well 38/18-1. In this well, the main salt interval (900 feet thick) is underlain by 45 feet of anhydrite-dolomite-anhydrite, beneath which there is a 60-feet interval of salt (inferred from the Schlumberger log). The latter thin salt has been interpreted as being of Z2 age, with the 45-feet interval above it being designated the Plattendolomit, which would make the main salt interval above to be of Z3 age. Its thickness, approaching 1000 feet, appears unusually great for this part of the North Sea. However, there are several places in both the Northern and Southern Salt Basins where wells, seismic data, or both, indicate an anomalously low position for the Z3 Plattendolomit within the total salt section, and hence an apparently abnormally thick sequence of later evaporites compared with those of Z2 age.

There may be several reasons for this situation, including flow of once thick Z2 halite into an adjacent pillow or diapir (unlikely in the present context), original depositional geometry, or greater subsidence locally during Z3 or later cycles than during the period of Z2 halite deposition. This topic is the subject of a separate paper (J. C. M. Taylor, in preparation). The evidence in the present case is inadequate to discriminate between the alternatives, but some aspects of the sequence 33 km further south, in Well 38/29-1 may favour late Zechstein subsidence.

In 38/29-1, as already mentioned, the basal "sandwich" is of "shelf" type. More than 150 m of Z1 anhydrite is succeeded by a typical shallow water Z2 Hauptdolomit containing oolites and algal pisolites. Yet after a further 100 m of impure anhydrite, which probably includes the equivalent of the Z3 Plattendolomit/Hauptanhydrit (*cf.* 38/18-1), the Zechstein is completed by 230 m of halite which must have been accomodated by subsidence relative to sea level during Z3 or later cycles.

Of the wells marked on the map in Fig. 2, salt is absent in the following: 38/1-1, 38/3-1, 38/16-1, 38/22-1, 38/25-1, 39/1-1, and 36/13-1.

Positive features at base Zechstein level

Any consideration of the possible nature of the features exemplified by Zones 1 and 2 in Figs. 4 and 5 must take into account the effects in the collapsed Mesozoic overburden between the top of the remnant Zechstein (TA) and the Late Cimmerian unconformity (CU). As noted previously, it has been pointed out (Jenyon, 1986) that already-compacted and indurated strata that subside during the removal of an underlying salt interval will indicate the shapes of any positive irregularities upon which they come to rest after complete salt removal, and that these subsidence drape effects will persist upwards, actually becoming wider the higher they are traced in the section.

There are several different types of initiating feature which may occur at the base of a salt interval to give rise to these subsidence drape effects. They include remnant salt or anhydrite pods, pre-existing fault scarps (or fault-line scarps), cuesta features due to erosion-resistant beds beneath an unconformity on which the salt may have been deposited, carbonate buildups (e.g. reefs), and in this area, Rotliegendes sand dunes and various igneous features.

In the examples under discussion, remnant salt pods can be discussed as a possibility, since one of these exists in Fig. 4 and is identifiable as such

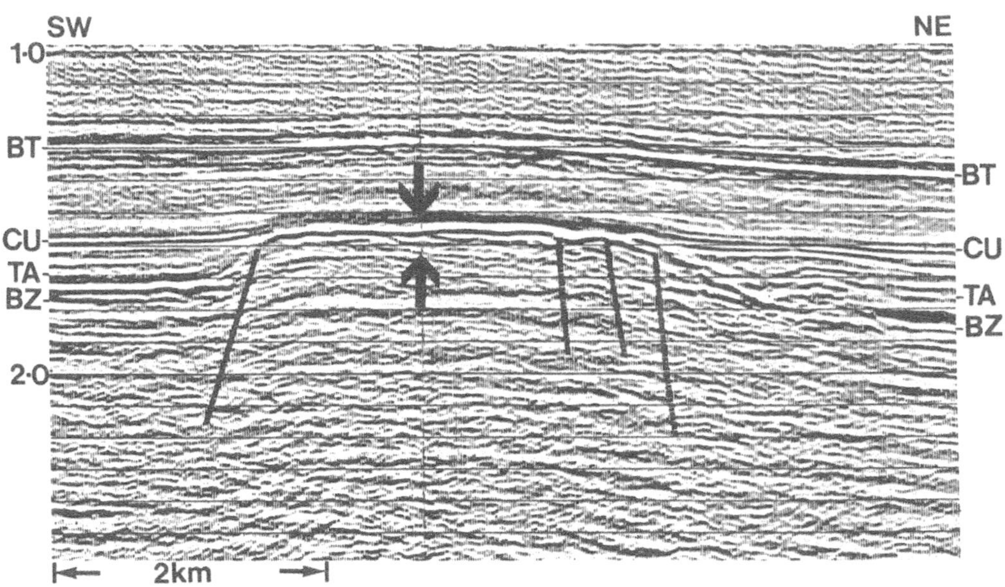

Fig. 10. Carboniferous horst block, with a 900 ft. (about 100 ms) capping of Zechstein anhydrite-dolomite-anhydrite section (between the arrows), as logged in a nearby well. No salt is present. Key as for Fig. 9. The onlapping relationship of seismic events in the TA-CU sequnce indicates that the main faulting occurred in Post-Permian, early Mesozoic times, with some movement continuing after the Late Cimmerian erosional phase. Location 'G' in Fig. 8. Courtesy of SS(E)L.

because the basal Zechstein events beneath it remain strong and continuous and show no velocity anomaly effects, as noted previously. Similarly, cuesta features are not involved, since the Pre-Permian stratal events are flat-lying. With estimated thicknesses of over 100 m, the features under investigation exceed known Rotliegendes dune hights by a factor of x 2 (see Glennie, 1984). Seismic velocities are probably too low for igneous intrusions to be involved, and characteristic reflection patterns of volcanic features are absent. The problem in this area is, therefore, to distinguish fault effects from those produced by possible carbonate buildup features or pods of anhydrite.

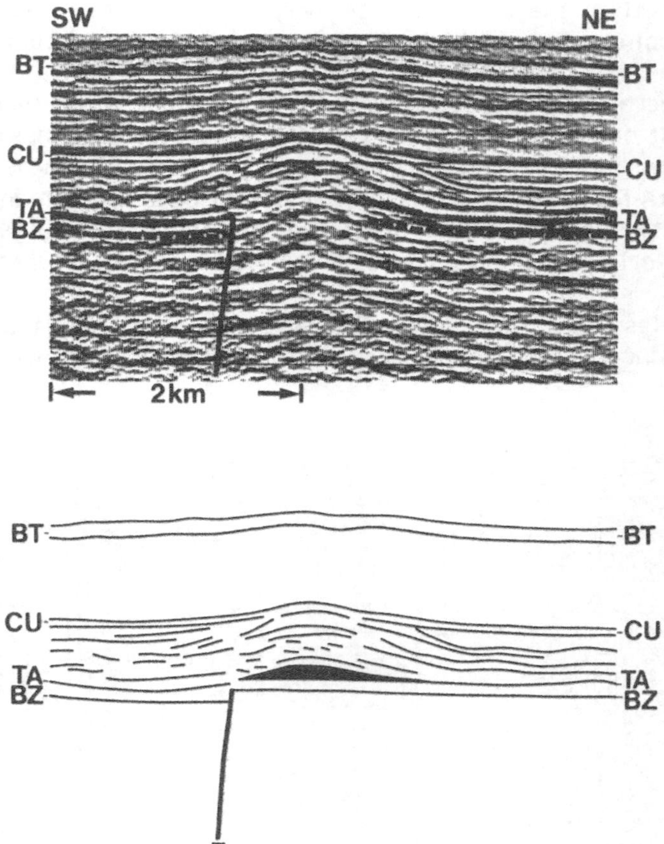

Fig. 11. Seismic section with sketch interpretation below. Unlike the previous example, the events in the TA-CU sequence here show definite signs of antiform drape over some positive feature on the upthrow side of the fault 'F' , rather than an onlapping relationship or monoclinal drape due to the presence of a simple fault scarp. Key as for Fig. 9. Location is Zone 2 in Fig. 5. Courtesy of SS(E)L.

It is known from drilling results that Carboniferous horst blocks project upwards through or into the Zechstein sequence in some locations, and an example of this effect is shown in Fig. 10 (location 'G' in Fig. 8). Well 38/16-1 was drilled into this horst a few kilometres away from the seismic profile seen, and after penetrating 900 feet of Zechstein anhydrite-dolomite-anhydrite section (about 100 ms two way time, as indicated by the arrows in Fig. 10), entered Lower Carboniferous (Visean) strata. The Zechstein sequence flanking the horst is indicated as BZ-TA, with the Late Cimmerian unconformity at CU and Base Tertiary at BT. The configuration of events in the TA-CU sequence suggests onlap against the fault scarps of the horst rather than drape over it.

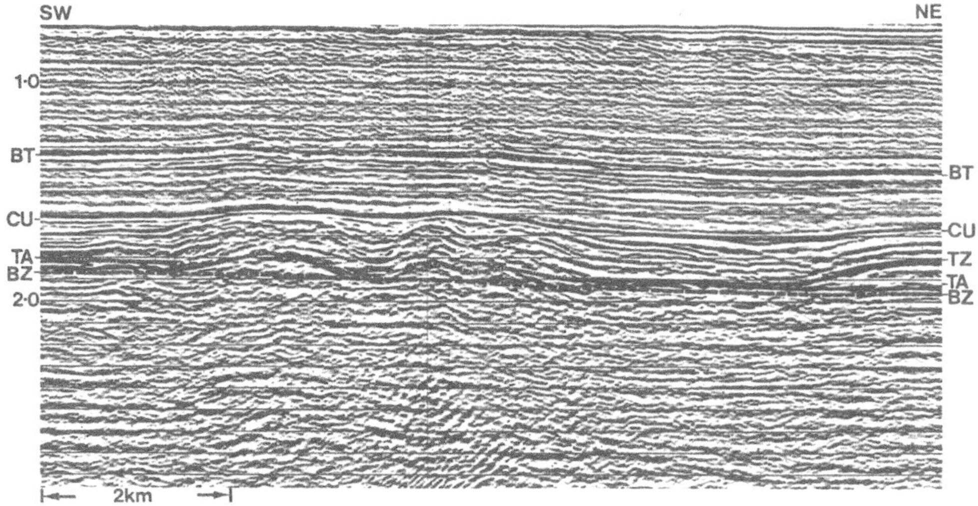

Fig. 12. Seismic example from Zone 1 of Fig. 5. Key as for Fig. 9. Courtesy of SS(E)L.

An interesting comparison can be made between this example of a proved horst feature, and the feature in Zone 2 of Fig. 5. An enlarged view of this is shown in Fig. 11, with the seismic section above, and a sketch interpretation below. Here the fault is seen affecting Base Zechstein on the southwest side of the feature. It is not possible from these data to state definitely that there is no fault at the east end of the feature, but it seems unlikely. In this case, the antiform drape feature seen in the TA-CU sequence indicates that there is a convex-upward structure present at basal Zechstein level which is initiating the observed subsidence drape in the Mesozoic sequence. This follows from the fact that if drape over a simple fault scarp were present, it would produce a monoclinal rather than an antiform structure. The interpretation shown gives a suggestion of the shape and size of a feature (solid black) which might produce the observed effects, with the BZ-TA interval "restored" - i.e. with any velocity "pull-up" removed.

The convex-upward structure shown has a time amplitude of about 40 ms (TWT), and assuming an appropriate velocity - say 5500 m/s - this results in a structure with a maximum thickness of about 110 metres. A carbonate buildup, or anhydrite pod, of this size and order of velocity, seems credible.

Turning the attention now to the features in Zone 1 of Fig. 5, an enlarged view of these is seen in Fig. 12. In this figure, the salt dissolution slope discussed earlier is seen at the NE end of the section, with the top Zechstein salt at TZ. Elsewhere on the section, the basal sandwich of Z1 and base Z2 anhydrites and carbonates is indicated by BZ-TA, with the remaining legend as for the previous figures. There is no good evidence for faulting related to the three antiform drape structures seen in the central part of the section, and the collapsed Mesozoic between the top of the Zechstein and CU is showing clear evidence of subsidence drape over convex-upward positive features at basal Zechstein level. There is some indication - particularly at the southwest end of the section - that the TA (top Zechstein here) event is onlapping the left-hand feature, which suggests that the initiating feature arises from near the base of the Z1, or is at any rate older than the base of the local salt. Subtracting the time-thickness of the BZ-TA interval from the total thickness of the larger feature at the southwest end of the section would leave an initiating structure of some 40 ms, while the thickness of the two smaller features to the right would be less - probably 20 ms (55 m) to 30 ms (80 m) at the velocity used previously. These figures would increase somewhat if the initiating features are indeed of Z1 rather than Z2 age.

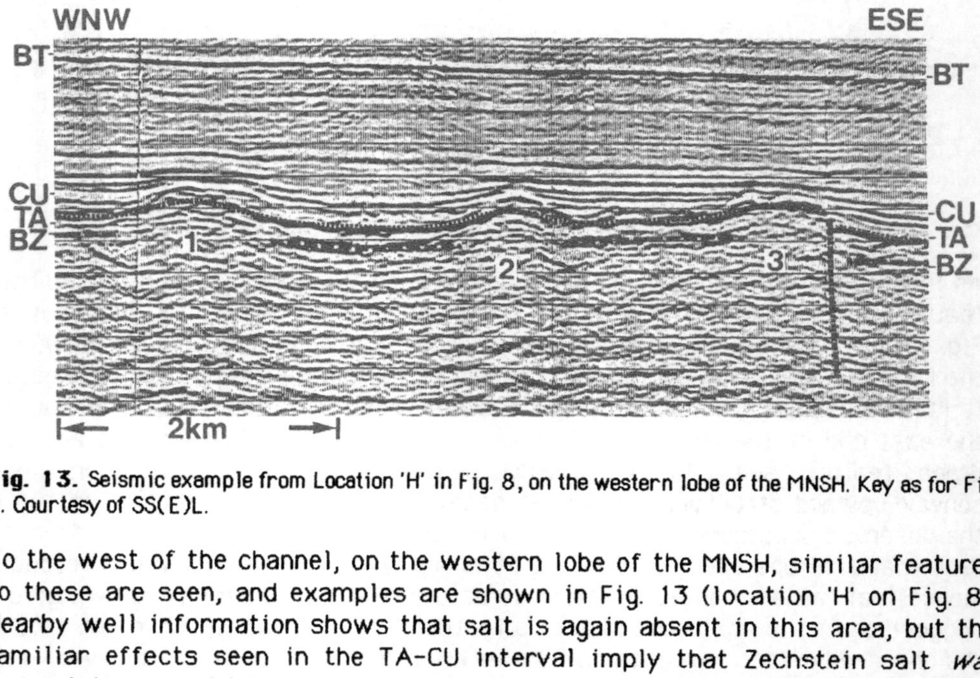

Fig. 13. Seismic example from Location 'H' in Fig. 8, on the western lobe of the MNSH. Key as for Fig. 9. Courtesy of SS(E)L.

To the west of the channel, on the western lobe of the MNSH, similar features to these are seen, and examples are shown in Fig. 13 (location 'H' on Fig. 8). Nearby well information shows that salt is again absent in this area, but the familiar effects seen in the TA-CU interval imply that Zechstein salt *was* present here, and has been removed by dissolution, leading to the observed collapse features. In this instance, the BZ-TA sequence is somewhat thicker (shelf type) than that seen in the previous examples, but the collapse effects and the drape over features numbered 1 and 2 are clear. On the other hand, only onlap appears to be present against feature 3, and this, together with the evidence of faulting, suggests that it is a small horst.

The stratigraphic situation, shape and size of these fatures is similar to those of the features on the eastern lobe of the MNSH discussed earlier. Although the evidence is not very clear, features 1 and 2 are almost certainly situated on the upthrow side of small faults, and the situation, of at least some of the features illustrated, on locally high points (in this case, fault-related), is of interest.

In the examples shown, the seismic line spacing in the areas involved is too wide to enable precise delineation of the plan shape of the features seen in cross section. However, effects observed on the adjacent lines in several cases suggest that the features seen are elongate in form, and are probably following the contours of the MNSH locally. Their nature cannot be determined with certainty, in the absence of drilling evidence; however, it is possible to make suggestions of a hypothetical nature by taking into account the facies displayed by local wells, and also by comparison with features in other areas which are located in analogous situations - i.e. on shelves or shallow water areas fringing evaporite basins.

Facies and thickness changes

The two wells intersected by the seismic profile BC in Fig. 5 display contrasting types of Zechstein sequence, as interpreted from the wireline logs and cuttings. They represent respectively "shelf" and "basin" facies. The crestal well on the left, 38/22-1, has the main distinguishing features of sequences occupying "shelf" positions in other parts of the Southern Zechstein Basin. That is, a moderately thick (about 125 m) anhydrite attributable to Z1 (the Werraanhydrit) provides a plinth on which a thinner (45 m) oolitic and pisolitic dolomite resembling the Z2 Hauptdolomit was deposited, evidently in predominantly shallower water. This is overlain by about 140 m of anhydrite (presumed shelf equivalent of the Z2 Basalanhydrit), followed by a thin dolomite which may represent the Plattendolomit, the sequence being completed by about 50 m of anhydrite of Z3 and/or later cycles. Broadly similar "shelf" sequences are shown by 38/16-1 and 38/25-1, also near the crest of the MNSH. Whether the evaporites formerly contained intervals of halite is not determinable from the well data, in the absence of cores which might have confirmed the presence or absence of characteristic collapse features.

Despite uncertainties about the interpretation of the later evaporite cycles, as already discussed, the flank well on the right of Fig. 5, 38/18-1, has an essentially "basinal" character, in that thick evaporites dominated by halite are underlain by a thin carbonate/anhydrite sandwich totalling only 61 m thick. In it, a typical basal Kupferschiefer, thin argillaceous Z1 carbonate, dolomitic Werraanhydrit, and brown laminated basin-facies Z2 carbonate can be recognised.

Thus between 38/22-1 and 38/18-1, before any leaching of evaporites, we would expect to have found "slope" facies in which the Z2 carbonate tapered northwards above a more rapidly thinning wedge of Z1 anhydrite, the overall loss in section being made good mainly by an increasing thickness of halite in Z2 and later cycles.

Oolite barrier shoals or algal buildups would be a possibility in the Z2 carbonates at the break in slope at the edge of the shelf. Z1 bryozoan-algal patch reefs could have been present, encased in the Werraanhydrit, though there is no special reason to expect them in the same part of the profile unless initiated over the upthrown side of basement faults. The Werraanhydrit itself, however, may have diminished in thickness irregularly rather than evenly, and instability at a steep shelf-edge may have led to gravity collapse of this unit, with or without the overlying Hauptdolomit, leading to a markedly hummocky topography. This would be particularly likely if the shelf-edge was in fact controlled by deep-seated faulting, as already considered. The irregularities, of whatever origin, would have become burried and concealed by later evaporite déposition, the top of the Zechstein being essentially horizontal at that time.

Later dissolution of halite would have led to subsidence of the overburden across an uneven substrate which would have arisen either through the depositional buildup of carbonates or anhydrite, or by collapse structures in these bodies. Equally, however, the observed relief could have been created by the irregular dissolution of the thick shelf anhydrite itself.

Comparison with other areas

In the Polish Zechstein Basin, algal carbonate barrier, pinnacle and atoll reefs have been identified in seismic data (Antonowicz & Knieszner, 1981, 1984) in the Z2 cycle, and have been confirmed by drilling. In the areas dealt with by the latter cited papers, the Zechstein is considerably thicker (by a factor of two to three times) than it is in the locations under discussion in this work; however, there are some similarities between the seismic examples from the MNSH area and those from the Polish Basin, where the buildup features are normally on localised highs.

In the German Zechstein, numerous examples of bryozoan-algal reefs, described by Smith (1985) as "complex mound-shaped bodies", are known both from surface exposures and from drilling results. Examples in Thuringia can be 3 km long by 1.5 km wide, and up to 60 m thick, although a subsurface example north of Münster reaches a thickness of over 100 m (Füchtbauer, 1980).

In Britain, Smith (1981a) has described a stromatolitic biostrome which overlies a barrier-type Z1 reef 100 m thick which runs parallel to, and just inland of, the coast of County Durham, northeast England, and also an area of bryozoan-algal patch reefs on the landward side of the barrier (Smith, 1981b). The stromatolitic biostrome *may* be of Z2 age, but is thought more likely to belong to Z1.

An interesting example for comparison also exists in the Otto Fiord Formation (of Late Mississippian to Early Pennsylvanian age) in the Sverdrup Basin of the Canadian Arctic Archipelago. The Formation is a major evaporite deposit composed of up to 600 m of rhythmically-alternating limestone and anhydrite, with some interbedded sandstones near the top. The area has been studied by Davies (1977), and his paper should be referred to for details of the complex depositional history of the area, with rhythmic changes from marine to restricted, hypersaline conditions (although no halite is present). Germane to the present discussion is Davies' observation that "... at several localities, limestone mounds up to 30 m thick and 350 m long are built on erosional "plinths" of anhydrite. The mounds have a crinoid-rich base, a main ...algal... facies of hypersaline aspect, and several marine limestones capping beds. The marine limestone phases of the mounds have thin marine limestone equivalents in the intermound setting, separated from each other by units of anhydrite.". Davies interprets the mounds as forming a record of several alternating cycles of marine limestone deposition and algal/anhydrite deposition under hypersaline conditions imposed by evaporitic drawdown.

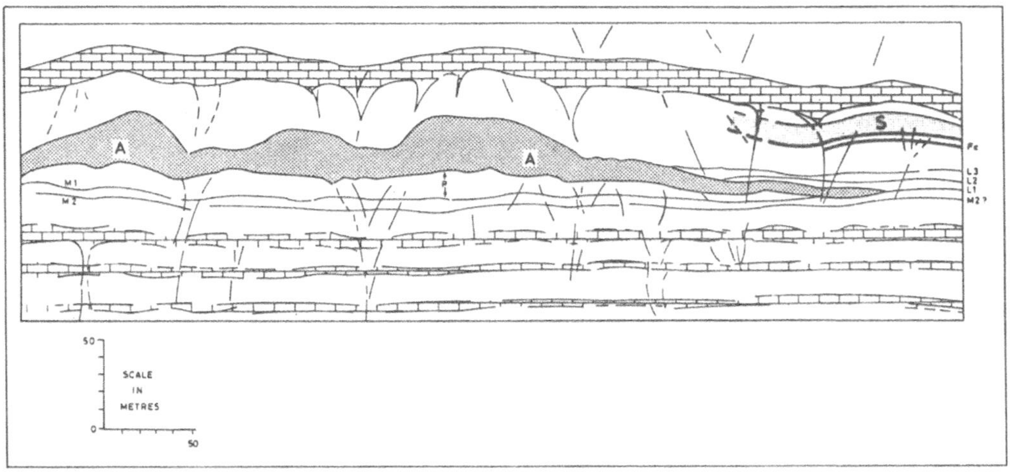

Fig. 14. Sketch from a photograph showing an algal mound (shaded A - A) in the Otto Fiord Formation, Sverdrup Basin. From Davies, 1977, courtesy of the American Association of Petroleum Geologists.

Although it may not be possible to suggest a close parallel between the depositional history of this area and the North Sea Zechstein, the morphological aspects of the Canadian algal mounds are of some interest, and their size may not be vastly different from the smaller features of those already discussed. In Fig. 14, a sketch from a photograph in Davies' paper is shown, with an algal mound shaded at A - A. The mound is up to 30 m thick, and rests on a plinth of anhydrite (P) about 14 m thick. Beneath it is seen a series of cyclic units of limestones (brick symbol) alternating with anhydrite after gypsum (shown blank). Thin limestones below (M1) and above (L1, L2, L3) the mound can be seen truncating against it. A sandstone unit (S) and a haematite layer (Fe) which are present in the intermound areas are also marked. The irregular series of antiform shapes forming the top of the mound should be noted and compared with, e.g., Fig. 12. In Fig. 15 is shown a schematic diagram of the structure of an algal mound of this type, showing limestone (brick symbol), anhydrite (chevron symbol), haematite layer (solid black), and sandstone (dot dashed) encasing the main algal facies of the mound. Although in this case no dissolution of salt, and overburden collapse is involved, the morphological similarities with the North Sea features is of interest.

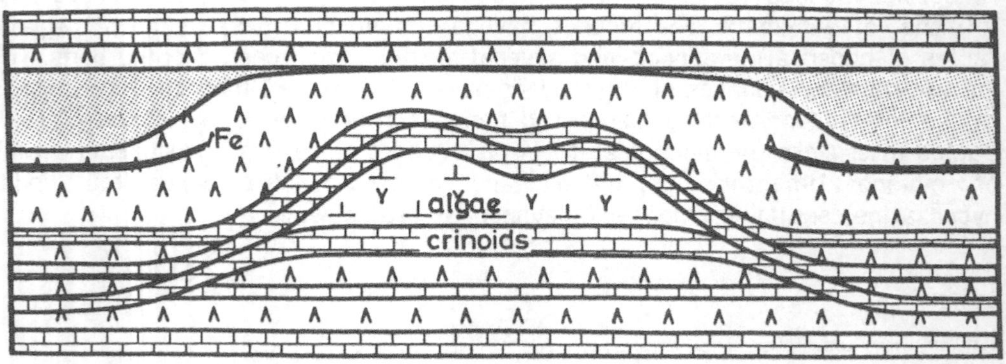

Fig. 15. Schematic of the structure of an algal mound of the type shown in the previous figure. From Davies, 1977, courtesy of the American Association of Petroleum Geologists.

Off-shelf features

Certain other structures of a quite different nature have been observed in seismic data in a fringing position just off the MNSH shelf area and down the slope into the Northern sub-basin. Two examples in adjacent locations on the same seismic line are seen in Fig. 16 (location 'I' in Fig. 8) and Fig. 17 (location 'J' in Fig. 8), with the features marked 'R'. From indications on adjacent lines, these ridge-like thickenings in the basal Zechstein carbonate/anhydrite units are again elongate parallel to the contours of the basinal slope, and could be 8 km or more long (in the direction perpendicular to the sections). The apparent basinal dip to the east-northeast is seen at Base Zechstein level (B) in both examples. Internal reflection events within these features seem to show that the main thickening occurs in the upper part of the interval, immediately beneath the Z2 halite which forms the major component of the Zechstein sequence (B-T). These features could be elongate fringing reefs, shoals, or possibly abrupt thickenings of the Z2 Basalanhydrit; features of the latter type are reported in the paper by Antonowicz & Knieszner (1981) cited previously, as occurring in the Polish area on the basinal slopes, forming thick lenses elongate parallel to the slope contours, and up to 100 m thick. Based on seismic velocities expected in this zone, in Figs .16 and 17 the thickness increase of the features 'R' could be anything from 100 to 200 m. Another example of somewhat different form, but at the same stratigraphic level, is seen in Fig. 18 (location 'K' in Fig. 8). Instead of more or less continuous stratal events running through this feature, as seen in the two previous examples, the Fig. 18 feature seems to have a core structure, which has suffucient acoustic impedance contrast with the surrounding material to cause it to show a closed lenticular body in cross section. This gives it a more reef-like appearance than the features in Figs. 16 and 17. It is possible that it has formed on the upthrow side of a small fault, as indicated.

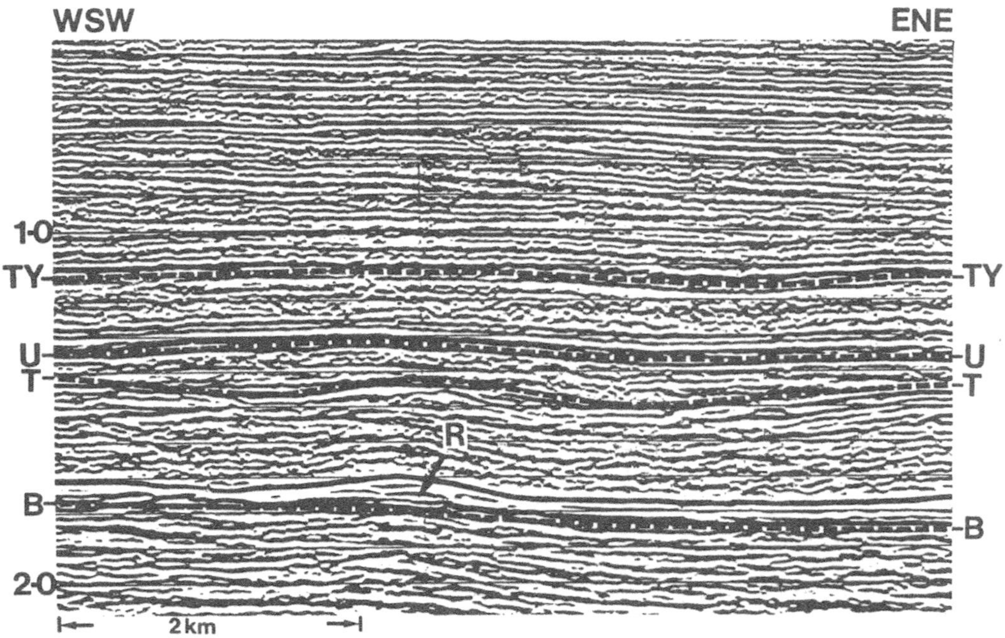

Fig. 16. Seismic example from location 'I' in Fig. 8, showing an off-shelf feature 'R'. B=Base Zechstein; T=Top Zechstein; U=Late Cimmerian Unconformity; TY=Base Tertiary. Courtesy of SS(E)L.

The difference in appearance between this, and the features in the two previous examples may be more apparent than real. One thing which all the features have in common is that there is no sign of underlying velocity anomaly related to any of them. This implies that despite the thickness increase, there is no time differential between seismic signal passing through them, and through adjacent locations. This means that the seismic velocity in the material of which they are formed differs little from the velocity in the overlying salt (about 4500 m/s). Such an observation does not tend to support the view that these features are reefs, unless anhydrite and dense carbonates are balanced by an appropriate interval of relatively high porosity - an interesting thought for hydrocarbon explorationists.

There is a possibility that, with the difference in direction of the seismic line in the Fig. 18 example (which is orthogonal to the line direction of Figs. 16 and

17), the core-like structure in the Fig. 18 feature is being seen in cross-section, whilst similar structures are being seen in longitudinal section in Figs. 16 and 17. If this were the case, it would mean that these features are only slightly elongate, and this slight elongation is perpendicular to the basin slope contours (i.e. in a direction *down* the basinal slope) which gives a very different situation to that suggested previously - i.e. elongate parellel to the basin slope contours, and 8 km or more long. It is not possible at present to explore this possibility further, due to the low density of seismic lines in the area.

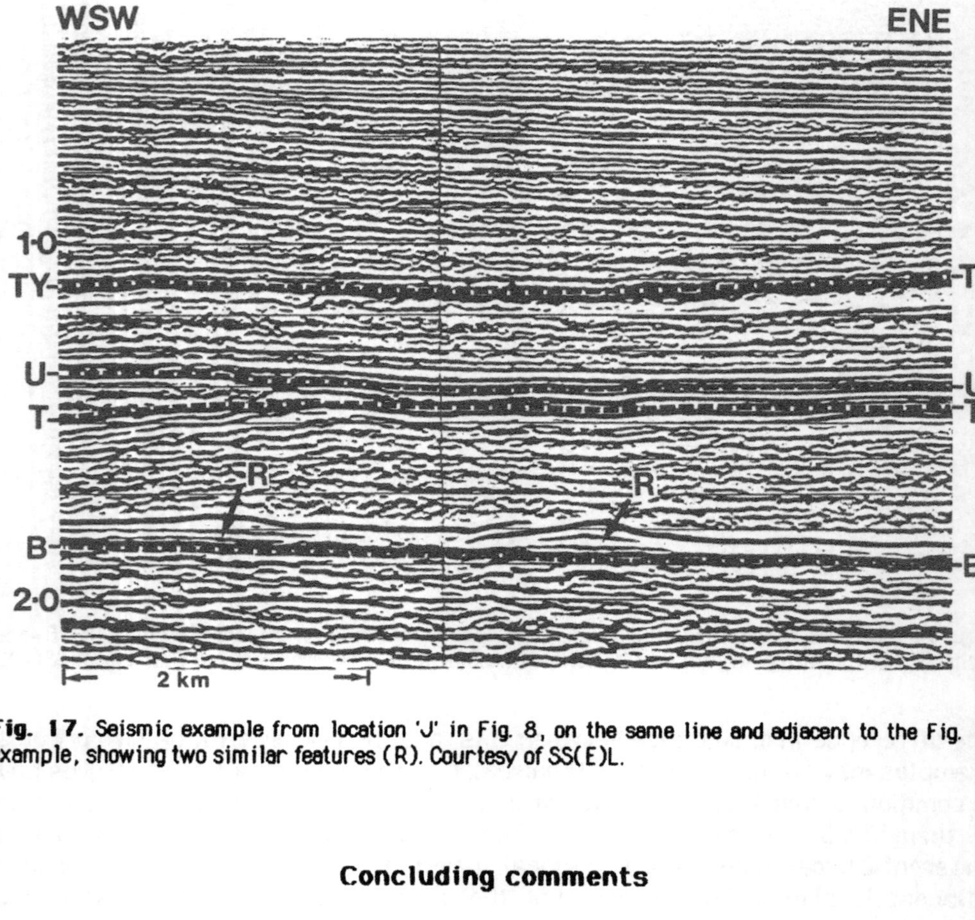

Fig. 17. Seismic example from location 'J' in Fig. 8, on the same line and adjacent to the Fig. 16 example, showing two similar features (R). Courtesy of SS(E)L.

Concluding comments

It is possible to explain many of the structural features in the Zechstein sequence over part of the Mid North Sea High by reference to a rather simple model. This invokes edge-(and possibly areal-) dissolution of Zechstein salt in the crestal region of the High, with intraformational undersaturated water in an aquifer either immediately overlying, or beneath, the Zechstein salt having dissolved and removed it. The edge of the remnant salt has retreated down both northern and southern flanks of the MNSH, to locations where it now terminates in salt-edge dissolution slopes. This process has led to subsidence of the overlying Mesozoic strata, mainly during Late Cimmerian erosion, but to a lesser extent later. An apparently isolated remnant patch of salt on the north flank of the MNSH may possibly be connected to an embayment in the south flank, in which case yet another breach in the High was present during the period of Zechstein deposition. The results of the study suggest that at least a thin development of the latest Zechstein Z2 salt, and/or possibly that of younger cycles, may have been deposited across the crest of the MNSH. If this were not so, then certainly only a very narrow belt, a few kilometres wide, along the crest of the High remained free of salt before the commencement of the postulated dissolution.

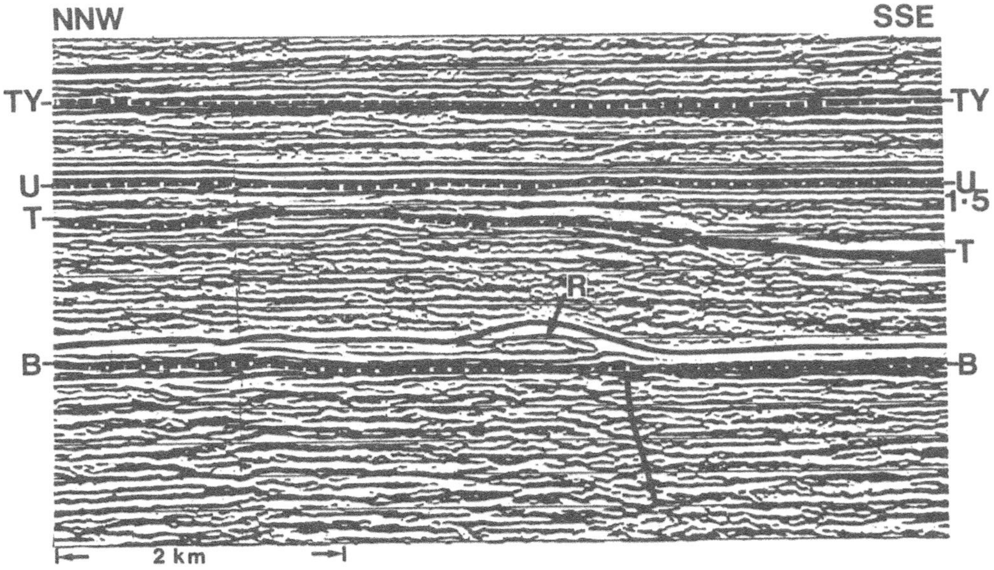

Fig. 18. Seismic example from location 'K' in Fig. 8. The thickened feature (R) seen in cross section appears to differ in internal structure to those in Figs. 16 and 17. Key as for Fig. 16. Courtesy of SS(E)L.

Subsidence drape of the collapsed Mesozoic units reveals the presence at basal Zechstein level of positive structures previously encased in salt. These structures *may* be of Z2 age, but are more likely to be Z1. Some of these are related to old faults, but in this case, velocity "pull-up" anomalies cannot be expected. From evidence of Carboniferous strata penetrated beneath the Zechstein in Well 38/18-1, and of Devonian rocks penetrated beneath the Zechstein in Well 38/29-1, the seismic interval velocities through these formations are about 3800 m/s. This is substantially lower than interval velocities in the Zechstein salt and the collapsed Mesozoic clastics (about 4500 m/s, as demonstrated earlier).

Other structures discussed may be carbonate buildups, from their shape, size, location, and seismic velocities. If so, they could represent fringing reefs of bryozoan-algal type which developed in the shallow-water zones along the contours of the MNSH in early Zechstein (Z1 or Z2) times. Alternatively, they could be composed of pods of the Z1 Werraanhydrit, either representing depositional lenses adjacent to the main, thick, shelf unit, or dissolution remnants of the latter.

Some other features observed further down the basinal slope seem to be elongate lenticular thickenings of basal Z2 evaporite or porous carbonate units, again (probably) following the structural contours of the High. These are present on the north flank of the MNSH, but have not, so far, been identified on the southern flank. Because of their broadly similar sizes, shapes, and distribution patterns, there is a natural temptation to regard all these features as being of similar origin. This may be unwise. Only when a number of them have been drilled will their true nature be clarified.

Acknowledgements

Thanks are expressed to the Directors of Seismograph Service (England) Limited for permission to submit this work for publication, and to use the high-quality seismic data examples. The authors are also grateful to Dr. Denys B. Smith for reading the manuscript and making many constructive suggestions.

References

Anderson, R. Y. & Kirkland, D. W., 1980. Dissolution of salt deposits by brine density flow. *Geology* 8: 66-69.

Antonowicz, L. & Kniesznar, L., 1981. Reef zones of the Main Dolomite set out on the basis of palaeogeomorphologic analysis and the results of modern seismic techniques. *Proc. Intern. Symp. Central Europ. Permian* 356-368. Warsaw.

Antonowicz, L. & Kniesznar, L., 1984. Zechstein reefs of the Main Dolomite in Poland, and their seismic recognition. *Acta Geol. Pol.*, 34: 81-94.

Davies, G.R., 1977. Carbonate-anhydrite facies relationships, Otto Fiord Formation (Mississippian-Pennsylvanian), Canadian Arctic Archipelago. *Am. Assoc. Petroleum Geologists, Studies in Geology* 5: 145-167.

Füchtbauer, H., 1980. Composition and diagenesis of a stromatolitic bryozoan bioherm in the Zechstein 1 (northwestern Germany). *Contr. Sedimentology* 9: 233-251.

Jenyon, M.K., 1986. *Salt Tectonics*. Elsevier Applied Science Publishers, 191 pp. Barking, U.K.

Jenyon, M.K., Cresswell, P.M. & Taylor, J.C.M., 1984. The nature of the connection between the Northern and Southern Zechstein Basins across the Mid North Sea High. *Marine & Petrol. Geol.*, 1: 355-363.

Smith, D.B., 1981a. The Magnesian Limestone (Upper Permian) reef complex of northeastern England. *SEPM Spec. Publ.*, 30: 161-186.

Smith, D.B., 1981b. Bryozoan-algal patch-reefs in the Upper Permian Lower Magnesian Limestone of Yorkshire, northeastern England. *SEPM Spec. Publ.*, 30: 187-202.

Smith, D.B., 1985. Zechstein reefs and associated facies. In: Taylor, J.C.M. (Ed.), *The Role of Evaporites in Hydrocarbon Exploration. Course Notes* 39, Joint Association for Petroleum Exploration Courses (U.K.), London.

Taylor, J.C.M., 1984. Upper Permian-Zechstein. In: Glennie, K.W. (Ed.) *Introduction to the Petroleum Geology of the North Sea*, 61-83. Blackwell, Oxford.

REGIONAL SALT MOVEMENT EFFECTS IN THE ENGLISH SOUTHERN ZECHSTEIN BASIN

Malcolm K. Jenyon
Seismograph Service (England) Ltd.
Holwood, Keston, Kent, U.K.

Abstract: The halokinetic model of basinal salt movement described by Trusheim (1960) for the North German Zechstein Basin, with large salt-wall diapirs occurring in the basin centre, and diapirs, pillows, and swells of decreasing size evolving towards the basin margins, has been used to explain salt movements in many other basins. The concept of halokinesis implies vertically upward salt movements in response to the isostatic readjustments taking place in the salt/overburden system. This model does not fit the situation observed in the Southern Zechstein sub-basin of the British North Sea at all. In the basin centre, where the thickest salt is present, there are no diapirs whatsoever, but only a few isolated salt swells of broad, shallow form. The very few salt-wall diapirs which occur in the general area are invariably associated with major regional faults in directions other than that of the main trend of the majority of salt pillows and swells. On the other hand, around the margins of the sub-basin, the seismic data reveal basin-edge diapiric effects which appear to be the remnants of major piercement, extrusion and subsidence events that took place largely during the Late Cimmerian erosional phase. These, together with other effects to be discussed, indicate that movement of the Zechstein salt in this area took the form of lateral salt flow centrifugally from the basin centre towards the margins.

Introduction

The Southern Zechstein sub-basin is bounded, except to the east, by old, stable highs. To the north is the Mid North Sea High; to the south the London-Brabant Ridge; whilst to the west, the Pennine High of mainland Britain effectively forms the boundary. Eastwards, the basin extends through North Sea waters into Germany, and beyond into Poland and the USSR. The basin margins have been well studied in the eastern areas where they lie on land, but have been less well investigated at the western end since here the transition from basin to marginal zones lies mainly offshore.

Lecture Notes in Earth Sciences, Vol. 10
T. M. Peryt (Ed.), The Zechstein Facies in Europe

The north-western margin of the sub-basin has been particularly neglected until fairly recently. Even during the period of extensive drilling operations in the Southern Basin in the late 1960's and early 1970's this area received scant attention; at that time, rightly or wrongly, the Carboniferous section which represents the main focus of interest as a gas source in the region was considered to be too shallow to have evolved any hydrocarbons. Only in recent years, with the discovery of gas at relatively shallow depths within the Upper Carboniferous in the adjacent areas, has interest revived. This interest has taken the form of rather extensive reconnaissance seismic surveys offshore, producing better seismic resolution than in the earlier period of exploration, and an increasing number of wells - although these have so far been confined still to the locally deeper parts of the basin. The southern basin margin is rather better known, due to the proximity of the main Gas Fields area; even here, however, there is little public-domain literature which includes discussion of regional salt movement.

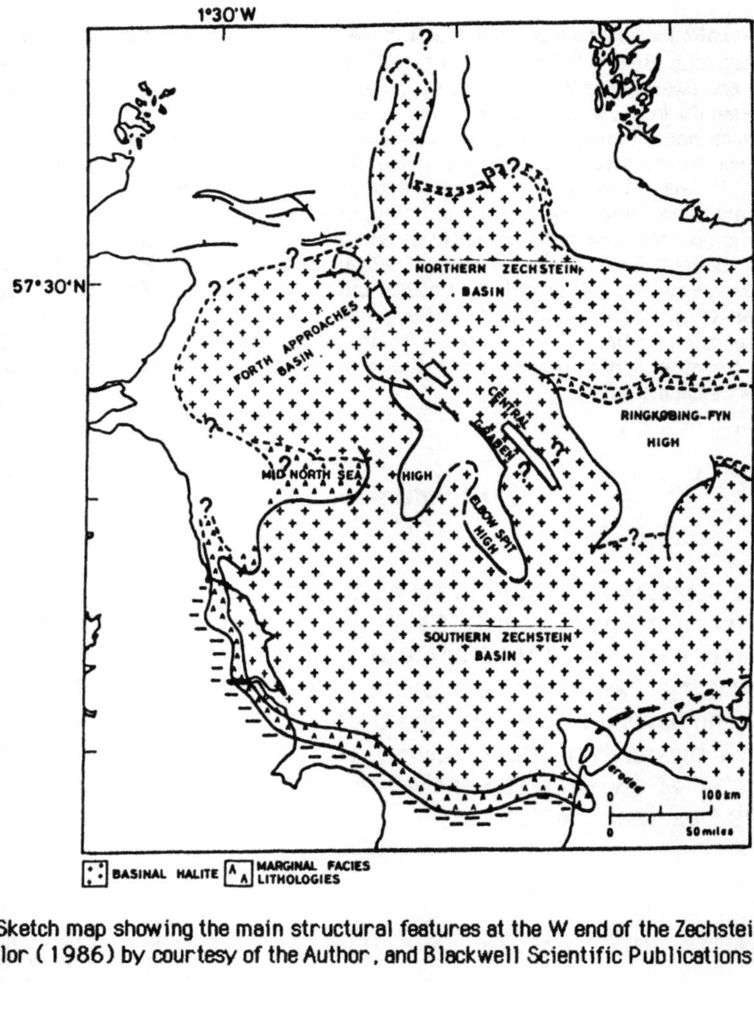

Fig. 1. Sketch map showing the main structural features at the W end of the Zechstein basin. Modified from Taylor (1986) by courtesy of the Author, and Blackwell Scientific Publications, U.K.

Seismic surveys of recent date have provided the opportunity for more detailed investigation of the sub-basin margins, and some results of these studies are presented here in a preliminary way. Suggestions are made regarding the palaeophysiography of the basin, the movement of the Zechstein salt, and the effects of the latter on the oberburden.

The local basinal configuration

The sketch map of Fig. 1 shows in simplified form the main structural features of the western end of the Zechstein Basin, with indications of the distribution of halite and marginal facies.

In Fig. 2A another sketch map shows the northern and NW part of the study area in more detail; the zoning by shaded boundaries seen in the sketch will be discussed at a later stage. The Mid North Sea High (MNSH) is traversed by a channel infilled with Z2 (Stassfurt) Halite; this channel has been the subject of a separate study (Jenyon *et al.*, 1984). Briefly, it is controlled by an old fault of early- or pre-Zechstein age which follows the arcuate, heavy-dashed line on the western side. There is no major fault of any kind on the eastern side. Both western and eastern sides of the channel are marked by edge-dissolution slopes (terminating the Z2 salt) dipping westwards and eastwards respectively. These are belived to be due to the flow of undersaturated water intraformationally down towards the channel position from higher areas of the MNSH on each side of the channel. The arcuate shepes in plan of these two dissolution slopes, indicated approximately by the heavy dashed lines, form a kind of "tide-mark" which indicates either subsidence of the channel area, or uplift of the two segments of the MNSH related to postulated deep-seated granite batholith emplacement within the MNSH on both sides of the channel (see also Donato *et al.*, 1973). Also in Fig. 2A, the approximate location of a fringing reef is shown on land in NE England, discussed in Smith (1981). Fig. 2B is a sketch map of part of the southern margin of the sub-basin, showing the position of a seismic example to be considered later.

Seismic zonation of the Zechstein

In the part of the area shown in Fig. 2B, the Zechstein interval is resting on relatively thin (Lower Permian) Rotliegendes rocks which are of mainly argillaceous lithology. The Rotliegendes thins to northwards, coming to a feather-edge at about 55°N, and is absent over the MNSH. There is a Rotliegendes depocentre to the ESE of this area, consisting mainly of some 300 meters of claystones.

In the Basinal Facies Zone, the total Zechstein interval includes a thickness, in this area, of some 800 metres of Z2, Z3, and Z4 halites, with the Z2 Stassfurt

Halite making up the bulk of this. The "Intermediate Zone" is not a separate facies; the Zechstein is still in basinal facies in this zone, although the total salt thickness is less, averaging some 450 to 500 metres. The reason for the demarcation of this Zone lies in the nature of its boundary with the Basinal

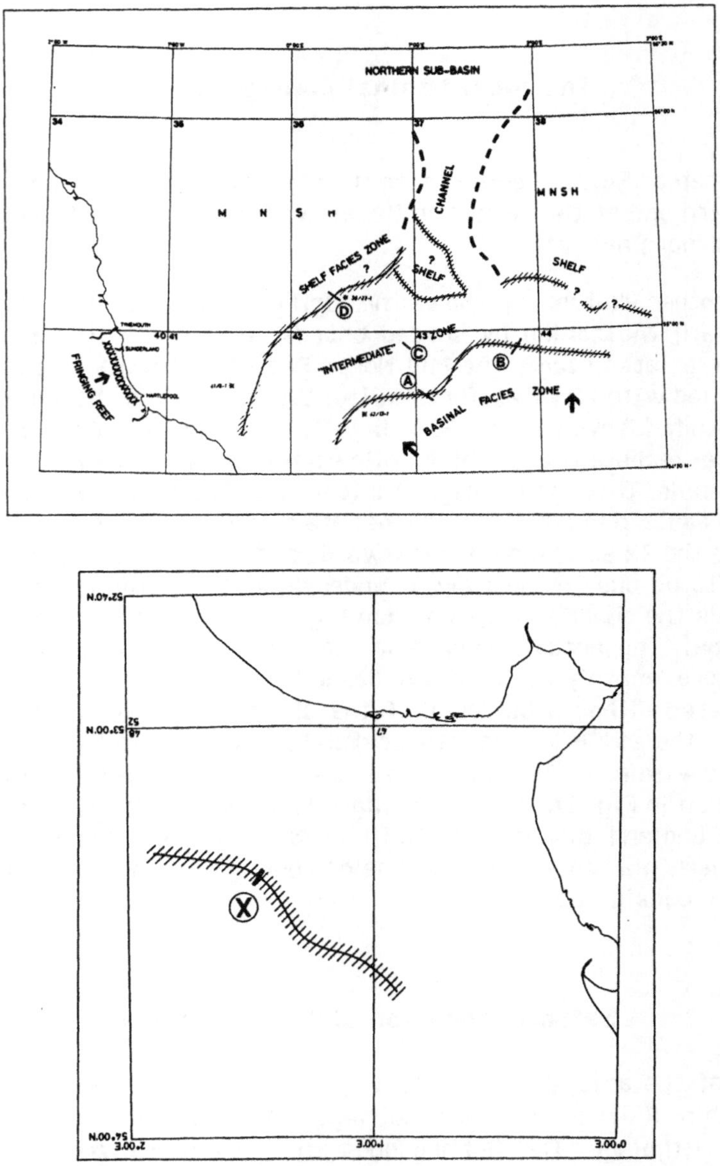

Fig. 2. (A) Sketch map showing features at N and NW margins of the western end of the southern Zechstein sub-basin. Hatched lines are the boundaries of zones showing different seismic characteristics. Heavy dashed lines show approximate boundaries of a channel of thick Z2 salt cutting across the MNSH. Large arrows in basinal zone indicate postulated directions of lateral, regional salt movement. Locations A, B, C and D refer to Figs. 3, 4, 5 and 6 respectively. **(B)** Sketch map showing part of the southern shelf/basin boundary (hatched line) of the southern Zechstein sub-basin. Fig. 8 at X.

Facies Zone. This boundary is marked by diapiric structures, although in many locations the original diapiric salt has been largely lost by extrusion at the depositional/erosional surface, and the positions of the former diapirs can be inferred only from resulting collapse features and other effects.

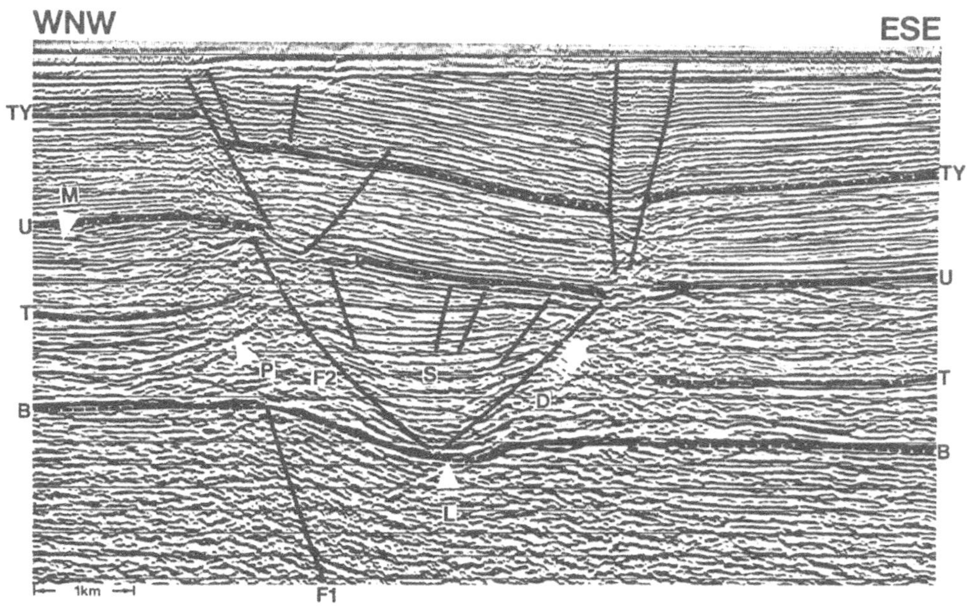

Fig. 3. Migrated seismic section from location A in Fig. 2A, which shows evidence of basin-edge diapiric effects. See text for explanation. Courtesy of Seismograph Service (England) Limited.

A portion of migrated seismic line, Fig. 3, from the location marked A on the sketch map in Fig. 2A, and running across the western boundary between the two zones, exhibits features of what has been referred to as "basin-edge diapirism". The suggested mechanisms associated with this phenomenon have been discussed in detail elsewhere (see Jenyon, 1985), but a brief description based on this figure is as follows.

Events B and T mark the base and top of the Zechstein interval, and U is the Late Cimmerian Unconformity, whilst TY is the Base Tertiary. F_1 marks the Intermediate Zone/Basinal Zone boundary fault, L is a velocity pull-down effect due to salt being thin or absent, and D is the original route of salt which moved upwards through the overburden into a primary edge-diapir and also extruded at the erosional surface during the Late Cimmerian phase. The wedge-shaped mass at S is a subsided prism of Mesozoic overburden which slid down the fault plane of F_2, a low-angled normal fault initiated by a tendency to rotation of the overburden during early salt withdrawal from beneath the prism S into the primary diapir via D. The primary diapir was removed by erosion or dissolution at the surface during the U period. P is the competent Plattendolomit event, and M indicates Mesozoic units showing subsidence/uplift effects.

The fault F_1 marks the boundary between the two zones, the Intermediate Zone being to the WNW and the Basinal Zone to the ESE. The Basinal Zechstein to the ESE is relatively thin because of the major salt withdrawal and subsidence which have taken place; it thickens rapidly off section to the ESE, and the commencement of this thickening can be seen. The amount of subsidence caused by the major withdrawal of salt up D can be gauged by the relative levels of horizons T, U, and TY across the feature.

The configuration of P to the left of fault F_2 should be noted. P is the seismic response to the competent Plattendolomit band of anhydrite/dolomite within the total salt interval, which separates the Z2 salt below from the Z3 salt above. The steep uplift of this event into the fault zone may be partly due to "splay" resulting from the edgeward movement of the basinal salt being dammed at this location - a matter which will be discussed more fully in relation to the next example - and possibly also partly due to differential overburden load pressure across the fault. The latter has a large downthrow to the ESE in this location, and the load differential across the fault may have led to salt flow from the downthrow side across to the upthrow side (see Jenyon, 1986a). There also seems to have been some limited flow of salt from the Intermediate Zone into the F_2 fault zone, as suggested by the subsidence effects at and adjacent to M.

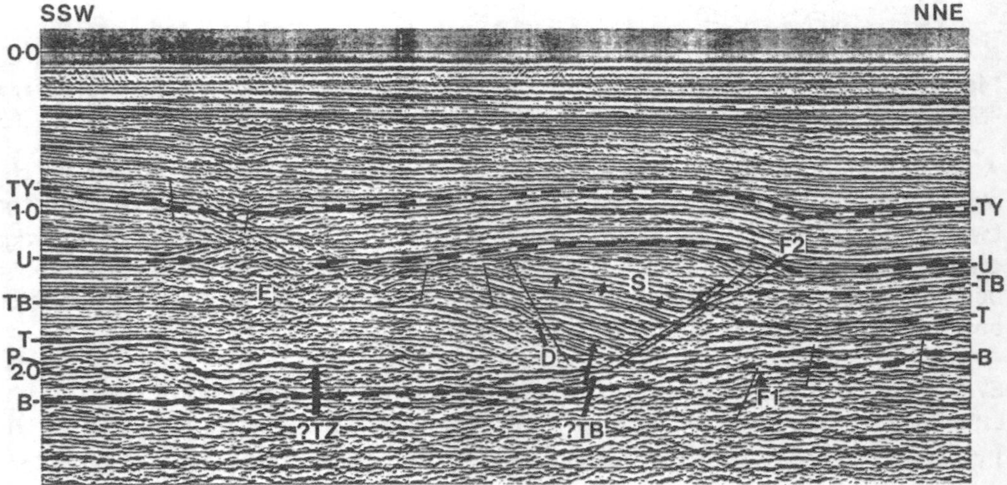

Fig. 4. Migrated seismic section from location B in Fig. 2A. Effects of basin-edge diapirism. See text for explanation. Courtesy of Seismograph Service (England) Limited.

The faulting above the level of U, of listric normal and antithetic type, is of much later date than that below the unconformity, having been caused by movement of salt and overburden subsidence during the later Mesozoic and the Tertiary.

Similar effects to those outlined above, can be traced all the way around the zone boundary in this area. At location B in the sketch map of Fig. 2A is a portion of migrated seismic section seen in Fig. 4, across the northern part of the zone boundary. This shows a general similarity to the previous example, and the key lettering remains largely the same. In Fig. 4, the Basinal Zechstein is present at the SSW end of the section, and the Intermediate Zone Zechstein at the NNE end. The movement of basinal salt laterally towards the margin has been inhibited at this zone boundary. Continuing lateral pressure by the moving salt has caused first a pillow, then a diapir to form parallel to the boundary (and perpendicular to this section). The salt has broken through the overburden along the route indicated by D, leading to the occurrence of the F_2 fault and subsidence of the prism of Mesozoic overburden S. This prism slid down the fault plane of F_2 to come to rest on basal Zechstein units. The bulk of the salt which pierced the overburden at D was extruded at the surface, and eventually removed during the Late Cimmerian erosional phase represented by unconformity U.

There are several factors which may effect the "damming" or inhibition of further lateral salt movement in such zones at basin margins. One general factor which must influence the course of events is the increasing viscosity of mobile salt due to decreasing load pressure and geothermal temperature towards a basin margin. Other factors are:

(i) The progressive constriction of the thinning salt interval towards the margin. The effect of this on the lateral movement of salt towards the margin must at some point exceed the load pressure of the thinning overburden. At this point, stress relief would take the form of elongate salt swell parallel to the basin contours; this could eventually evolve into a piercement.

(ii) Any down-to-basin faulting affecting the base of the salt interval would act as a focus for upward movement in the circumstances of (i) above, with either physical upward deflection of the moving salt by the upthrow side of the fault, or differential overburden pressure across the fault zone leading to the formation of a salt swell/pillow, and eventually a diapir.

(iii) Another possible factor, for which suggestion I am indebted to Dr. J. C. M. Taylor, involves the steep slope of the top of the Z2 carbonates at the edge of the shelf. In the examples shown here, and in others in the Polish area as shown by Antonowicz and Knieszner (1984), this reflecting horizon shows a remarkably steep slope - possibly of 10° or more - and at outcrop and in offshore cores, evidence of considerable instability (slumping, sliding, fracturing) has been observed at this interface even when substantially lower inferred depositional dips were involved. There is thus a strong possibility that the bedding and other planes of weakness in the carbonates (which both overlie and underlie anhydrites) lie at a variety of angles to, and in places intersect, the boundary with the overlying Z2 halite. In these circumstances, the injection of halite back into the Z2 carbonates and even the Werraanhydrit seems a distinct possibility. This injection, together with the pressure of the moving salt against the rather steep depositional slope and the tendency of the latter to deflect the salt upwards, may be another factor operating in the damming mechanism.

(iv) In some areas, apart from the progressive constriction as mentioned in (i) above, there may occur a facies change at the margins of some basins, with basinal salt interdigitating with marginal facies lithologies over a relatively short distance; alternatively, a more rapid constriction of the salt interval may occur due to some palaeogeographic feature such as an old fault scarp, or a fringing carbonate buildup, which may define the basin/shelf boundary. Such features, alone, or combined with some other factors mentioned, may provide the necessary inhibition to further marginward flow of the salt which initiates the basin-edge diapiric phenomena under discussion.

In Fig. 4, additional lettering indicates TB, the (Triassic) Top Bacton horizon, both in the adjacent section, and within the subsided prism S, where its position is estimated. Feature E is a secondary salt diapir which occurred during the U-TY interval. Although some late salt movement has taken place, with accompanying subsidence effects, as is indicated by the shapes of horizons U and TY above the prism S, it was clearly less than in the previous example; no strongly-marked late listric and antithetic faulting has developed in the shallow section at this location.

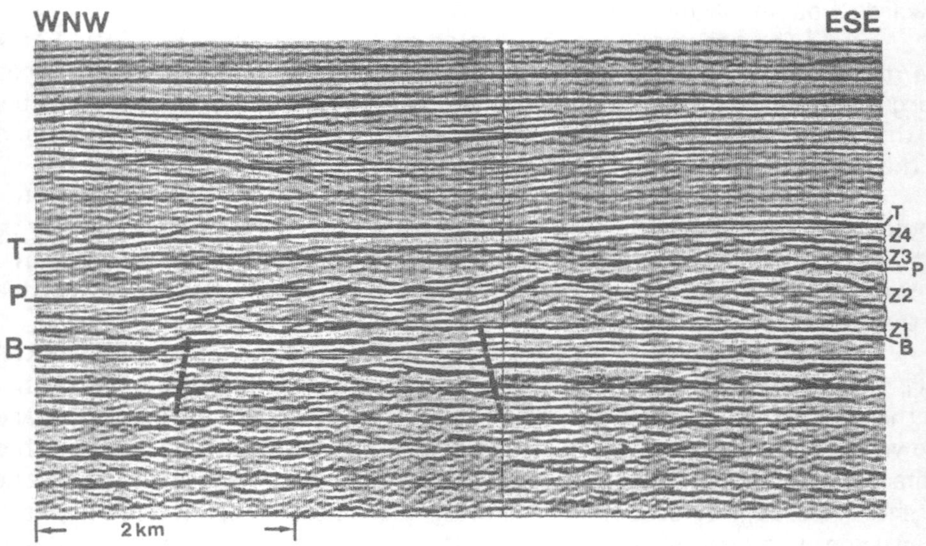

Fig. 5. Migrated seismic section from location C in Fig. 2A. Deformation and disintegration of the Plattendolomit competent band P-p due to lateral salt movement within the B-T interval.

The Intermediate Zone Zéchstein here is thinner than that in Fig. 3,and the proportion of the marginal facies units to salt in the thickness of the total interval is greater. The damming effect which inhibited further edgeward movement of salt and led to the diapirism may, as was suggested above, be the result of some interdigitation of basinal halite with marginal facies units, or may be due to a combination of the constriction of the thinning salt interval, the presence of a base-salt fault, and to the non-mobile condition of the shelf halite.

In the Fig. 3 example, the Intermediate Zone section is thicker, with a greater proportion of salt of basinal facies within the column. It would appear most likely that in this case the damming effect which led to diapirism was caused by the salt on the upthrow side of the boundary fault, F_1, being more viscous than (at least) the lower part of the salt interval on the downthrow side which is likely to have been more mobile. This greater mobility would be brought about by the differential loading with respect to the lower overburden/salt loading on the upthrow side, and to higher geothermal temperature.

Evidence for lateral, rather than vertical (halokinetic) movement of the Zechstein salt in this area is shown by the seismic event which is the response to the competent Plattendolomit band - the basal unit of the Z3 evaporite cycle lying within the total Z2/Z3 salt interval. In Fig. 5 is seen a part of a migrated section from location C, in Fig. 2A. Horizons B and T are the base and top of the Zechstein respectively, and P-p is the strong Plattendolomit event. Following event P across the section, it can be seen that considerable deformation and disintegration of Plattendolomit band has occurred, even to the extent of shortening by overlap of fragmented sheets of the band in places. At the same time, the Top Zechstein T shows no deformation which could be in any way related to the effects seen in the Plattendolomit event; it is clear that the latter must be the result of purely lateral movement within the salt interval.

The migrated seismic section in Fig. 6 is from location D in the Fig. 2A map, and shows effects seen at the Shelf Facies Zone/Intermediate Zone boundary. These effects are belived to be analogous,although not identical to, the effects discussed earlier at locations A and B on the map, since at both boundaries they are due to lateral, centrifugal salt movement. It is not clear whether the effects at the location D are contemporaneous with, predate, or post-date those at the boundary at A and B. The "front" of marginward salt movement would have reached the A/B boundary earlier than that at D, but the faulting affecting the base salt at the A/B boundary may have occurred, or been reactivated quite late, after the salt movement front had reached the location at D. The salt and overburden thicknesses here at the time of the Late Cimmerian erosional phase (unconformity at U) were much less than at the Basin boundary at A and B. As in previous figures, B and T mark the base and top of the Zechstein, with Base Tertiary at TY. There is a deep fault at F_1 which is located approximately at the Zone boundary and may control the latter, at least locally, and a shallow fault system (F_2 with an antithetic fault). The configuration of event A within the Zechstein may indicate splay of marginal facies bedding due to the lateral salt movement towards the shelf.

The interpretation given is that at this location, laterally moving salt met resistance to further flow at the shelf boundary, and pierced the probably thin overburden at the location of fault F_2. The salt then would have been extruded at the surface during Late Cimmerian times, and the fracture system of F_2 developed contemporaneously with overburden subsidence related to salt withdrawal. During later deposition, F_2 evolved as a growth fault, with initiation of the antithetic fault and graben structure now present.

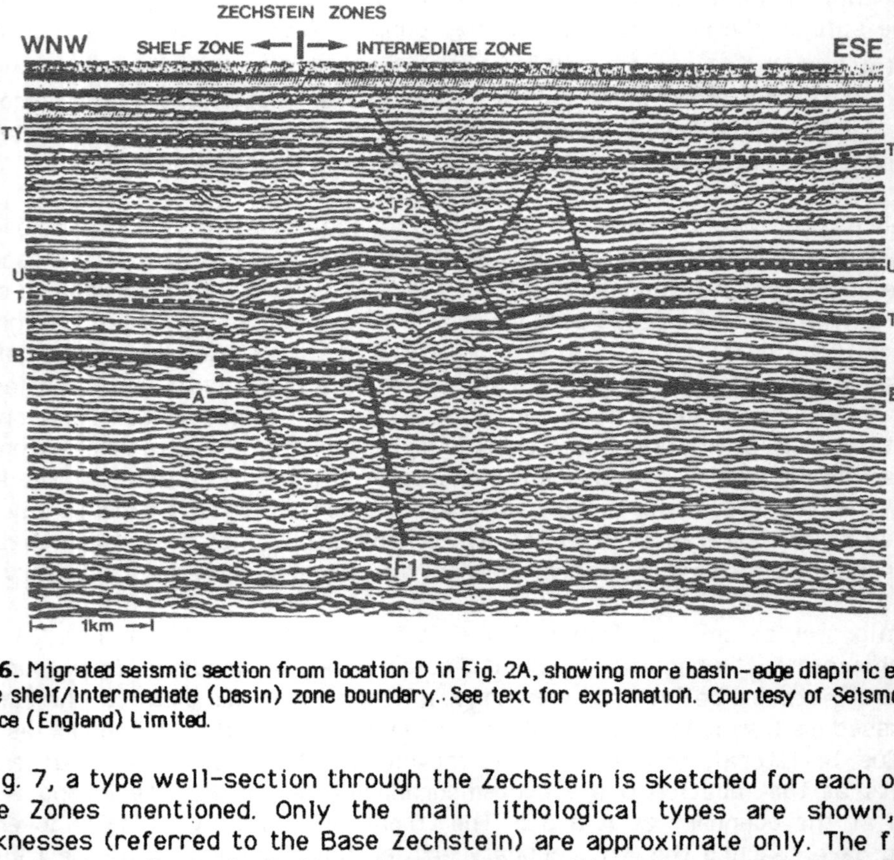

Fig. 6. Migrated seismic section from location D in Fig. 2A, showing more basin-edge diapiric effects, at the shelf/intermediate (basin) zone boundary. See text for explanation. Courtesy of Seismograph Service (England) Limited.

In Fig. 7, a type well-section through the Zechstein is sketched for each of the three Zones mentioned. Only the main lithological types are shown, and thicknesses (referred to the Base Zechstein) are approximate only. The figure serves to show the main features in each zone, with very thick salt in the Basin Zone, thinner salt with more marginal facies units in the Intermediate Zone, and very thin total salt in the Shelf Zone. In this instance, the well section from the Shelf Zone shows a complete absence of Z2 salt; in fact, elsewhere on the outer part of the Shelf Zone there may still be a thin development of Z2 salt present. The location of these three wells is indicated in the sketch map of Fig. 2A.

Fig. 8 shows a seismic example from the southern margin of the sub-basin, the location of which is marked X on the sketch map of Fig. 2B. It will be seen that there are some close similarities between this example and those shown previously from the northern and the NW margins of the sub-basin.

In particular, comparisons with the example of Fig. 3, in which the section orientation is similar - i.e., the shelf direction is to the left and the basinal direction to the right - show the similarity of the principal features. This is such that the mechanisms involved are clearly related in both locations, to the extent that the description and key lettering of Fig. 3 can be used almost as they stand to describe the Fig. 8 example, although the two locations are separated by several hundreds of kilometres. Note particularly the shape of the events marked P in both examples, interpreted as being due to the bedding splay resulting from pressure of salt moving from the basinal zone on the right towards the shelf zone on the left.

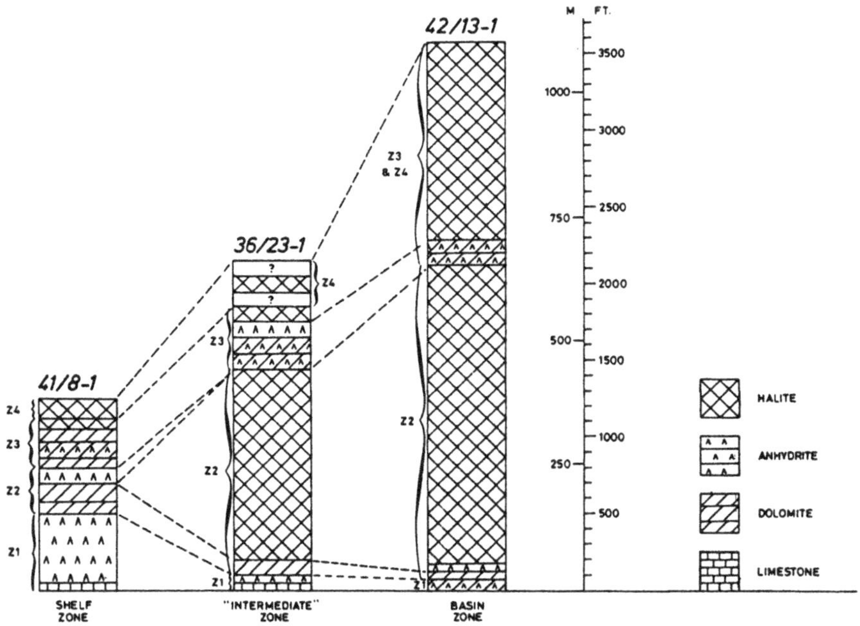

Fig. 7. Type well-sections for each of the three zones discussed in the NW area (Fig. 2A map). Only major lithologies shown. The well locations are marked on the Fig. 2A sketch map.

In Fig. 8, the Z2 Stassfurt salt is within the seismic interval A-B,which is seen to be thicker, and thickening, to the NNE into the basin. On the SSW side of the section, interval A-B is substantially thinner, and may consist mainly of marginal facies lithologies with little or no salt towards the left-hand end of the section. The relatively rapid constriction of the interval shelfward over a ahort distance probably initiated a salt swell and eventually a piercement in the position of the fault zone D, with continuing lateral shelfward salt movement. The piercement stage may have been penecontemporaneous with the occurrence, or reactivation, of strike-slip fault F_3. Salt flow into the diapir from beneath the overburden segment now represented by S resulted in the rotational subsidence of the latter, causing the fracturing which led to the development of the gravity fault F_2. Subsequently, the overburden prism S, now unsupported on both sides, slid down the F_2 fault zone with some later rotation which may have been caused by the torque from salt flow into the diapir up route D.

Salt which flowed upwards through the pierced overburden along zone D was extruded at the surface, and probably removed during the Late Cimmerian erosion (the Unconformity now being off the top of the section in this location). B and T mark the base and top of the Zechstein interval, whilst A is interpreted as the seismic response to the Plattendolomit equivalent here. The Top Bacton horizon is marked TB, and the Top Triassic is at TT. The explanation offered here is effectively the same as that put forward for the development of the previous examples of Figs. 3, 4, and 6.

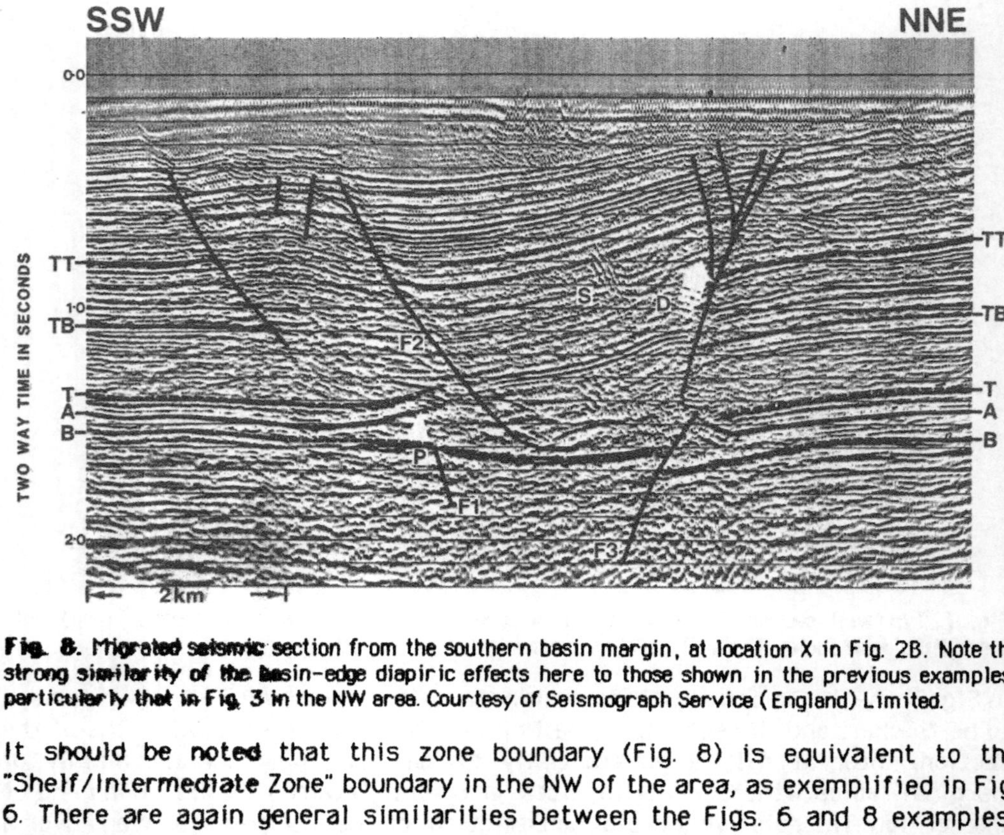

Fig. 8. Migrated seismic section from the southern basin margin, at location X in Fig. 2B. Note the strong similarity of the basin-edge diapiric effects here to those shown in the previous examples, particularly that in Fig. 3 in the NW area. Courtesy of Seismograph Service (England) Limited.

It should be noted that this zone boundary (Fig. 8) is equivalent to the "Shelf/Intermediate Zone" boundary in the NW of the area, as exemplified in Fig. 6. There are again general similarities between the Figs. 6 and 8 examples, although the salt thicknesses, details of the faulting, and also probably the overburden thickness at the time of the diapirism, are different.

It is not known if there is any equivalent at this southern basin margin of the "Intermediate Zone/Basinal Facies Zone" boundary as seen at the NW margin, since the available seismic coverage into the basin stops in the area of Fig. 8. The answer to this question probably depends upon whether any equivalent down-to-basin faulting occurs to the north of the Fig. 8 example. However, from the seismic evidence it seems quite probable that at least one, and possibly two, zones of basin-edge diapiric effects may be present around the N, W, and S sub-basin margins.

Distribution of salt structures

It should be clear that the areas discussed in this paper are peripheral to the main area of formation of salt structures - mainly pillows and swells - in the basin centre. It is interesting to recall that Brunstrom and Walmsley (1969) in their pioneer paper on the Permian evaporites of the North Sea, drew attention to the fact that although the area of maximum salt movement, including some diapirism, occupies approximately the central area of the Southern Basin, it does *not* coincide with the area of the thickest Zechstein salt in its primary development, which occurs somewhat to the north of the central area and is devoid of diapirs in spite of having an overburden thickness in excess of 2500 metres.

The reason for this seems undoubtedly to be that the area of maximum salt movement is coincident with a belt of major reactivated regional faults, the reactivation having occurred over a period which included, at its peak, the Late Cimmerian movements and erosional phase, based on seismic observations in the area. Attention has been drawn to the close relationship in many localities between salt diapirism and faulting (see, for example, Jenyon, 1986a, 1986b).

In the area of the thickest salt mentioned above, only a few broad, shallow, isolated salt swells are observed. The lack of any major salt structures in this area of thickest salt, even under the large thickness of overburden present, and adjacent to another area in which major fault-related salt structures including diapirs have developed, is of some significance. It brings into question current ideas related to "halokinesis" (Trusheim, 1960), and the rheological behaviour of salt rock; this subject is beyond the terms of reference of the paper, but has been discussed by the writer elsewhere (Jenyon, 1986b).

It is apposite to note that at least two areas of complete salt depletion due to lateral salt flow are present in the central area of the sub-basin, one of these being illustrated by the seismic example in Fig. 9. Such zones where there is a total absence of salt in the central area are the only examples which occur purely due to lateral salt flow, although other zones of depletion occur in more peripheral areas which are the result of subsidence related to diapirism and extrusion of salt at the depositional/erosional surface.

The evidence assembled here supports the view that regional salt movement in the southern Zechstein sub-basin has teken the form of lateral, centrifugal flow from the central part of the basin towards the northern, western, and southern margins. Upon this general marginwards flow have been superimposed other effects, mainly those produced by major fault movements. Some areas in the basin centre have been totally depleted of salt, whilst basin-edge effects resulting from the lateral pressure of salt moving towards the margins from the basin centre can be observed in seismic data. It would be of interest to know if similar depletion areas along the trough of the Zechstein Basin, and edge-diapiric effects along its margins, are identifiable eastwards from the North Sea in the German, Polish, and USSR parts of the Basin.

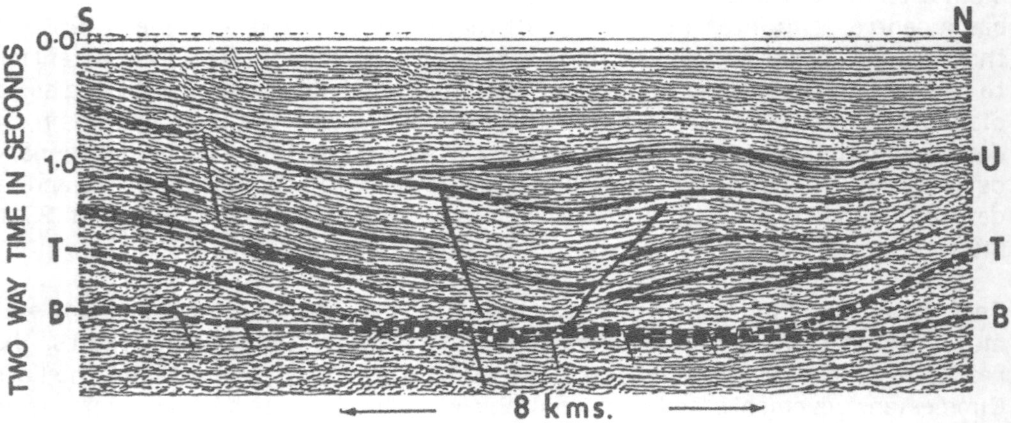

Fig. 9. Migrated seismic section from a basin-central location, showing complete salt depletion over a zone 8 km wide in the section, due to lateral salt movement. The long axis of this depletion zone runs perpendicular to the section for a distance which results in an area of about 200 square kilometres over which the salt is absent from the Zechstein interval B-T. Other, similar zones exist in the axial trough zone of the basin, and suggest the provenance of the salt which exerted lateral pressure at the basin margins resulting in edge diapiric effects. Courtesy of Seismograph Service (England) Limited.

Concluding comments

The zones of deformation and salt piercement which are represented by the boundaries between shelf and basinal facies, as indicated in the seismic examples presented, contrast sharply with the Zechstein within the basinal zones (particularly in the N and NW where the seismic coverage is more extensive), where in general only quite gentle salt swells occur. Absence of any widespread development of large salt pillows or diapirs away from the zone boundaries (except in the zone of clearly fault-related structures in the basin), together with the evidence of deformation and disintegration, in many places, of the competent Plattendolomit band within the total salt interval, clearly indicates lateral movement of the Zechstein salt.

The marked effects at the Intermediate/Basin Facies Zone boundary, and on a smaller scale, those at the Shelf Facies/Intermediate Zone boundary, suggest that this lateral flow was directed centrifugally away from the basin centre towards the northern, western, and southern margins of the basin. Other effects have been superimposed locally on this general movement by, for example, reactivation by strike-slip movements of older faults, to determine fault-related trends of the salt swells and features which do exist in the basin.

The prime cause of this edgeward salt movement cannot be established unequivocally at this stage. Once the salt became mobile, the overburden load pressure differential between the basin centre and margins must have exerted the main controlling influence. As far as the prime, initiatory factor is concerned, however, there may have been other considerations. The salt at the basin centre may have been already in a metastable state when the Late Cimmerian reactivation of older faults took place; some force involved may have acted as a triggering mechanism, starting the basin centre salt into motion towards the margins. The Late Cimmerian movements may themselves have been initiated by a tectonic pulse related to lithospheric plate movement, or a change in the rate of such movement (pulsation tectonics).

There two other points of interest which are suggested by the seismic evidence. The first of these is that, whilst the overall controlling factor of the regional salt movement, once initiated, may be regarded as "halokinetic" - i.e. gravitationally controlled due to load stress differential between basin centre and margins - at a local level, salt swells, pillows and piercements at zone boundaries around the margins are the result of lateral salt movement, and are thus not "halokinetic" in origin. The second point is that if the mechanisms suggested for the basin-edge diapiric effects are accepted, they imply forceful upward injection of salt through the overburden. Such a process has been discounted by those who favour the "downbuilding" hypothesis first advanced by Barton (1933), and who believe that all positive salt features, both piercement and non-piercement, can be explained by the downbuilding process (see, for example, Woodbury *et al.*, 1980). It is believed that salt structures produced directly by lateral movement within a salt interval, as discussed in this paper, provide an important field of study which is somewhat neglected.

Acknowledgements

Thanks are expressed to the Directors of Seismograph Service (England) Limited for permission to use the seismic examples which illustrate this paper. The author is also grateful to Dr. J. C. M. Taylor for comments and suggestions on the first draft of the work.

References

Antonowicz, L. & Knieszner, L., 1984. Zechstein reefs of the Main Dolomite in Poland and their seismic recognition. *Acta Geol. Polon.*, 34: 81-94.

Barton, D. C., 1933. Mechanism of formation of salt domes with special reference to Gulf Coast salt domes of Texas and Louisiana. *Am. Assoc. Petroleum Geologists Bull.*, 17: 1025-1083.

Brunstrom, R. G. W. & Walmsley, P. J., 1969. Permian evaporites in North Sea Basin. *Am. Assoc. Petroleum Geologists Bull.*, 53: 870-883.

Donato, J. A., Martindale, W. & Tully, M. C., 1983. Buried granites within the Mid North Sea High. *J. geol. Soc. London* 140: 825-837.

Jenyon, M. K., 1985. Basin-edge diapirism and updip salt flow in the Zechstein of the Southern North Sea. *Am. Assoc. Petroleum Geologists Bull.*, 69: 53-64.

Jenyon, M. K., 1986a. Some consequences of faulting in the presence of a salt rock interval. *Jour. Petroleum Geology* 9: 29-52.

Jenyon, M. K., 1986b. *Salt Tectonics* Elsevier Applied Science Publ., 191 pp. Barking, U.K.

Jenyon, M. K., Cresswell, P. M. & Taylor, J. C. M., 1984. Nature of the connection between the northern and southern Zechstein Basins across the Mid North Sea High. *Marine & Petr. Geol.*, 1: 355-363.

Smith, D. B., 1981. The Magesian Limestone (Upper Permian) reef complex of Northeastern England. *SEPM Spec. Publ.*, 30: 161-186.

Taylor, J. C. M., 1986. Late Permian-Zechstein. In: Glennie, K.W. (Ed.) *Introduction to the Petroleum Geology of the North Sea*, 87-111. Blackwell Scientific Publications, Oxford.

Trusheim, F., 1960. Mechanism of salt migration in northern Germany. *Am. Assoc. Petroleum Geologists Bull.*, 44: 1519-1540.

Woodbury, H. O., Murray, I. B. & Osborne, R. E., 1980. Diapirs and their relation to hydrocarbon accumulations. *Can. Soc. Petrol. Geol. Memoir* 6: 119-142.

FACIES AND GEOCHEMICAL ASPECTS OF THE DOLOMITE-ANHYDRITE TRANSITION ZONE (ZECHSTEIN 1-2) IN THE BATUM-13 WELL, NORTHERN JUTLAND, DENMARK: A KEY TO THE EVOLUTION OF THE NORWEGIAN-DANISH BASIN

Martin Sønderholm
Geological Survey of Greenland
Øster Voldgade 10
DK-1350 Copenhagen K
Denmark

Abstract: The Dolomite-Anhydrite Transition Zone is an important marker horizon separating the thick successions of grey halite of the Zechstein 1 and 2 cycles in the central parts of the Norwegian-Danish Basin. Within this unit, six main facies have been recognised in the Batum-13 well in northern Jutland. These can be grouped into three facies associations which, due to their physical log characteristics can be traced over large parts of the basin. The three facies associations reflect major events in the evolution of the basin, which are closely related to the evolution seen in the southern part of the Zechstein Basin as reflected by the Stassfurt carbonates (Ca2) and the Basal Anhydrite (A2). Hence, facies association 1 consisting of bedded massive anhydrite and carbonate conglomerates, records the transgression heralding the Zechstein 2 cycle, whereby the brines of the basin went through a recessive phase (A1r). Facies association 2 is dominated by dark bituminous carbonate rhythmites with nodules of anhydrite and marks the beginning of the Zechstein 2 cycle; it reflects the maximum transgression of the Zechstein sea (Ca2). Continued rising salinities probably led to the formation of a sulphate platform as reflected by the resedimented sulphates in the bottom of facies association 3 (A2). Sulphates, now seen as distorted bedded mosaic anhydrite in the upper part of facies association 3, were subsequently precipitated over the entire basin. Ultimately, halite precipitation (Na2) returned to the basin.

Introduction

Four evaporite cycles reaching a total thickness of up to 1600-2000 m, have been recognised within the Zechstein deposits in the Norwegian-Danish Basin (Olsen, 1978; Ormaasen *et al.*, 1980; Jacobsen, 1984; Taylor, 1984). In general, the evolution of these cycles closely resembles the evolution of the four major

Lecture Notes in Earth Sciences, Vol. 10
T. M. Peryt (Ed.), The Zechstein Facies in Europe

cycles in the southern part of the Zechstein Basin. However, the Zechstein 1 cycle in the central parts of the Norwegian-Danish Basin contains an approximately 400 m thick sequence of grey halite (Richter-Bernburg, 1955a, b; Sannemann *et al.*, 1978).

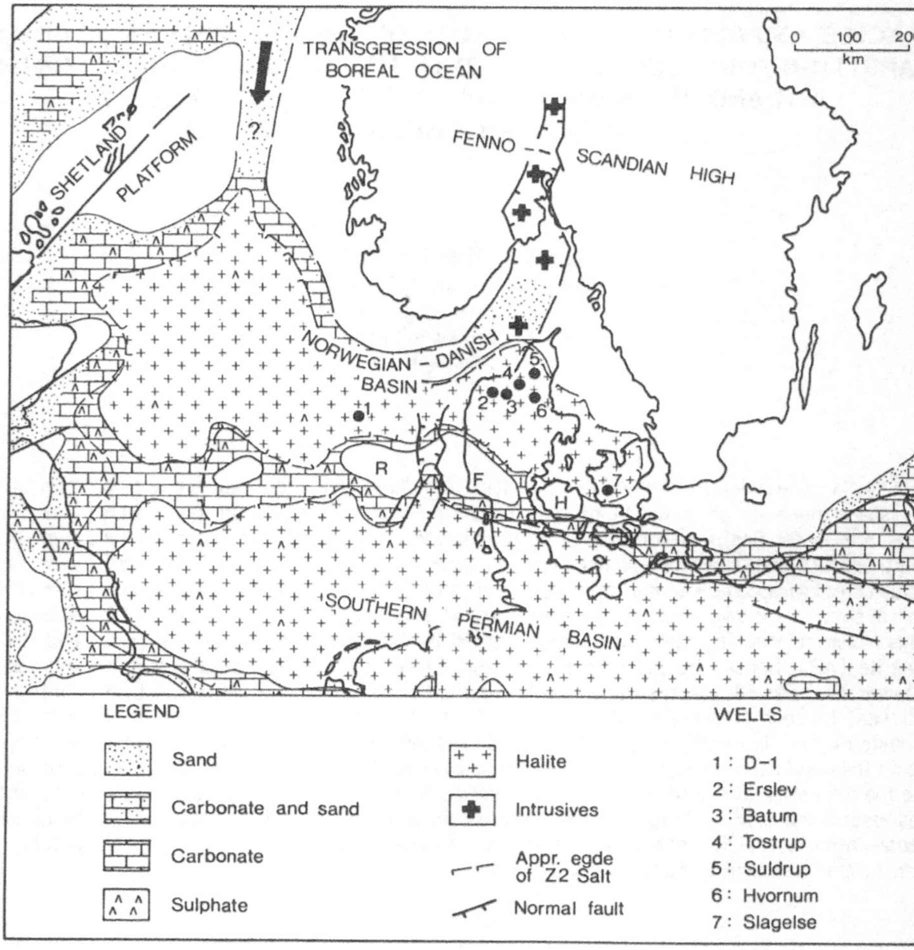

Fig. 1. Generalised map of Zechstein basins. Wells containing the Dolomite-Anhydrite Transition Zone in the Norwegian-Danish Basin are indicated. Map based on Ziegler (1982).

This paper deals with a conspicuous unit of carbonates and anhydrite, the so-called Dolomite-Anhydrite Transition Zone, which forms an important marker horizon in the basin, separating the thick successions of grey halite of the Zechstein 1 and 2 cycles. Lithostratigraphically, this unit closely correlates with the Stassfurt carbonates and Basal Anhydrite in the southern part of the Zechstein Basin (A1r, Ca2, A2; Richter-Bernburg, 1962). There, it has been possible to relate the different developments of the carbonates to a regional depositional model. Based on lithology, sedimentology and thickness, the Stassfurt carbonates have been subdivided into the 30-100 m thick Haupt-

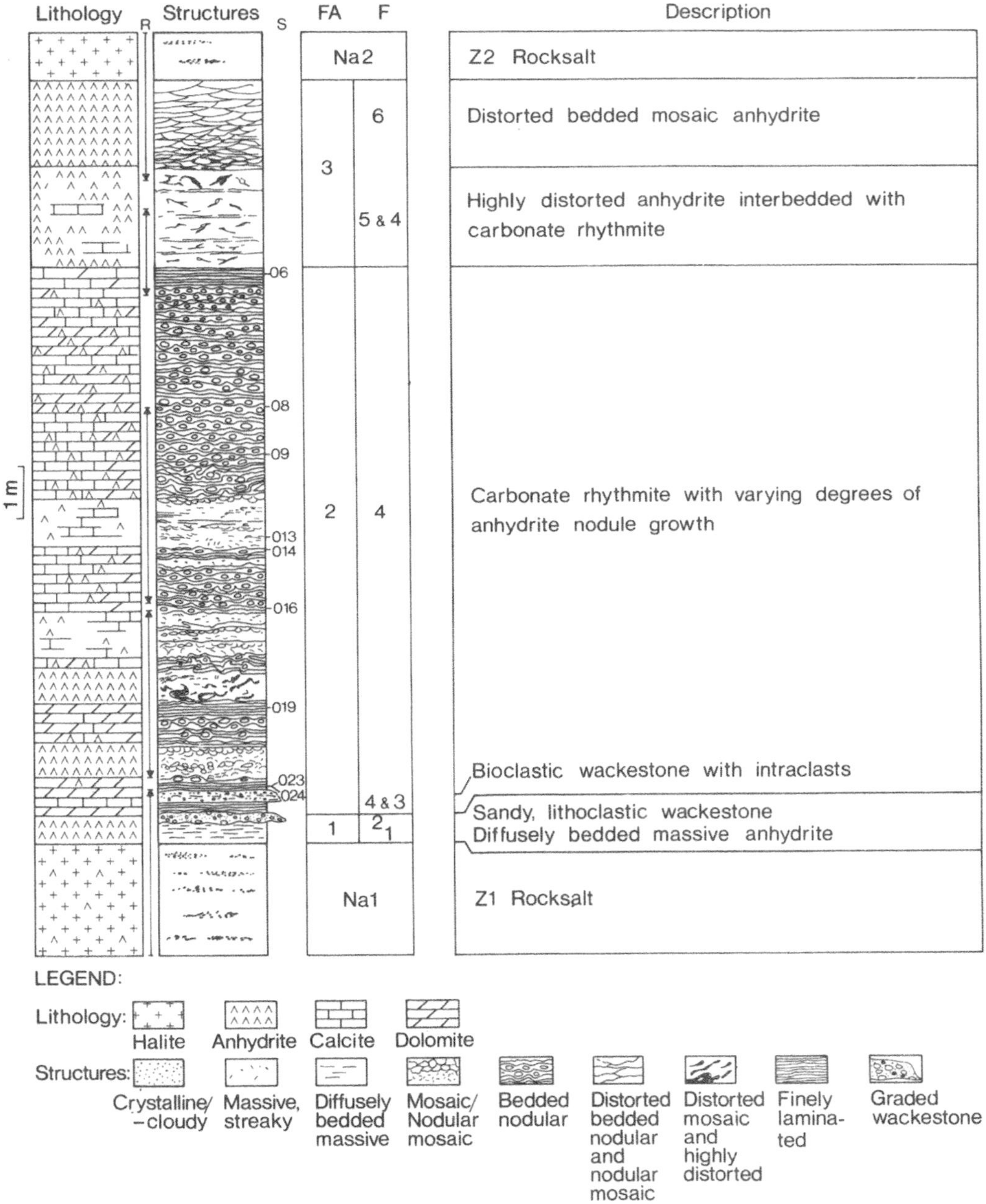

Fig. 2. Log of the Dolomite-Anhydrite Transition Zone in the Batum-13 well (419.9-437.8 m). Thicknesses of units are calculated from original steep dips of beds. FA and F indicate facies associations and facies, respectively. Recovery (R) and sample numbers (S) used in this paper are indicated.

dolomit (shallow water, mainly basin margin facies), the 10->100 m thick Stinkdolomit and Stinkkalk (basin-slope facies) and the less than 10 m thick deep water basin-centre facies of the Stinkschiefer (Richter-Bernburg, 1955a; Füchtbauer, 1964, 1968).

This study concentrates on the equivalents of the Stinkschiefer and the Basal Anhydrite, since well-preserved core material only exists from these facies. Although the Dolomite-Anhydrite Transition Zone has been encountered in a number of wells (Fig. 1), the description of facies is based upon the Batum-13 well, since this is the only well where a nearly complete succession is preserved (Fig. 2).

Facies description

Within the Dolomite-Anhydrite Transition Zone in the Batum-13 well, six main facies have been recognised. These can be grouped into three facies associations which, due to their physical log characteristics, can be traced over large parts of the basin.

In the following description of facies, the structures and textures of the anhydrite are described according to the scheme of Maiklem *et al.* (1969), while the classification of Randazzo & Zachos (1984) is used for the dolomite fabrics.

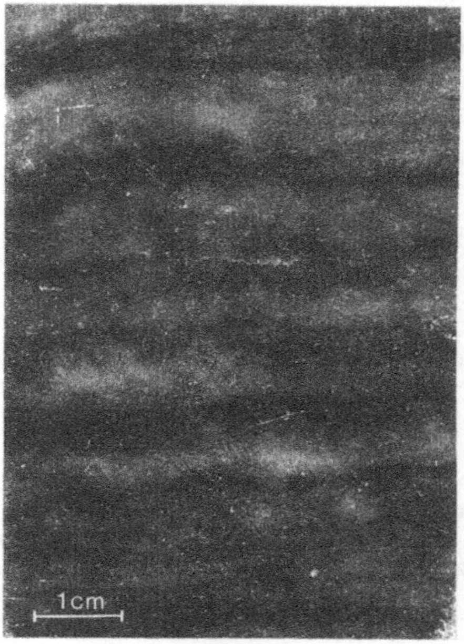

Fig. 3. Facies 1: Diffusely bedded massive anhydrite. Polished slab.

Facies 1: Bedded massive anhydrite

This facies consists of bluish to grey, diffusely bedded, massive anhydrite. The lamination is wavy, parallel, 1-5 mm thick (Fig. 3). The anhydrite crystals are sub- to euhedral, less than 60 µm long or lath-shaped, subhedral up to 500 µm long. The texture is sub- to aligned felted. A cellular fabric is weakly developed due to the lack of carbonate matrix. It consists of cells 1-2 mm in diameter with a subfelted, sometimes fibroradiate core surrounded by partly tangentially arranged anhydrite laths.

Interpretation. The parallel lamination and the lack of matrix indicate deposition below wave-base in the subphotic zone (Richter-Bernburg, 1955b; Davies & Ludlam, 1973; Dean *et al.*, 1975; Schreiber *et al.*, 1976; Schlager & Bolz, 1977; Dean & Anderson, 1978; Crawford & Dunham, 1982). The lamination is interpreted to be a product of changing growth or precipitation rates governed by fluctuating salinities. These can be due to climatic variations, often interpreted as seasonally controlled (Richter-Bernburg, 1955b; Anderson *et al.*, 1972; Davies & Ludlam, 1973). Although it is impossible from geochemical data alone to deduce whether the $CaSO_4$ was precipitated as gypsum or anhydrite (Kushnir, 1982a), some constraints on the concentration of the brine precipitating the sulphate can be given from the strontium content of the anhydrite.

At high brine concentrations (seawater concentration factor C.F > 10) primary anhydrite is coprecipitated with halite (Butler, 1969), and the theoretical strontium content of the anhydrite can be calculated from the formula:

$$(Sr/Ca)_A = (Sr/Ca)_B \times D_{Sr} \quad (1)$$

where the subscripts A and B stand for anhydrite and brine, respectively. The partition coefficient for strontium into anhydrite (D_{Sr}), is dependent on temperature (*cf.* Kushnir, 1982a). A temperature estimate can be given by the oxygen isotope data (Fig. 4) obtained from the carbonate rhythmites (facies 4) using the formula:

$$t\,(^\circ C) = 16.9 - 4.2(\partial_c - \partial_w) + 0.13(\partial_c - \partial_w)^2 \quad (2)$$

where ∂_c is $\partial^{18}O$ (PDB) for calcite and ∂_w is $\partial^{18}O$ (SMOW) for the brine (Epstein *et al.*, 1953). Calcite, however, is found both as an early diagenetic phase and as late diagenetic dedolomite (*cf.* Fig. 5) and hence $\partial^{18}O$-values vary considerably. As the dolomite values are much more stable, the ∂_c value can be substituted by:

$$\partial_c = \partial^{18}O_{(dolomite)} - \partial^{18}O_{(dolomite-calcite)} \quad (3)$$

assuming that the two phases are cogenetic. If the brines were derived from sea-water having the same isotopic composition as present day sea-water, a ∂_w-value of 2-4 ‰ could be used for the brines (*cf.* Bein & Land, 1982).

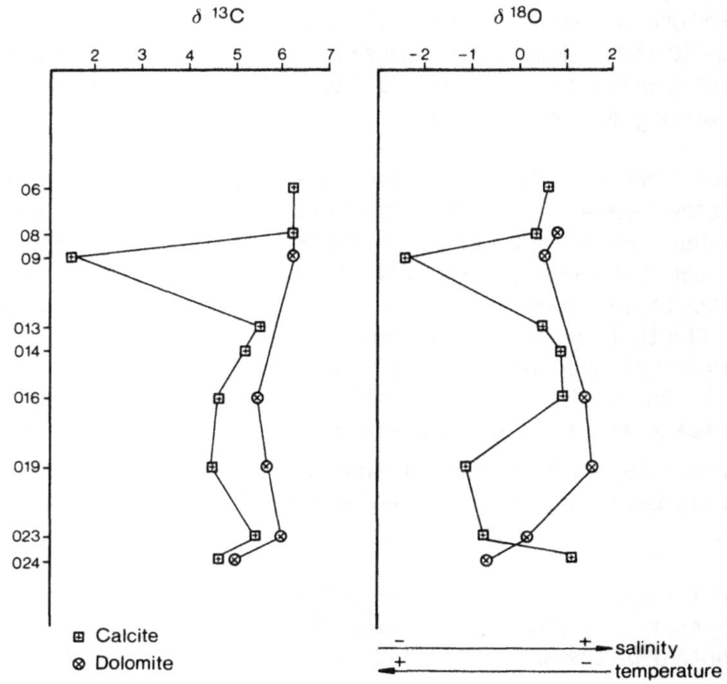

Fig. 4. Stratigraphic plot of $\partial^{13}C$ and $\partial^{18}O$-values from facies 4. Covariation of $\partial^{13}C$ values in the calcite and dolomite suggest cogenesis between the early diagenetic calcite and dolomite. The low $\partial^{13}C$-value in the calcite of sample 09 is attributed to the influence of diagenetic calcite formed from pore waters enriched in light carbon due to thermally induced decarboxylation of organic material (*cf.* Irwin *et al.*, 1977). Calcite is found both as an early diagenetic phase and as late diagenetic dedolomite resulting in a considerable variation in $\partial^{18}O$-values of the calcite. The determination of the fractionation coefficient $\partial^{18}O_{(dolomite-calcite)}$ = 0.6‰ is therefore only based on samples 08, 016, and 023 where influences of the late diagenetic calcite phase seems minimal. This value is lower than the 2-4‰ given by Land (1980). Based on values from the upper part of the sequence, where salinities are constant, $\partial^{18}O_{(dolomite)}$ is 1.1‰.

Based on the values given in Fig. 4, a temperature estimate of 23-33 °C can be obtained from (2), which is close to other determinations in similar settings (Bein & Land, 1982; Kushnir, 1982b). Hence, a value of 0.35 is chosen for D_{Sr} (Kushnir, 1982a). At C.F. >10 the $(Sr/Ca)_B$ is equal to 0.09 (Kushnir, 1982b; Fig. 2), which, following formula (1), gives $(Sr/Ca)_A$ value of 0.032. This is equivalent to a strontium content of 3500 ppm in the anhydrite. The actual measured strontium content of the anhydrite is 1500 ppm (Sønderholm, 1984), indicating that the anhydrite could not be a primary phase coprecipitated with halite at C.F. values >10. At lower C.F. values (between 3.5 and 10), the strontium content of the suphate will be lower. Gypsum, however, will be the stable sulphate phase (Butler, 1969; Kushnir, 1982b) and the complicated chemical conditions related to the transformation of gypsum to anhydrite preclude a safe estimate of the concentration of the original precipitating brine (*cf.* Kushnir, 1980, 1982a).

The cellular texture could reflect a gypsum precursor and similar textures have been described as a diagenetic product of the replacement of gypsum by anhydrite (West, 1964). This complete recrystallization has totally obliterated any primary structure and texture. Hence, the gypsum could either have grown on the sediment surface (*cf.* Schreiber *et al.*, 1976), or have been precipitated within the water column (*cf.* Schmalz, 1969), while some of the thicker beds could have been deposited by gravity flows (*cf.* Schlager & Bolz, 1977).

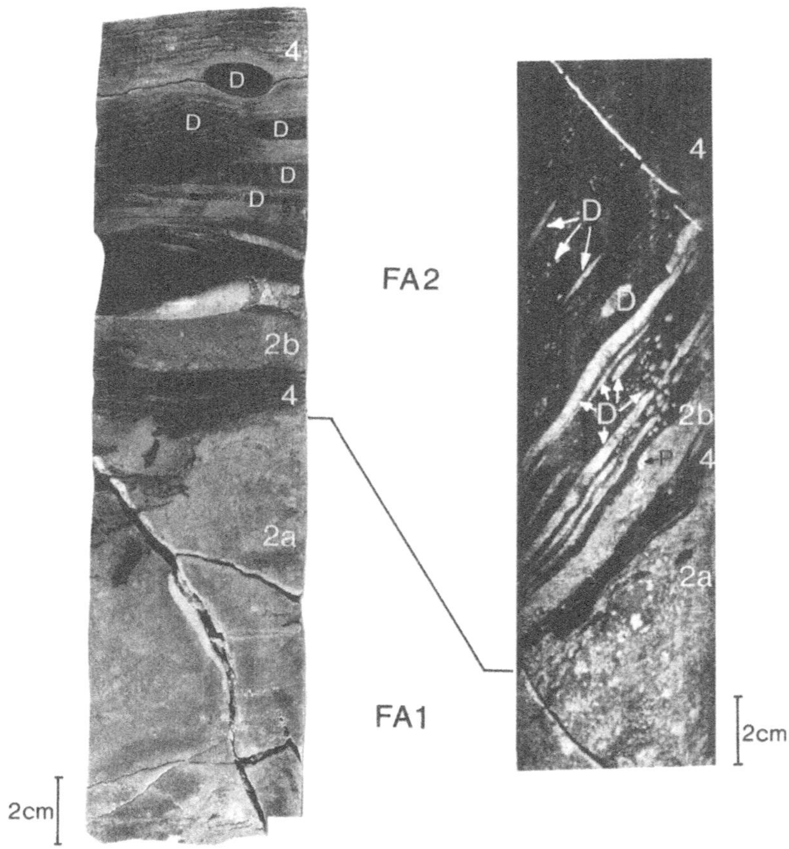

Fig. 5. Boundary between facies association 1 (FA1) and facies association 2 (FA2) defining the transition between the Zechstein 1 and 2 cycles. 2: Lithoclastic wackestone (facies 2). A poorly developed coarse-tail grading and protruding intraclasts (p) suggest deposition by debris flow. The irregular upper surface of 2a is interpreted to be due to water-escape or hydroplastic flow. 4: Carbonate rhythmite (facies 4) with nodules (D) consisting of calcified dolomite (dedolomite).

Facies 2: Lithoclastic wackestone

This facies occurs in 1-10 cm thick light grey to grey beds. Upper and lower contacts of beds are sharp and non-erosive, but irregular due to flame and load structures. Beds are non-graded or exhibit a poorly developed coarse-tail grading and protruding clasts are seen in some beds (Figs. 5 & 6). The matrix consists of dolomitic mud with an equigranular, microxenotopic, peloidal fabric and poorly sorted (20-200 µm) subangular to rounded siliciclastic material in amounts varying up to 50% of the total. Intraclasts are well rounded to angular (Fig. 5), ranging in size from 0.2 to 10 mm, and show no preferred orientation. Most of the clasts consist of the same material as the carbonate matrix, although some clasts show an alternation of carbonate-rich and siliciclastic laminae.

Interpretation. The sedimentary structures and fabrics indicate deposition by debris flows (Middleton & Hampton, 1976). The irregular top of the beds is interpreted as being due to liquefied or hydroplastic flow (Lowe, 1976, 1982). The pale dolomites and the high content of siliciclastic material suggest a source area within facies equivalent to the Hauptdolomit (Ca2d). This is considered to have been deposited in oxic littoral environments which, as suggested by the lack of fossils, were probably highly saline.

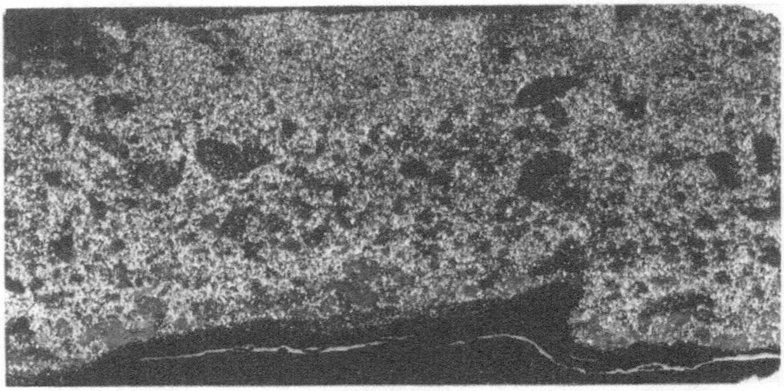

Fig. 6. Facies 2: Lithoclastic wackestone showing poorly developed coarse tail grading. Thin section from 2b in Fig. 5.

Facies 3: Sandy bioclastic wackestone with intraclasts

This facies consists of light grey-brown beds, 2-6 cm thick. The individual beds have irregular bases due to flame and load structures. The lower main part of the beds consists of massive wackestone containing randomly oriented intraclasts which sometimes can be concentrated in horizons. The upper boundary of this massive part of the bed is irregular and exhibits protruding intraclasts. The upper part of the beds is an approximately 1 cm thick dolomitic interval with diffuse discontinuous lamination (Fig. 7).

The matrix consists of microsparitic calcite with a clotted peloidal structure, and minor amounts (<10%) of subangular to rounded grains of siliciclastic material.

The bioclastic material is totally dominated by shells and shell fragments of which indeterminate bivalves and ostracods constitute 2/3 and the rest is formed by foraminifers. The shells either lack internal sediment (spar-filled) or are filled with matrix. Occasionally, geopetal surfaces showing that the shells are not in their original position, can be observed (Fig. 8). The foraminifers have not been identified with any certainty due to poor preservation, but two types, a totally dominating glomospiriid and a very rare nodosariid, have been observed (Fig. 9).

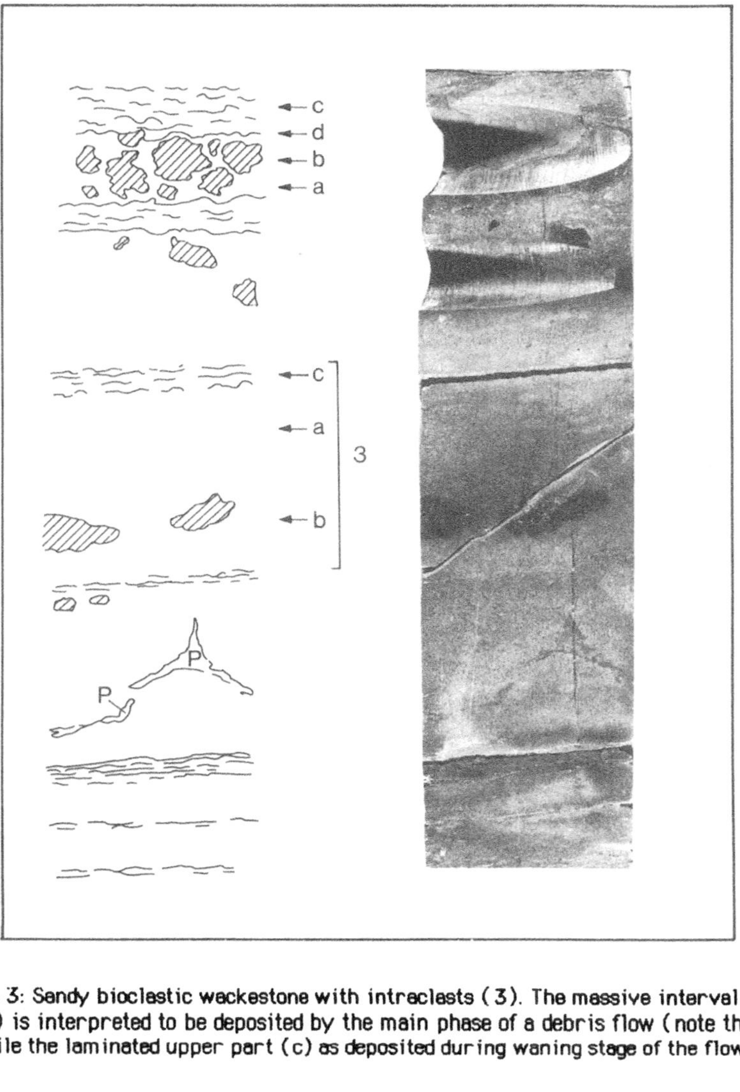

Fig. 7. Facies 3: Sandy bioclastic wackestone with intraclasts (3). The massive interval (a) with the intraclasts (b) is interpreted to be deposited by the main phase of a debris flow (note the protruding clast at d), while the laminated upper part (c) as deposited during waning stage of the flow. P – pyrite.

Intraclasts are very irregular in shape and vary in size from less than 1 mm to up to 3 cm. The larger clasts consist of aggregates of smaller clasts (Fig. 8). Both, individual and aggregated clasts are micritised in an outer zone 50-100 µm wide, probably due to boring algae.

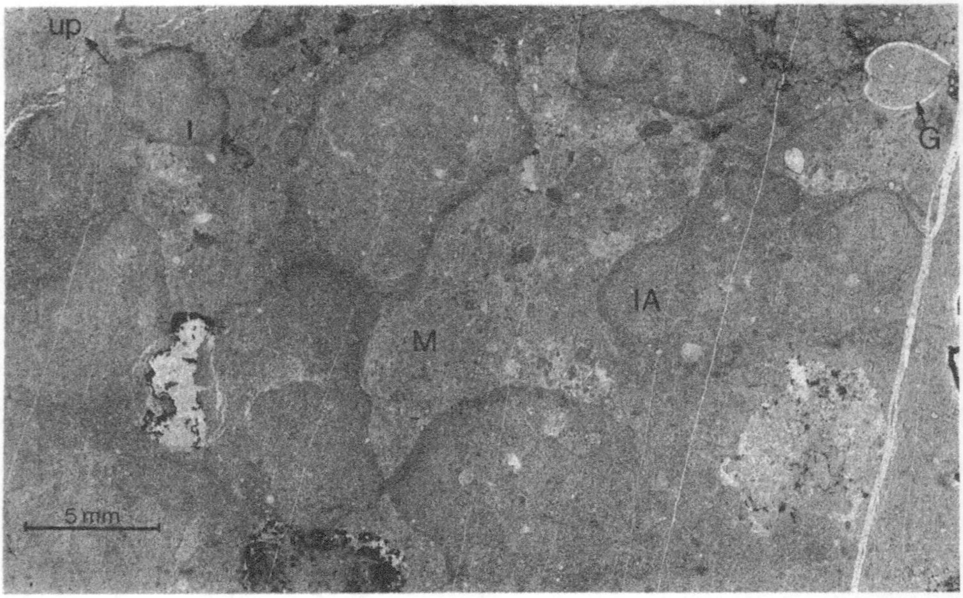

Fig. 8. Facies 3, thin section. The intraclasts are found both as individuals (I) and aggregates (IA) in a matrix (M) having the same composition as the intraclasts. The surface of the intraclasts is micritised, probably by boring algae. The geopetal surface (G) in the shell in the upper right corner shows that the shell is not in its original position.

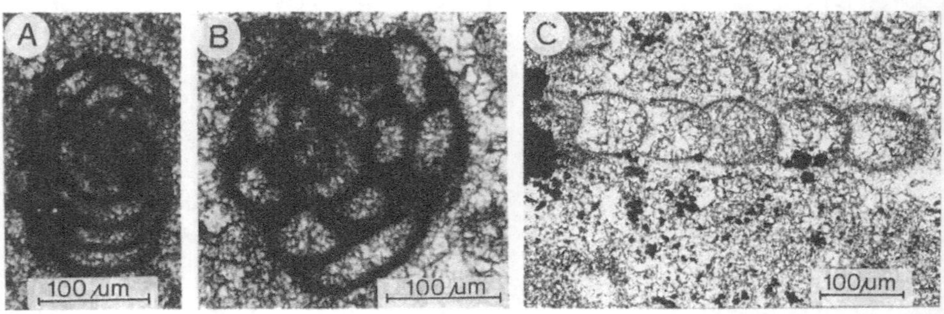

Fig. 9. Foraminifers from facies 3. A&B: glomospiriid, C: nodosariid.

Interpretation The boring by algae of the whole surface of the intraclasts, the internal sediment in the shells and the pale colour suggest that the sediments originated in a high-energy, oxic environment in the littoral zone. The low diversity of the fauna could indicate raised salinities.

These sediments were subsequently carried into the central part of the basin by subaqueous debris-flows, as indicated by the sedimentary features observed in the massive part of the beds. The concentration of clasts in certain horizons in the beds is attributed to differential shear-stress within the flow, and the laminated upper part of the beds is interpreted to be due to the waning-stage of the flow (Middleton & Hampton, 1976; Lowe, 1976, 1982).

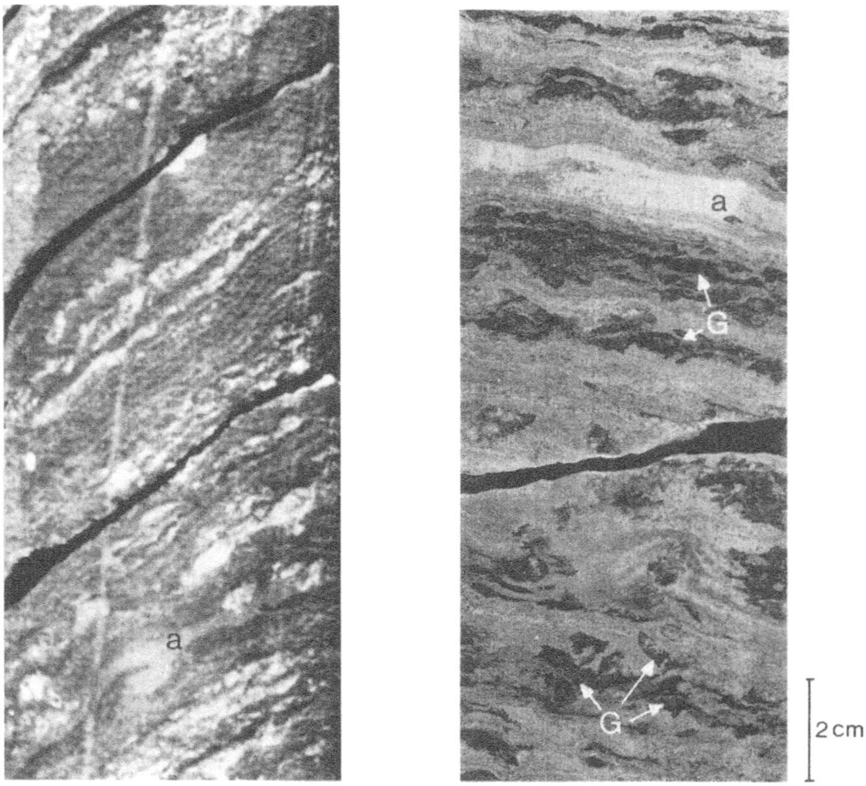

Fig. 10. Facies 4: Carbonate rhythmite with anhydrite nodules (bedded nodular anhydrite). In the polished slab, many angular, discoidal pseudomorphs after gypsum are seen (G). The fine lamination is interpreted to be of possible varvic origin, while the thicker beds with very irregular lower and upper boundaries (a) are suggested to be deposited by resedimentation processes.

Facies 4: Carbonate rhythmite with anhydrite nodules

This facies consists of light brown to black, partly bituminous, laminated, dolomitic mudstone with varying degrees of growth of displacive anhydrite nodules (Figs. 10 & 13). The mudstone contains generally less than 5% clastic material, consisting of angular to rounded clay minerals and silt-grade quartz and feldspar. Organic content is low, generally between 0.1 and 0.3%, but may range up to 1.2% (up to 1.7% in the Hvornum-2/64 well; Sønderholm, 1984).

The lamination in the mudstone is very fine (between 0.1 and 0.5 mm), continuous and parallel. It exhibits a rhythmic alternation of darker laminae, which are often dolomitised, containing amorphous organic material and pyrite, and paler laminae consisting of calcite (Fig. 12). The lamination, however, is often more or less disturbed, mainly due to the growth of displacive anhydrite nodules (Fig. 12), but occasional thin beds subjected to synsedimentary slumping have been observed (Fig. 11). Bioturbation has not been recorded.

Fig. 11. Deformed bedding, probably due to penecontemporaneous slumping.

Fig. 12. Facies 4, thin section, crossed nicols. The pale laminae mainly consist of calcite, while the darker consist of dolomite. The nodules are pseudomorphs of discoidal or prismatic gypsum. Most nodules are replaced by anhydrite with a felted or subfelted texture (a), while some nodules exhibit a blocky texture (b).

Rarely occurring pale laminae, from 5 to 10 mm thick with very low organic content, have irregular bases and wavy tops (Fig. 10). The calcite of the pale laminae is microsparitic, crystal sizes range from 10 to 50 µm but are generally smaller than 30 µm. The dolomite of the dark laminae displays a microxenotopic sutured mosaic fabric, indicating a single-staged,

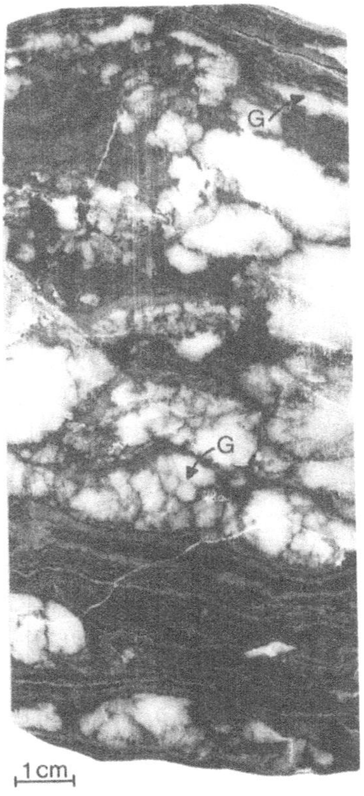

Fig. 13. Facies 4. Bedded nodular mosaic anhydrite. The larger nodules are intergrowths of smaller nodules, which in some cases (G) still are seen as pseudomorphs after pseudo-hexagonal gypsum twins (*cf.* Hardie & Eugster, 1971, fig. 3; Dronkert, 1977, fig. 5). Polished slab.

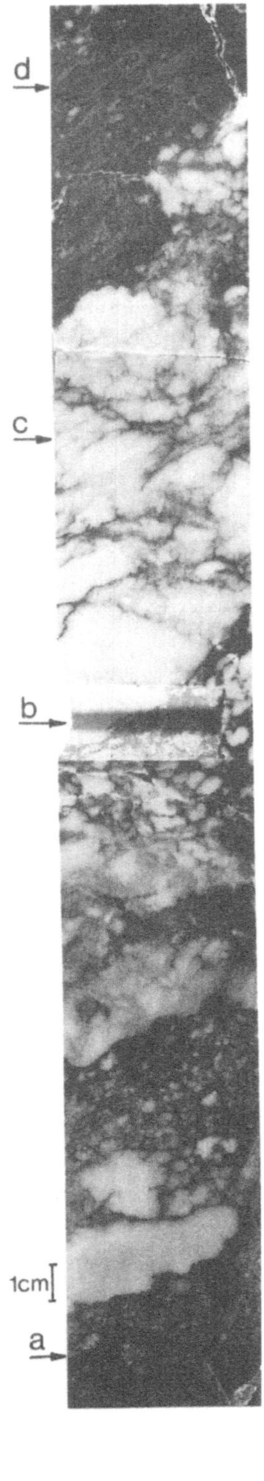

Fig. 14. Facies 4. Carbonate-anhydrite sequence; in the bottom carbonate rhythmite with few and small nodules (a) with an upwards gradual transition into nodular mosaic and mosaic anhydrite (b), eventually entherolitic folded. The folds show constant vergence in the core (c). From this point, the degree of anhydrite nodule growth decreases upwards, and the sequence is capped by carbonate rhythmite (d). Polished slab.

homogeneous, penecontemporaneous dolomitisation of an original mudstone (Randazzo & Zachos, 1984). Penecontemporaneous diagenetic growth of displacive anhydrite nodules is recorded by the distribution of the lamination and by the presence of syndepositional micro-thrusting above the nodules.

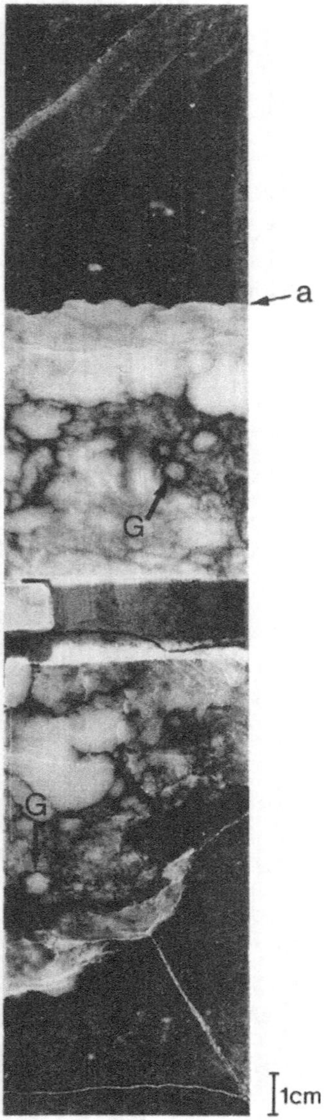

Fig. 15. Facies 4. Carbonate-anhydrite sequence as in Fig. 13. This sequence, however, is capped by massive anhydrite followed by carbonate rhythmite. The boundary between the anhydrite and the carbonate (a) is interpreted to be the sediment surface at the time of anhydrite growth. G is a pseudomorph after a pseudo-hexagonal gypsum twin.

The displacive anhydrite occurs in two general ways with a gradational transition between these (Fig. 13). The anhydrite is partly seen as bedded nodules, with sizes up to 30 mm, completely or partially separated by matrix (nodular to nodular mosaic anhydrite) and partly as larger, coalesced masses (mosaic to massive anhydrite).

The **bedded nodular anhydrite** is seen either as nodules less than 5 mm in diameter with slightly rounded prismatic or discoidal shape and with an orientation subparallel to bedding (Figs. 10 & 12) or as irregular masses (5 to 30 mm) of partly intergrown smaller nodules (Figs. 14 & 15).

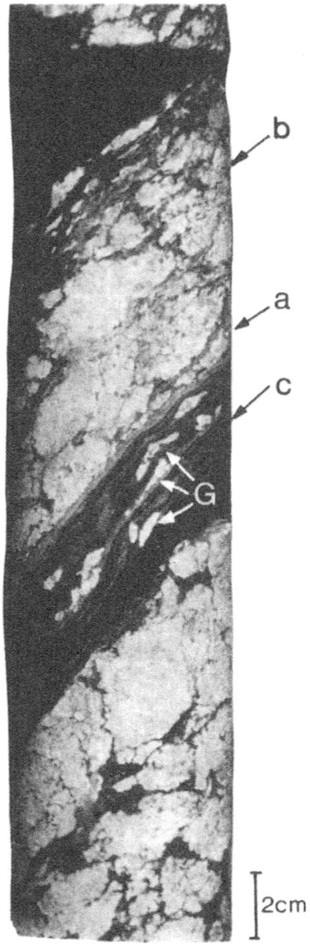

Fig. 16. Facies 4. Carbonate-anhydrite sequence displaying an abrupt start with a rather massive anhydrite (a) and an upwards decrease in nodule growth (b). Within the carbonate rhythmite (c) many pseudomorphs after discoidal gypsum are seen (G).

The more **massive mosaic anhydrite** is found in sequences 10 to 40 cm thick. These, when ideally developed, consist of often strongly deformed carbonate rhythmite with a few, small anhydrite nodules, which pass upward, due to increased growth of anhydrite nodules, through nodular mosaic into mosaic anhydrite, eventually entherolitic folded (Shearman, 1966), or nearly massive anhydrite (Figs. 14, 15 & 16). The ideal sequence is capped by nodular mosaic anhydrite which, through nodular anhydrite returns, to deformed carbonate rhythmite (Fig. 14). This ideal sequence, however, is rarely fully developed since some sequences have sharp basal contact between the underlying rhythmite and the mosaic anhydrite (Fig. 16) while others have a sharp upper contact between the mosaic anhydrite and the overlying rhythmite (Fig. 15).

In small nodules which have not coalesced, the core shows mainly a subbelted texture, while the rim is often more coarsely crystalline felted to parallel felted. Some nodules exhibit a blocky texture. The nodules contain abundant small inclusions of matrix (Fig. 12). In the larger coalesced nodules the outer part is parallel felted, while the anhydrite in the central part is arranged in a weakly developed cellular fabric formed by a subfelted core surrounded by tangentially arranged felted anhydrite (see facies 1). Occasional fragments or streaks of matrix are seen along the rims of the cells.

SAMPLES	LECO/ROCK EVAL							GASCHROMATOGRAPHY	
	TOC %	T_{max} °C	S1	S2	S3	HI	OI	Ali/Aro/Het	Pri/Phy
Batum-13 (03)	0.11							59/29/13	1.08
(06)	0.25		0.11	(0.05)	0.23				
(011)	0.09								
(019)	0.28								
(023)	1.23	(435)	0.92	(0.97)	0.52	(97)	(52)	49/36/15	1.36
(11)	0.82	(434)	0.71	(0.64)	0.47	(64)	(47)	60/32/8	1.32
Hvornum-2/64	1.66	439	2.49	2.60	0.50	260	(50)	41/54/5	1.56
Erslev-2								57/41/2	1.31

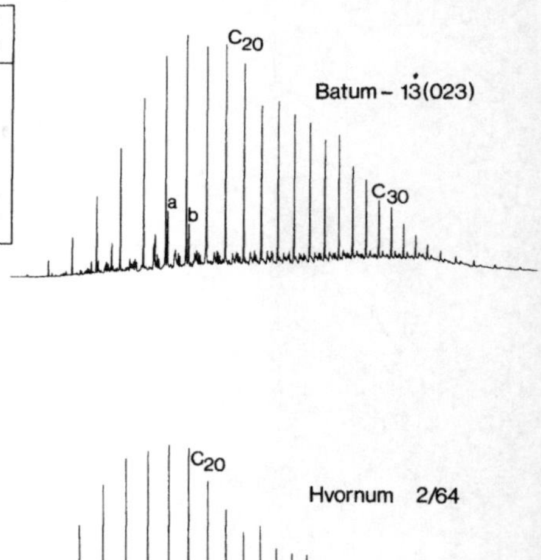

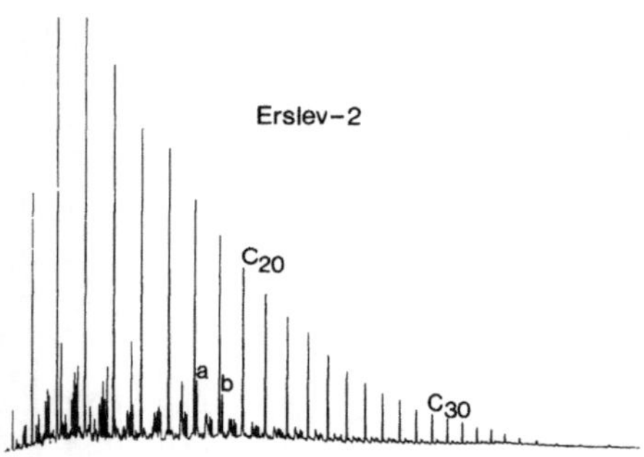

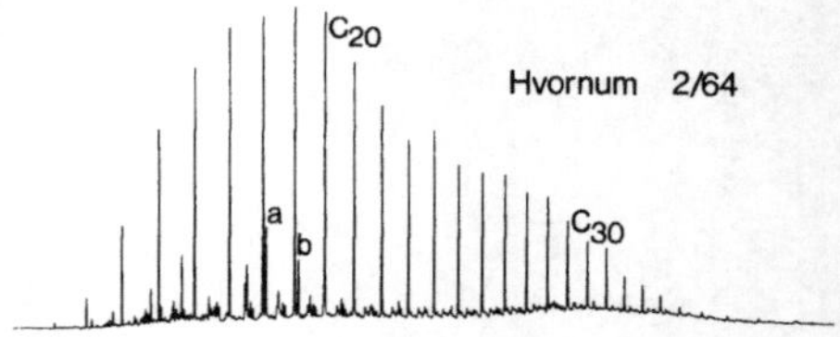

Fig. 17. Gas chromatograms of the aliphates extracted from the organic material. Table gives main results of the LECO/ROCK EVAL analysis (*cf.* Epistalie *et al.*, 1977). Ali/Aro/Het is the percentages of aliphates, aromates and heterocompounds in the soluble part of the organic material. Pri, (a): pristane. Phy, (b): phytane. The great similarity in the fine-structure of the chromatographs under the n-alkanes higher than nC_{21} shows that the composition of the organic matter is nearly identical in the different wells.

Interpretation. The interlamination of calcite and dolomite has been suggested to be of algal origin (Gebelein & Hoffman, 1973). However, no direct evidence for algal lamination (*cf.* Aitken, 1967) has been recorded in facies 4.

The original continuous parallel lamination suggests deposition from suspension below wave-base, probably caused by periodic chemical precipitation. The lamination can be interpreted as being varvic, where changes in salinity, plankton productivity and/or oxygen content are due to secular variations (Richter-Bernburg, 1955b, Seibold, 1958, Anderson *et al.*, 1972, Davies & Ludlam, 1973, Kushnir, 1981).

Analyses of organic material of facies 4 (Sønderholm, 1984) are very similar to those of the Stinkschiefer (Ca2st) in the German part of the Zechstein Basin (Botz *et al.*,1981). The origin of the organic matter is difficult to ascertain on account of the maturity, poor quality (type II/III to III kerogen) and general scarcity. Gas chromatograms (Fig. 17), however, suggest a mixed origin with contribution from both a "terrestrial" and planktonic algal source (Sønderholm, 1984).

The lack of bioturbation and fossils, the content of pyrite and organic material with a low pristane/phytane ratio (Fig. 17, Welte & Waples, 1973) suggest deposition in low-oxic reducing environments and/or higher salinities than normal.

Due to their irregular bases and their very low organic content, the thicker interbeds are interpreted to be resedimented micritic material from more oxic, probably shallow water environments. The irregular basal surface could either be due to erosion or water-escape triggered by sudden deposition of sediment on an unconsolidated sediment surface. Nothing conclusive can be said about the resedimentation process due to the lack of internal structures within the beds caused by the uniform grain size.

The early diagenetic nodules record an original penecontemporaneous growth of primary gypsum crystals within the sediment. This is illustrated by the small nodules (less than 5 mm in diameter) which clearly are pseudomorphs after prismatic or discoidal gypsum (Figs. 12 & 16). Gypsum crystals of this aspect have earlier been described from subaqueous evaporites, which have not undergone diagenetic alteration (Hardie & Eugster, 1971, Dronkert, 1977, 1978, Schreiber *et al.*, 1976).

The gradual transition between the different types of nodules, from small isolated replacements of gypsum crystals to larger, coalesced masses indicates a genetic relation between these two forms. Holliday (1968) has suggested that the small gypsum crystals acted as nucleation cores for early diagenetic growth of primary anhydrite. The cellular structure seen in the mosaic anhydrite probably reflects the earlier gypsum crystals (see facies 1).

From this evidence, it is suggested that the different types of nodule development reflect the salinity within the pore waters of the sediment. Crystallisation rates are slow when the brines are only slightly oversaturated with respect to $CaSO_4$ and crystallisation rates will be faster, resulting in the

Fig. 18. Facies 5: Highly distorted anhydrite interbedded with carbonate rhythmite (facies 4). Note the erosive lower boundary of the anhydrite (a) and the conformable deposition of carbonate on the irregular upper surface of the anhydrite (b)

precipitation of gypsum as small crystals within the nodules. The sequences capped by nearly massive anhydrite (Fig. 15) do not show erosion of the anhydrite and could therefore define a primary sedimentary surface. The other sequences probably developed slightly deeper in the sediment column.

Facies 5: Highly distorted anhydrite

This facies consists of grey, highly distorted anhydrite with streaks and fragments of dark brown carbonate and typically occurs in beds 3-5 cm thick. The bedding is defined by strongly deformed and discontinuous streaks of carbonate which sometimes can be seen to consist of facies 4. Lower and upper boundaries are mostly diffuse and very irregular, although erosive bases are occasionally seen. The lamination in the carbonate is conformable with the irregular upper boundary of the anhydrite (Fig. 18). Fragments of brown carbonate, ranging in size from single crystals to aggregates up to 2 mm, and strongly deformed streaks from less than 1 mm up to 3 mm thick are found with the anhydrite (Fig. 18).

The texture of the anhydrite is microcrystalline. More than 90% of the crystals are less than 10µm, and equidimensional, but dispersed, laths up to 400 µm long are also seen.

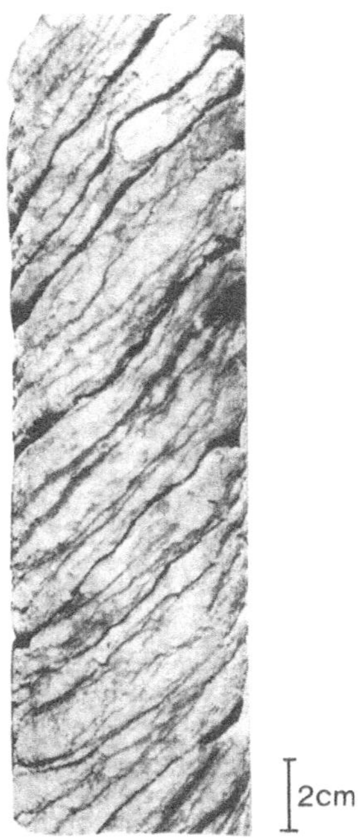

Fig. 19. Facies 6: Distorted bedded anhydrite.

Interpretation. The primary sediment on the site of deposition is the carbonate-rhythmite (facies 4) which has been partly incorporated in the anhydrite during incoherent slumping. Deposition of the anhydrite during slumping is suggested by the strongly deformed (folded) structures of the carbonate streaks, the erosive lower boundary of the anhydrite, the microcrystalline texture and the conformable deposition of the carbonate-rhythmite on the irregular upper surface of the anhydrite beds (Schlager & Bolz, 1977, Kendall, 1978, Rupke, 1978).

The texture of the anhydrite is considered to be formed by crushing during slumping of originally larger, lath-shaped crystals. Similar textures have been observed in anhydritic turbidites (Schlager & Bolz, 1977).

Facies 6: Distorted bedded mosaic anhydrite (Flaseranhydrit)

This facies consists mainly of pale grey to bluish grey distorted bedded mosaic anhydrite (Fig. 19). The bedding is discontinuous, non-parallel wavy. Due to variations in the content of carbonate matrix, however, both massive anhydrite and nodular mosaic anhydrite can be found. The anhydrite nodules are lensoid, from 3 to more than 50 mm long and from 1 to 10 mm thick, and an upwards increase in the size of nodules has been observed. The nodules are arranged in beds which are 1 to 2 nodules thick. These are separated by discontinuous laminae and streaks up to 3 mm thick consisting of calcite reminiscent of the rhythmite of facies 4.

The texture of the anhydrite is sub-felted. The smaller anhydrite crystals are concentrated along the margin of the nodules, while the larger tend to occur in the centre. The lath-shaped crystals can be found both aligned parallel to bedding and as radially arranged clusters (Fig. 20).

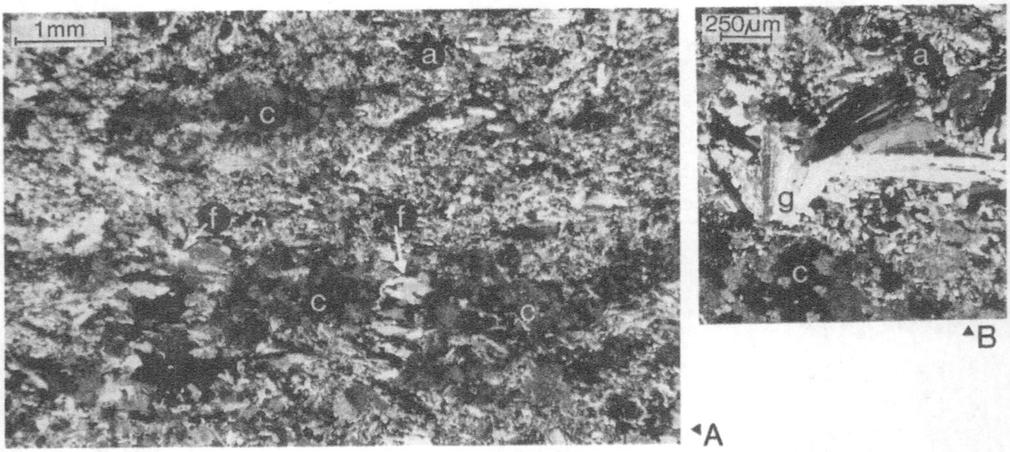

Fig. 20. Photomicrograph of facies 6, crossed nicols. **(A)** Boudinaged calcite laminae (c) where the fractures (f) between the boudins are filled with anhydrite laths. The anhydrite nodules (a) are more coarsely crystalline in the centre (upper part of picture) than close to the calcite. **(B)** Possible gypsum pseudomorph (g).

Calcite is found as discontinuous laminae and streaks defining the bedding and as streaks and fragments between the nodules. The size of the calcite crystals is mainly around 50 µm, and they are rich in inclusions probably consisting of clay. The streaks are boudinaged, and the individual boudins are separated by large, up to 600 µm, lath-shaped anhydrite crystals which are parallel to bedding (Fig. 20).

Interpretation. The formation of the mosaic structure has been ascribed to shearing (Mossop, 1979), which could be induced during halokinetic movements, or compaction of anhydrite nodules in an impermeable matrix (Shearman & Fuller, 1969). The discotinuous wavy bedding has further been interpreted as a primary sedimentary structure (Richter-Bernburg, 1955b, Schreiber *et al.*, 1976). However, these interpretations do not fit the observed textural and structural requirements.

Dean *et al.* (1975) suggested that the bedded mosaic structure is the end-product of a recrystallisation process which begins with sporadic bulges in a carbonate-sulphate laminite. The texture of the anhydrite in the nodules is described as being felted, while it is microcrystalline in the original laminite. This difference could explain the observed distribution in crystal size in the nodules of facies 6. Hence, the small crystals along the margin of the nodules represent an original laminite while the large, lath-shaped crystals in the centre represent the recrystallisation phase. Compaction and boudinage of calcite laminae will be a consequence of the process (Fig. 21).

Thus, the original sediment is interpreted as carbonate-sulphate rhythmite. The primary $CaSO_4$ was probably gypsum, since rare pseudomorphs after swallow-tail gypsum twins have been observed (Fig. 20). The texture of the recrystallisation phase is typical of diagenetic anhydrite according to Holliday (1968).

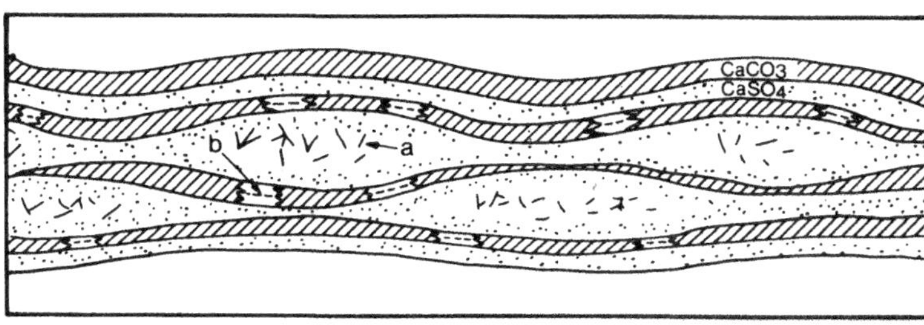

Fig. 21. Schematic representation of the formation of distorted bedded anhydrite. Due to diagenetic growth of anhydrite laths (a) within the sulphate laminae in the original carbonate-sulphate laminite, the bedding displays bulges. Continued growth causes disruption (boudinage) of calcite laminae, and the interstices between boudins are filled with coarse anhydrite laths (b).

Facies sequence

Due to very scarce amounts of collected core material, it has not been possible to evolve a model for the evolution of the Norwegian-Danish Basin during the transition from Zechstein 1 to Zechstein 2 evaporite cycle solely based on the knowledge obtained from the drilled wells. It is presumed, however, that the processes involved in both the northern and southern parts of the Zechstein

Basin are closely related (Jacobsen & Larsen, 1980, Smith, 1980, Jacobsen, 1984). Hence, parallels to the southern part of the Zechstein Basin are drawn in the interpretation of the facies associations, using established models for the genesis and distribution of the various lithofacies obtained from many detailed studies of the Zechstein deposits from these areas (*cf.* Richter-Bernburg, 1955b, 1985, Füchtbauer, 1968, Schlager & Bolz, 1977, Sannemann *et al.*, 1978, Colter & Reed, 1980, Taylor, 1984).

The sequence in the Dolomite-Anhydrite Transition Zone can be grouped into three facies associations which, due to their physical log characteristics, can be traced over large parts of the Norwegian-Danish Basin. These facies associations are related to major steps in the evolution of the basin during this time.

The Zechstein 1 cycle was terminated by the formation of a renewed connection between Greenland and Norway to the Boreal Ocean (Fig. 1), probably due to glacio(?)-eustatic sea level rises (Smith, 1980, Taylor, 1984). During this rapid sea level rise of up to 200 to 250 m whereby depths of up to 400 to 500 m were reached in the basin centre (Sannemann *et al.*, 1978, Smith, 1979, 1980), the brines in the basin went through a recessive sulphate phase marking the end of the Zechstein 1 evaporite cycle (A1r, *cf.* Richter-Bernburg, 1955a). This phase in the evolution of the Norwegian-Danish Basin is represented by facies association 1.

Facies association 1: Recessive sulphate phase (A1r)

This association comprises a basal unit formed by diffusely bedded anhydrite (facies 1) capped by limestone conglomerates of facies 2 (Figs. 2 & 5). Facies 1 records the initial dilution of the Zechstein brines, causing a shift from halite to sulphate precipitation. It has been suggested that facies 1 originated as an insoluble residue reflecting a sudden incursion of fresh marine waters accompanied by dissolution of some of the Zechstein 1 rock salt (Jacobsen, 1984). However, both sedimentological and geochemical evidence seem to indicate primary gypsum deposition in low energy environments suggesting a progressive, although rapid, sea level rise. This probably caused density stratification of the water column as the less saline marine waters overlay the residual brines of the Zechstein 1 cycle; the latter protecting the Zechstein 1 halite from dissolution.

Fig. 22. Genetic scheme of the Dolomite-Anhydrite Transition Zone in the Norwegian-Danish Basin, based on analogy to the southern Zechstein Basin (*cf.* Füchtbauer, 1968, fig. 4, Sannemann *et al.*, 1978, fig. 7). NaCl, $CaSO_4$ and $CaCO_3$ indicate brines satureated with respect to these compounds. A1r: Recessive Deckanhydrit, Ca2st: Stinkschiefer, Ca2std: Stinkdolomit, Ca2d: Hauptdolomit, A2: Basal Anhydrite, Na2: Zechstein 2 halite (after Richter-Bernburg, 1955a). RFH: Ringkøbing-Fyn High. In (a) the saturation boundaries of the transgressive fresh marine water lens at different (t_1 and t_2) are indicated. Signatures on logs as for Fig. 2. See text for description of stages a–f in the scheme.

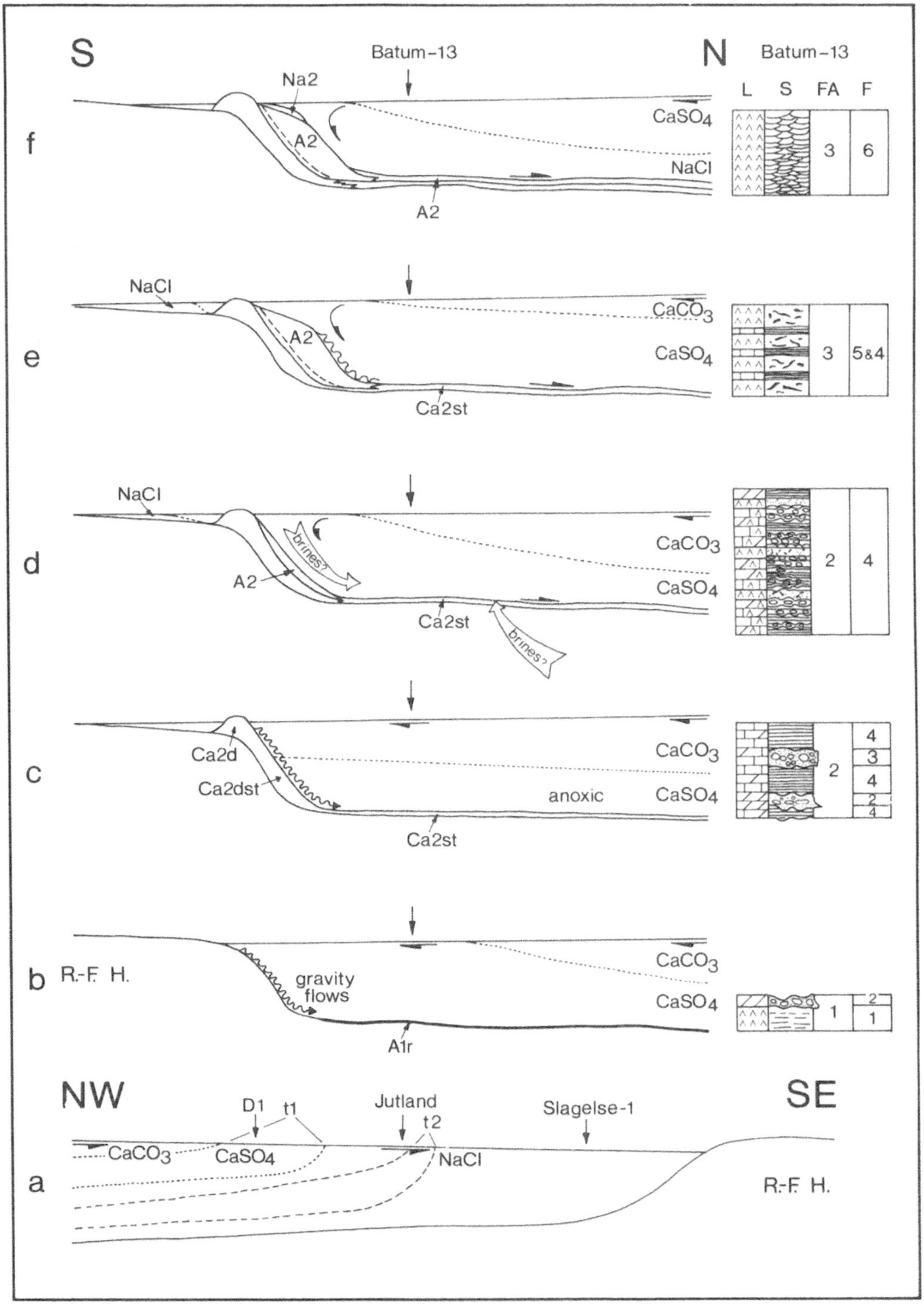

S
Batum-13
N
Batum-13
L
S
FA
F
Na2
CaSO4
A2
NaCl
A2
3
6
f
NaCl
CaCO3
A2
CaSO4
3
5&4
e
Ca2st
NaCl
CaCO3
brines?
d
A2
CaSO4
2
4
Ca2st
brines?
CaCO3
Ca2d
c
Ca2dst
anoxic
CaSO4
2
4
3
4
2
4
Ca2st
CaCO3
b
R.-F. H.
gravity flows
CaSO4
1
2
1
A1r
NW
SE
D1
t1
Jutland
t2
Slagelse-1
CaCO3
CaSO4
NaCl
a
R.-F. H.

After deposition of facies 1, pale carbonates and siliciclastics of shallow marine origin were transported into the deeper parts of the basin by debris flows (facies 2). This facies probably records decreasing salinities in the basin and increased instability along the basin margin due to continuing sea level rises (Fig. 22b).

The thickness trend observed on the geophysical logs in facies association 1, where the thickest developments occur in the western part of the basin (Fig. 23), could be explained by the combined effects of density stratification and the formation of a lateral salinity gradient causing fractionated sedimentation. The density stratification probably evolved as a response to the inundation of the residual brines by a transgressive fresh marine water lens (Fig. 22a). This fresh marine water lens, however, will show lateral salinity gradient due to evaporation, causing the lowest salinities to be found closest to the seaway to the Boreal Ocean. This process leads to fractionated sedimentation of the different evaporitic phases (Richter-Bernburg, 1955b). It is envisaged that the brines were diluted faster than the lateral salinity gradient formed by evaporation could evolve, causing the recessive sulphate phase to have its longest duration close to the seaway (Fig. 22a).

The maximum sea level rise in the Zechstein Basin was reached in a relatively short time (Smith, 1979) and subsequent sedimentation during Zechstein 2 was governed by the balance between the amount of supplied fresh sea water and the evaporation rate (Richter-Bernburg, 1955b). These processes caused the basin to evolve from a carbonate precipitating phase, through a sulphate phase into a halite precipitating phase, represented by facies associations 2 and 3 of the Dolomite-Anhydrite Transition Zone in the Norwegian-Danish Basin.

Facies association 2: Carbonate precipitating evaporite phase (Ca2st)

This association is dominated by the carbonate rhythmites of facies 4, which were probably deposited as varves under low-oxic to anoxic bottom conditions. Minor amounts of pale carbonate conglomerates (facies 2 and 3) occur near the base of the association, suggesting that oxic, high energy conditions with lowered salinities prevailed along the rim of the basin.

The bottom and top of the association are marked by very dark mudstones with a high organic content (1.2% and 0.3% respectively), reflected by the conspicuous peaks seen in the Gamma Ray logs (Fig. 23). These intervals with raised organic content showing sedimentation under reducing conditions (low pristane/phytane ratio) are probably related to density-layering of the water column caused by transgressive events (*cf.* Schmalz, 1969, see Fig. 22c). This

---

Fig. 23. Log characteristics of facies associations across the Norwegian-Danish Basin. Gamma-Ray (upper) and Formation Density (FDC) logs (lower) from selected wells. Scale bar indicates 3 m drilled thickness. Thicknesses are calculated based on information on dips of beds.

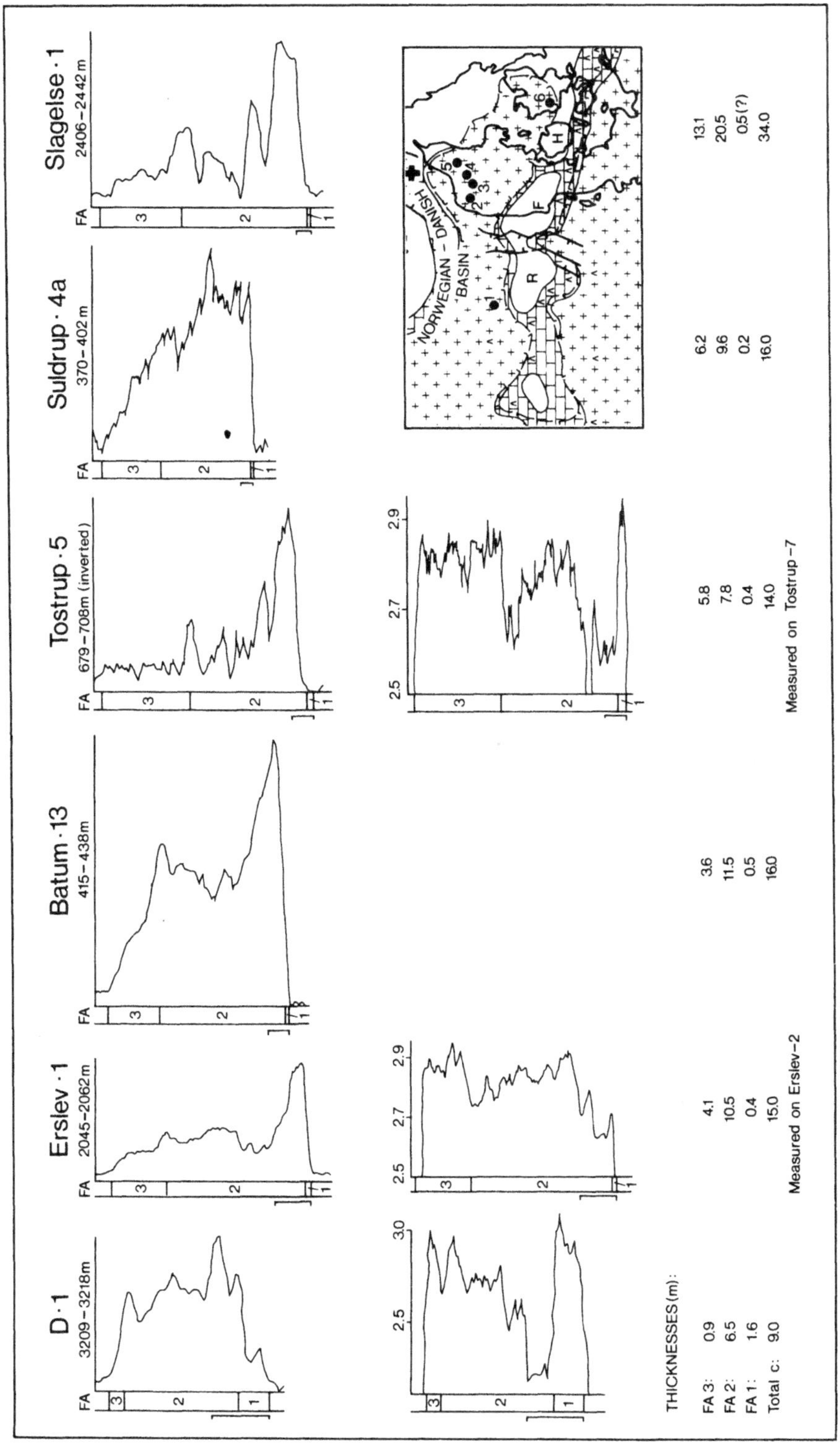
D·1
3209–3218m
Erslev·1
2045–2062m
Batum·13
415–438m
Tostrup·5
679–708m (inverted)
Suldrup·4a
370–402m
Slagelse·1
2406–2442m
FA
NORWEGIAN – DANISH BASIN
R
F
H
Measured on Erslev-2
Measured on Tostrup-7
THICKNESSES (m):
FA 3: 0.9 4.1 3.6 5.8 6.2 13.1
FA 2: 6.5 10.5 11.5 7.8 9.6 20.5
FA 1: 1.6 0.4 0.5 0.4 0.2 0.5(?)
Total c: 9.0 15.0 16.0 14.0 16.0 34.0

is particularly clear in the bottom of the association which marks the maximum sea level during Zechstein 2, since carbonate conglomerates with clasts originally deposited under more normal shallow marine conditions are found here (facies 3, see Fig. 22c).

The early diagenetic growth of gypsum within the sediments is probably related to dolomitisation of the lime muds by magnesium-rich brines according to the equation:

$$2CaCO_3+Mg^{2+}+SO_4^{2-} = CaMg(CO_3)_2+CaSO_4$$

The brines possibly originated either as a result of precipitation of $CaSO_4$ along the basin margins, from where they sank to the bottom in the basin centre (*cf.* Taylor & Colter, 1975) or by expulsion from the underlying rock-salt sequence (Fig. 22d).

Facies association 3: Sulphate sedimentation phase (A2)

Facies association 2 is terminated by the income of resedimented sulphate (facies 5) which, interbedded with the carbonate rhythmites of facies 4, forms the lower part of facies association 3. This initial event suggests that rapidly rising salinities in the basin caused a change from carbonate to sulphate precipitation along the rim of the basin, probably leading to the formation of a prograding sulphate platform (Richter-Bernburg, 1955b, Sannemann *et al.*, 1978, Clark, 1980; see Fig. 22e).

Facies 5 is also found in the Erslev-2 well (Sønderholm, 1984) and geophysical logs from other wells suggest that facies 5 can be found in large areas of the basin. Only the most westerly areas close to Central Graben may lack this phase, as suggested by the thin development of facies association 3 in the D-1 well (Fig. 23). However, in the central parts of the basin, carbonates were still precipitated for some time as indicated by the interbedding of carbonate rhythmites (Fig. 22e). According to this, the sedimentation of the Basal Anhydrite (A2) started along the margins of the basin and not, as reported from some parts of the Zechstein Basin (*cf* Sannemann *et al.*, 1978, Clark, 1980, Smith, 1980) contemporaneously in both shallow and deep water environments.

The upper part of facies association 3, consisting of distorted bedded mosaic anhydrite (facies 6), reflects continued rising salinities in the basin leading to the precipitation of carbonate-sulphate rhythmites in the central part of the basin, probably deposited as annual varves (*cf.* Richter-Bernburg, 1955b, Anderson & Kirkland, 1966, Anderson *et al.*, 1972, Dean *et al.*, 1975, Taylor, 1980). The rising salinities towards the termination of facies association 3 and the change to halite precipitation (Na2) are probably reflected by the upwards increase in the size of anhydrite nodules seen in facies 6.

Acknowledgements

This paper is based on a cand. scient. thesis completed at the University of Copenhagen under supervision of Niels Oluf Jørgensen to whom I am greatly indebted. Some of the work was carried out at the Geological Survey of Denmark (DGU) as a part of the Salt Research Project EFP-81, supported by the Ministry of Energy, file no. 221-80, 1981-04-23, and I am very grateful to Fritz Lyngsie Jacobsen and Johannes Fabricius (DGU) for their help and discussions. Dansk Olie- & Gasproduktion A/S has permitted the reproduction of their logs. I thank Lars Clemmensen who provided support from the International Association of Sedimentologists to participate in the 4th IAS Regional Meeting in Split, where an early account of the work was presented. Furthermore, I wish to thank Bjørn Burchardt, Henrik Olsen, Lars Stemmerik and Poul Henrik Due for fruitful discussions of various aspects of the manuscript. Anne-Marie Madsen, Bente Thomas, Birgitte Larsen, Carlos & Irma Torres, Jakob Lautrup and Jan Aagård are thanked for technical assistance. I am grateful to John K. Warren and John S. Peel for their critical comments on the manuscript. The paper is published with permission of the directors of the Geological Survey of Denmark and the Geological Survey of Greenland.

References

Aitken, J.D. 1967. Classification and environmental significance of cryptalgal limestones and dolomites, with illustrations from the Cambrian and Ordovician of southwestern Alberta. *Jour. Sed. Petrology* 37: 1163-1178.

Anderson, R.Y. & Kirkland, D.W. 1966. Intrabasin varve correlation. *Geol. Soc. Amer. Bull.*, 77: 241-255.

Anderson, R.Y., Dean, W.E., Kirkland, D.W. & Snider, H.W. 1972. Permian Castile varved evaporite sequence, West Texas and New Mexico. *Geol. Soc. Amer. Bull.*, 83: 59-86.

Bein, A. & Land, L.S. 1982. San Andreas carbonates in the Texas Panhandle: Sedimentation and diagenesis associated with magnesium-calcium-chloride brines. Bureau of Econ. Geol., Univ. Texas, Austin. Report of Investigations 121, 48p.

Botz, R., Hiltmann, W., Schoell, M., Teschner, M. & Wehner, H. 1982. Kriterien und Bewertung des Zechstein-Stinkschiefers im hinblick auf sein Erdöl- und Erdgaspotential. *Geol. Jahrb.*, D/47: 113-132.

Butler, G.P. 1969. Modern evaporite deposition and geochemistry of coexisting brines, the sabkha, Trucial Coast, Arabian Gulf. *Jour. Sed. Petrology* 39: 70-89.

Clark, D.N. 1980. The sedimentology of the Zechstein 2 Carbonate Formation of Eastern Drenthe, The Netherlands. *Contr. Sedimentology* 9: 131-165.

Colter, V.S. & Reed, G.E. 1980. Zechstein 2 Fordon Evaporites of the Atwick No. 1 borehole, surrounding areas of N.E. England and the adjacent southern North Sea. *Contr. Sedimentology* 9: 115-130.

Crawford, G.A. & Dunham, J.B. 1982. Evaporite sedimentation in the Permian Yates Formation, Central Basin Platform, Andrews County, West Texas. *Core Workshop* 3: 238-275.

Davies, G.R. & Anderson, R.Y. 1978. Salinity cycles: evidence for subaqueous deposition of Castile Formation and lower part of Salado Formation, Delaware Basin, Texas and New Mexico. *New Mexico Bureau of Mines and Mineral Resources, Circ.* 159: 15-20.

Dean, W. E., Davies, G.R. & Anderson, R.Y. 1975. Sedimentological significance of nodular and laminated anhydrite. *Geology* 7: 367-372.

Dronkert, H. 1977. The evaporites of the Sorbas Basin. *Publ. Inst. Invest. Geol. Barcelona* 32: 55-76.

Dronkert, H. 1978. Late Miocene evaporites in the Sorbas Basin and adjoining areas. *Mem. Soc. Geol. Ital.*, 16: 341-361.

Epistalie, J., Madec, M., Tissot, B., Mennig, J.J. & Leplat, P. 1977. Source rock characterization, method for petroleum explortation. *Proc. 9th Annual Offshore Technology Conference, Houston.* 439-444.

Epstein, S., Bucksbaum, R., Lowenstam, H.A. & Urey, H.C. 1953. Revised carbon-water isotope temperature scale. *Geol. Soc. Amer. Bull.*, 64: 1315-1326.

Füchtbauer, H. 1964. Fazies, Porosität und Gasinhalt der Karbonatgesteine des norddeutschen Zechsteins. *Z. dt. geol. Ges.*, 114: 484-531.

Füchtbauer, H. 1968. Carbonate sedimentation and subsidence in the Zechstein Basin (Northern Germany). In: Müller, G. & Friedman, G. M. (Eds.): *Recent Developments in Carbonate Sedimentology in Central Europe*, 196-204. Springer, Berlin.

Gebelein, C.D. & Hoffman, P. 1973. Algal origin of dolomite laminations in stromatolitic limestone. *Jour. Sed. Petrology* 43: 603-613.

Hardie, L.A. & Eugster, H.P. 1971. The depositional environment of marine evaporites: a case for shallow clastic accumulation. *Sedimentology* 16: 187-220.

Holliday, D.W. 1968. Early diagenesis in Middle Carboniferous nodular anhydrite of Spitsbergen. *Proc. Yorks. Geol. Soc.*, 36: 277-292.

Irwin, H., Curtis, C. & Coleman, M. 1977. Isotopic evidence for source of diagenetic carbonates formed during burial of organic-rich sediments. *Nature* 269: 209-213.

Jacobsen, F.L. 1984. Lithostratigraphy of the Zechstein Salts in the Norwegian-Danish Basin. *Danm. Geol. Unders.*, Ser. C, 1/2: 7-66.

Jacobsen, F.L. & Larsen, J.G. 1982. Permian. *Danm. Geol. Unders*, Ser. B., 8: 28-32.

Kendall, A.C. 1978. Subaqueous evaporites. *Geoscience Canada* 5: 124-139.

Kushnir, J. 1980. The coprecipitation of strontium, magnesium, sodium, potassium and chloride ions with gypsum. An experimental study. *Geochim. Cosmochim. Acta* 44: 1471-1482.

Kushnir, J. 1981. Formation and early diagenesis of varved evaporite sediments in a coastal hypersaline pool. *Jour. Sed. Petrology* 51: 1193-1203.

Kushnir, J. 1982a. The partitioning of seawater cations during the transformation of gypsum to anhydrite. *Geochim. Cosmochim. Acta* 46: 433-446.

Kushnir, J. 1982b. The composition and origin of brines during the Messinian dessication event in the

Mediterranean Basin as deduced from concentrations of ions coprecipitated with gypsum and anhydrite. *Chemical Geol.*, 35: 333-350.

Land, L.S. 1980. The isotopic and trace element geochemistry of dolomite: the state of the art. *SEPM Spec. Publ.*, 28: 87-110.

Lowe, D.R. 1982. Sediment gravity flows: II. Depositional models with special reference to the deposits of high-density turbidity currents. *Jour. Sed. Petrology* 52: 279-297.

Maiklem, W.R., Bebout, D.G. & Glaister, R.P. 1969. Classification of anhydrite - a practical approach. *Bull. Can. Petrol. Geol.*, 17: 194-233.

Middleton, G.V. & Hampton, M.A. 1976. Subaqueous sediment transport and deposition by sediment gravity flows. In: Stanley, D.J. & Swift, D.J.P. (Eds.): *Marine Sediment Transport and Environmental Management*, 197-218. John Wiley & Sons Inc. New York.

Mossop, G.D. 1979. The evaporites of the Ordovician Baumann Fiord Formation, Ellesmere Island, Arctic Canada. *Bull. Geol. Surv. Canada* 298, 52 p.

Olsen, R.C. 1978. Lithology. Well 17/4-1. Norwegian Petroleum Directorate (Oljedirektoratet), *NPD Paper* 14, 26 p.

Ormaasen, D.E., Hamar, G.P., Jakobsson, K.H. & Skarpnes, O. 1980. Permo-Triassic correlations in the North Sea area north of the Central Highs. In: *The Sedimentation of the North Sea Reservoir Rocks.* Norsk Petroleumsforening (NPF) meeting, Geilo 11.-14. May 1980.

Randazzo, A.F. & Zachos, L.G. 1984. Classification and description of dolomitic fabrics of rocks from the Floridan aquifer, USA. *Sediment. Geology* 37: 151-162.

Richter-Bernburg, G. 1955a. Stratigraphische Gliederung des deutschen Zechstein. *Z. dt. geol. Ges.*, 105: 843-854.

Richter-Bernburg, G. 1955b. Über salinare Sedimentation. *Z. dt. geol. Ges.*, 105: 593-645.

Richter-Bernburg, G. 1962. Geologischer Bericht über Ergebnis der bisherigen und Planung der weiteren Exploration auf Kalisalze in Nord-Jutland. In: *Kaliboringerne ved Suldrup 1959-1961*, 1: 108-114. Rapport udarbejdet af Egnsudviklingsrådets Boreudvalg i samarbejde med Saltudvalget og Danmarks geologiske Undersøgelse. København.

Richter-Bernburg, G. 1985. Zechstein-Anhydrite, Fazies und Genese. *Geol. Jahrb.*, A/85, 82 pp.

Rupke, N.A. 1978. Deep clastic seas. In: Reading, H.G. (ed): *Sedimentary Environments and Facies*, 372-415. Blackwell, Oxford.

Sannemann, D., Zimdars, J. & Plein, E. 1978. Der basale Zechstein (A2-T1) zwischen Weser und Ems. *Z. dt. geol. Ges.*, 129: 33-69.

Schlager, W. & Bolz, H. 1977. Clastic accumulation of sulphate evaporites in deep water. *Jour. Sed. Petrology* 47: 600-609.

Schmalz, R.F. 1969. Deep-water evaporite deposition: A genetic model. *Am. Assoc. Petroleum Geologists Bull.*, 53: 798-823.

Schreiber, B.C., Friedman, G.M., Decima, A. & Schreiber, E. 1976. Depositional environments of Upper Miocene (Messinian) evaporite deposits of the Sicilian Basin. *Sedimentology* 23: 729-760.

Seibold, E. 1958. Jahreslagen in Sedimenten der mittleren Adria. *Geol. Rundschau* 47: 100-117.

Shearman, D.J. 1966. Origin of marine evaporites by diagenesis. *Trans. Inst. Min. Metall.*, B 75: 208-215.

Shearman, D.J. & Fuller, J.G. 1969. Anhydrite diagenesis, calcitization, and organic laminites, Winnipegosis Formation, Middle Devonian, Saskatchewan. *Bull. Can. Petrol. Geol.*, 17: 496-525.

Smith, D.B. 1979. Rapid marine transgressions and regressions of the Upper Permian Zechstein Sea. *J. geol. Soc. London*, 136: 155-156.

Smith, D.B. 1980. The evolution of the English Zechstein Basin. *Contr. Sedimentology* 9: 7-43.

Sønderholm, M. 1984. *Dolomit-Anhydrit Overgangszonen (Zechstein-1/Zechstein-2) i det Dansk-Norske Bassin: Sedimentologi og geokemi.* Unpubl. thesis, Univ. of Copenhagen, 223 p.

Taylor, J.C.M. 1980. Origin of the Werraanhydrit in the U.K. Southern North Sea - a reappraisal. *Contr. Sedimentology* 9: 91-113.

Taylor, J.C.M. 1984. Late Permian-Zechstein. In: Glennie, K.W. (Ed.): *Introduction to the Petroleum Geology of the North Sea*, 61-83. Blackwell, Oxford.

Taylor, J.C.M. & Colter, V. S. 1975. Zechstein of the English Sector of the Southern North Sea Basin. In: Woodland, A.W. (Ed.): *Petroleum and the Continental Shelf of North West Europe* 1: 249-263. Appl. Sci. Publ.Ltd., Barking.

Welte, D.H. & Waples, D. 1973. Über die Bevorzugung geradezahliger n-Alkane in Sedimentgesteinen. *Naturwissenschaften* 60: 516-517.

West, I.M. 1964. Evaporite diagenesis in the Lower Purbeck beds of Dorset. *Proc. Yorks. Geol. Soc.*, 34: 315-330.

Ziegler, P.A. 1982. *Geological Atlas of Western and Central Europe.* Shell International Petroleum Maatschappij B.V., 130 p., 40 encl.

SEDIMENTOLOGY AND FACIES DEVELOPMENT OF THE STASSFURT MAIN DOLOMITE IN SOME WELLS OF THE SOUTH OLDENBURG REGION (WESER-EMS AREA, NW GERMANY)

Silvin Mausfeld and Heinrich Zankl
Institut für Geologie und Paläontologie
der Universität Marburg, Lahnberge,
D-3550 Marburg, F. R. Germany

Abstract: In the cored sections of 7 wells of the South Oldenburg region 6 main facies types may be distinguished: (1) Basin-slope facies, (2) shallow-marine facies with low turbulence occurring in three palaeogeographic positions: shallow open marine, lagoonal and sheltered bay, (3) oncoid facies, (4) ooid sand facies, (5) intertidal facies, and (6) supratidal facies. The coated grains of the sediments of the high energy environments are mainly formed by microbiological activities. Facies association corresponds to a prograding "distally steepened ramp" (Read, 1985). The history of deposition is characterized by the development of two distinct phases: (1) a phase characterized by an overall shallowing upward sequence; (2) a phase dominated by the construction of a barrier-lagoon complex. The phases are probably separated by a slight drop of the sealevel.

Introduction

During the last few decades the Stassfurt-cycle carbonates (Ca2) have received growing geologic attention due to the increasing hydrocarbon exploration activities (Richter-Bernburg, 1955a, b; Füchtbauer, 1964, 1968, 1972; Sannemann *et al*, 1978; Clark, 1980). These investigations gave answer to the basic stratigraphic, palaeogeographic and facies-related questions of the Ca2. However, case studies dealing with facies relations and facies development of the Ca2-carbonates of a distinct area have rarely been published so far. The intention of this paper is to present such a case study for a gas field in South Oldenburg. The location of this gas field is shown in Fig. 1.
The lithology and the sedimentological structures of the cored sections of seven wells have been investigated for the study.

Lecture Notes in Earth Sciences, Vol. 10
T.M. Peryt (Ed.), The Zechstein Facies in Europe

The sediments consist of dolomites and secondary limestones with six main facies types to be distinguished in the cores, which are briefly characterized in the following sections.

The association of facies types in the gas field as inferred from the cores and wire-line data is shown in Fig. 2.

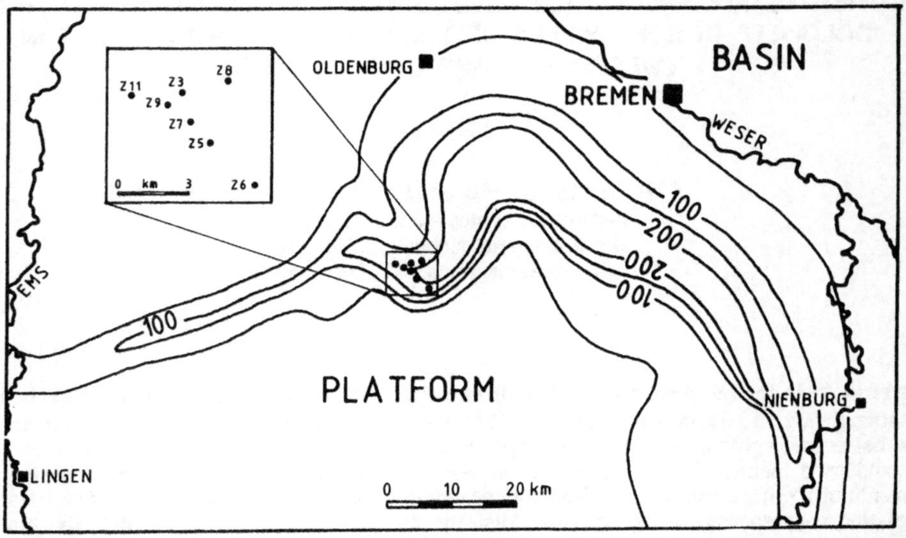

Fig. 1. Map showing the location of the study area and the thickness of the Ca2 sediments (in metres) in the Weser-Ems area. The black spots represent the studied wells (after Sannemann *et al.*, 1978).

Basin-slope facies

This facies type forms one of the thickest units in the studied area. It consists of dark mudstones and packstones with small peloids (50-100 µm). The sediments are distinctly evenly laminated; the thickness of the laminae varies between less than a mm to a few centimetres. Interbedding of cm-thick, lighter layers with darker mm-laminated layers is common. The lamination tends to be more undulating in the upper parts of the facies where the thickness of the laminae increases.

The content of biogenic remains is low. Apart of a few ostracods and forams, lumachelle beds formed mainly of thin-shelled pelecypods occasionally occur.

These sediments may show a variety of early deformation features:
(1) Early diagenetic transformation of the even bedding to a lenticular or flaser bedding.
(2) Formation of non-tectonic fissures, restricted to sedimentary layers only some cm to dm thick. These fissures are steeply inclined and often two shear planes are developed. Both compressional and delatational structures may be recognized.

(3) Plastic deformation with sedimentary folds, more rarely boudinage and development of relief at the sedimentary surface.
(4) Brecciation of somewhat lithified sediments. The clasts of these breccias show subrounded and angular forms. The breccias show a varying amount of matrix in the interstices.

The lenticular to flaser type bedding may have originated by the downslope creep of unconsolidated muds inducing the formation of small subhorizontal shear planes in the sediments.

Synsedimentary fissures may be produced in somewhat more cohesive sediments as a compressional feature at the downslope end of a glide and by dilatation in the upper parts of a glide or slump.

Plastic deformation and brecciation of the sediments are often connected to rotational slides. During the downslope movement of a sediment mass, growing deformation of the sediment may lead to a complete brecciation of initially consolidated sediments (Cook & Mullins, 1983).

During the deposition of this facies a quiet low-energy environment below normal wave-base is indicated by a relatively thin even lamination. The processes of gravitational mass transport interrupting this quiet-water sedimentation were initiated by an instability of the deposited sediments on an inclined depositional slope. Rare arenitic layers may be regarded as turbidite deposits although grading is rare and Bouma sequences are absent. The more undulating lamination of the upper part of the facies might reflect storm effects.

The scarcity of biogenic relicts and the absence of bioturbation may be caused by an anaerobic environment in a density-layered water body (Schmalz, 1969).

Shallow marine facies with low turbulence

This facies type consists mainly of packstones with fine peloids and mudstones and shows a wide range of sedimentary structures. Flaser bedding, wavy bedding, and small-scale ripple lamination are common. The bedding is more discontinuous and irregular compared with the basin-slope facies and the beds show greater thickness.

The facies is differentiated in three shallow marine subenvironments, all having low turbulence as a common feature, yet differing from one another by

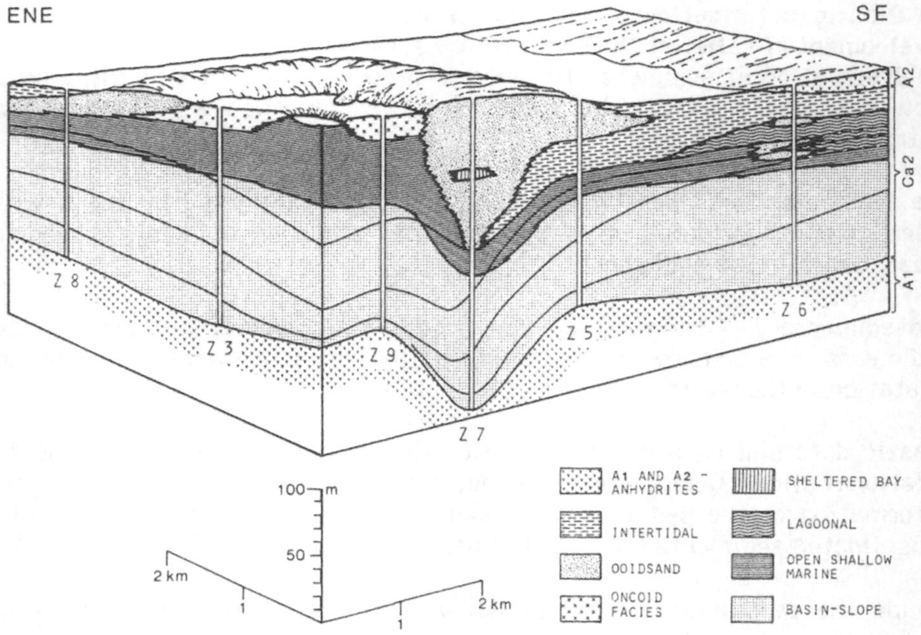

Fig. 2. Diagram showing the facies associations in the studied area. Vertical double lines represent the positions of the wells. The morphology and facies association on top of the sketch is an inferred picture of late Ca2 palaeogeography of the area.

their position within the stratigraphic sequence and by some sedimentary structures.

The small current ripple lamination and the wavy or flaser bedding developed in each subenvironment indicate the presence of low velocity currents, not strong enough to winnow the carbonate mud. This weak current activity, together with the stratigraphic positions of the subfacies suggest a shallow marine environment of formation for all three subfacies. Differences within subfacies are controlled by their different palaeogeographic positions.

The *open shallow marine subfacies* develops concordantly from the basin-slope facies and it forms the thickest unit of the three subfacies (20-45 m). Moreover, it contains some oolitic packstones and shows relative abundance of fossil remains consisting of ostracods, forams (*Nodosaria*-type and less commonly *Ammodiscus*-type forms) and pelecypod shells.

The *lagoonal subfacies* is about 18 m thick and develops from high energy shallow marine oolites by diminishing agitation. The biogenic content of these sediments is low. Sedimentary deformation features, such as brecciation, contorted and disrupted layers, are present.

Though indicating a local depositional slope these features are not thought to have originated in a basin slope environment. The overall stratigraphic position between two shallow marine oolites, the palaeogeographic position of the sediments behind an oolitic barrier and the type of bedding exhibited do not favour a deep water origin of the sediments.

The *third subfacies* is found intercalated within the high-energy oolite facies of the Z7-well. Its dark sediments are rather thin (about 5 m) and although unconformably overlie algal boundstones and grainstones, grade conformably into the coarse intraclastic sediments. This subfacies differs from the one found in other subenvironments by the occurrence of algal mats, which produced "rip-up clasts" during rare high energy events. It contains some forams and pelecypod shells and shows bioturbation. The sedimentary structures and the lack of shrinking features and fenestral pores suggest this facies was continually submerged and probably was formed in a very shallow sheltered bay.

Oncoid facies

This facies consists of an about 20 m thick succession of grainstones with coarse (1-15 mm) and fine (0.1-1 mm) coated grains, small lumps and peloids between 30 and 150 µm in diameter, and some pelecypod shells.

The sedimentary structures consist of a rather indistinct cross bedding with some sets exceeding 5 cm in height and, less frequently, roughly horizontally bedded sections.

The sediments show a small-scale interbedding of loosely packed grainstone layers with coarse coated grains and layers of more densely packed grainstones with coated grains of finer sizes, peloids and small aggregates.

This facies type may develop through an increasing frequency of coarse grainstone layers intercalated in shallow open marine peloid packstones or mudstones. The transition to supratidal "Basalanhydrit" facies is marked by thin, finer grained, possibly intertidal sediments.

Description of the coated grains

They consist of a nucleus and a surrounding cortex. The nucleus may be formed by a peloid, an aggregate grain or more rarely by a bioclast. Often the origin of the nucleus is obscure. The cortex is made of an interlamination of 4-10 µm thick micritic laminae and microsparitic laminae ranging in thickness between 5 and 30 µm.

In the smoother laminated cortices this fine lamination resembles the cortex-lamination of the Ca2 ooids. More commonly the generally concentric lamination of the cortex displays small discontinuities and irregularities, which become more frequent near the outer rim of the cortex.

Agglutinated peloids, coated grains, and even layers of small peloids, which are frequently glued to the cortices, may be incorporated by overgrowing of the cortex-lamination. The cortices of the bigger particles are rather thick but the smaller grains may exhibit just a thin cortex-lamination around a nucleus of varying origin.

Genesis of the coated grains

Füchtbauer (1964) interpreted the coated grains of the Ca2 carbonates as oncoids, but even in excellently preserved particles no signs of algal filaments have been observed, and spongiostromate fabric is lacking. Clark (1980) questioned this interpretation and suggested that the irregularly coated grains were "vadose pisolites", which were formed in a meteoric vadose environment during early diagenesis of the sediments by precipitation of laminated coats around a grain or a group of grains. Peryt (1983a, fig. 6b, e, c) published photographs of similar grains as examples for "vadoids" - coated grains which originated in a vadose environment.

Peryt (1983a) listed and discussed a set of characteristic features described in the literature on vadoid-bearing sediments of various age and settings.

In addition to the fabric any interpretation as vadoids depends on (1) the observed geologic profile (James, 1972; Arakel, 1982), (2) the kind of crystal fabric, and (3) the initial mineralogy (Chafetz & Butler, 1980).

Table 1 shows that the coated grains of this facies do not appear to be products of a meteoric vadose environment. An alternative interpretation of these grains as marine vadose products (Esteban & Pray, 1983) is favoured by the presence of primary sedimentary structures, but the sediments still lack a set of characterictic features (see Table 1), e.g., early brecciation, inverse grading, and *in situ* coating of the particles. Some grains are asymmetrically elongated in vertical directions but might equally be produced as an effect of sectioning of irregular grains.

No hints of a peritidal formation of the sediments could be detected in the cores (eg., brecciation by development of tepee structures). Only the overall stratigraphic position in the upper parts of the Ca2 sediments near the transition to supratidal "Basalanhydrit" conditions might correspond to the postulated position of marine vadose formations.

Given these findings a vadoid origin of the coated grains described seems not very probable.

Table 1. Comparison between the more important features of meteoric and marine vadoids, as discussed in the literature, and the features of the grainstones of the Ca2 oncoid facies. xxx = abundant, xx = common, x = rare.

Criteria	Meteoric vadoids	Marine vadoids	Ca2 grains
Multiple stage coated grains (Esteban & Pray, 1983; Arakel, 1982)	xxx	xxx	xxx
Development of a distinct soil profile (James, 1972; Arakel, 1982)	xxx	x	-
Laminated horizons (James, 1972; Arakel, 1982; Read, 1976)	xxx	x	-
Massive calcrete (Arakel, 1982)	xxx	-	-
Secondary mud-supported fabric (Read, 1976)	xxx	-	-
Brecciation (Read, 1976; Arakel, 1982)	xxx	xxx	-
Solution unconformities (Read, 1976)	xx	-	-
Thickness of sediments > several m (Arakel, 1982)	x	xxx	xxx
Sedimentary structures	x	xxx	xxx
Brecciation and recoating of the particles (Esteban &Pray, 1983; Bernoulli & Wagner, 1971)	xxx	xxx	-
Inverse grading (Esteban &Pray, 1983)	xxx	xxx	-
Polygonal fitting (Dunham, 1969)	xx	xxx	x
Vertical elongation of the grains (Dunham, 1969)	xx	xx	x
Vadose crusts (Scholle & Kinsman, 1974; Peryt, 1981a)	xx	xx	-
Rhombic sparry calcite (Chafetz & Butler, 1980)	xx	-	-

On the other hand the coated grains do not coincide with the "classical" oncoid model. The ooid-like cortex-lamination in more smoothly laminated cortices corresponds neither to the structures of recent oncoids (Schäfer & Stapf, 1978; Monty, 1967) nor to features of fossil porostromate and spongiostromate oncoids (Peryt, 1981b).

This ooid-like microstructure of the cortex might best be interpreted according to the findings of Fabricius (1977), who suggested that ooids are formed by the activity of non-calcifying cyanobactèria and stated that "the transition between ooids and oncoids is quite likely, since both are formed organically".

In their regional study of Ca2 carbonate facies Sannemann *et al.* (1978) found, "that there are many transitions between classical ooids and oncoids via irregularily formed grains".

The irregular coated grains of the Ca2 are therefore more aptly adressed as such transitions between ooids and oncoids. This cyanobacterial formation of the coated grains must have taken place in an environment of intermittent water agitation which transported the particles and winnowed the carbonate mud. During more quiet periods the agglutination of peloids and coated grains to the particles occurred while during more agitated periods the formation of the cortex-lamination was favoured. Continued higher agitation led to the development of more smoothly laminated more ooid-like cortices while moderate and more intermittent agitation caused the formation of irregular and multiple-coated grains.

Ooid sand facies

Grainstones and rudstones with ooids and aggregate grains form the bulk of this facies. The sediments are predominantly cross-bedded (angular and tangential), with heights of the sets ranging about 5 cm, rarely reaching 25 cm.

Two types of sediments may be distinguished:

(1) Moderate to well sorted ooid sand with only sparse aggregate grains with beds ranging in thickness from 3 to 20 m.

Fig. 3. (a) Oolitic grainstone from the Z6 well with dark, algal layers of densely packed, small micritic particles interbedded with more loosely packed layers of larger, recrystallized ooids. The centres of these ooids are leached and sometimes display internal sediment. A rather dense packing of the particles results from vadose compaction. **(b)** Fairly well sorted oolitic grainstone from the Z5 well with leached ooids. Completely leached ooids indicate that leaching postdates initial rim cementation. Cements now consist of subhedral to euhedral sparry dolomite. Some ooids are deformed after cementation and leaching (arrow). **(c)** Loosely packed oolitic grainstone from the Z7 well consisting of differently sized ooids, lumps and intraclasts. Grainstone displays interbedding of poorly sorted layers with a high percentage of smaller particles and better sorted, coarse layers with winnowed fines. **(d)** Poorly sorted rudstone with aggregate grains and ooids. Many of the aggregate grains show somewhat irregularly shaped compound subgrains. Sediment strongly compacted.
Length of black and white bar: 2 mm.

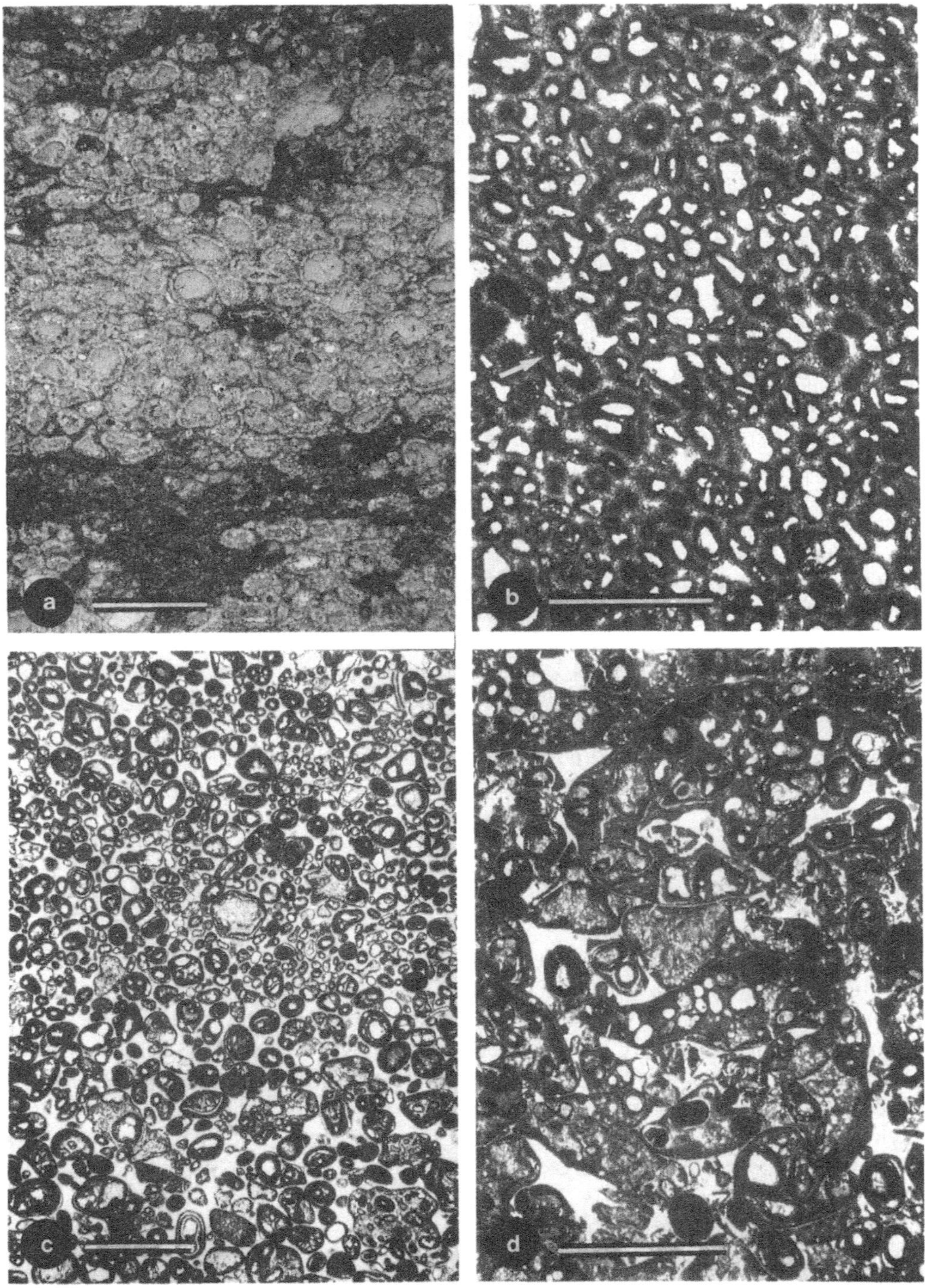
a
b
c
d

The ooid size varies according to the local level of energy (Bathurst, 1975). In only moderately agitated environments (as in the Z8 grainstones) small ooids (100-500 µm) are formed, while in higher turbulent environments ooids with sizes up to more than 1 mm (rarely to 4 mm) may be developed in the Z5 and Z6 wells (Fig. 3a, b). Within the foreset lamination the oolite of the Z6 well shows a fine interbedding of loosely packed ooids and densely packed layers of micritized ooids indicating algal activity (Fig. 3a).

(2) The second type of oolitic sediment mainly occurs in the Z7 well and to a much less extent at the top of the Z5 core sections. This type is characterized by a high percentage of coarse particles in the oolitic sediments, which normally vary in size between 0.5 and 5 mm, but may exceed 35 mm (Figs. 3b, c & 4a).

In the Z7 well a succession of more than 100 m of this carbonate sands is developed with some interruptions of relatively thin sediments of less agitated facies. In the cored section of this well, which does not transsect the whole oolitic succession, the coarse grains can either be admixed to the ooid sands to form poorly sorted sediments or, more common, in the upper part of the succession, form discrete coarse rudstone layers some dm thick (Fig. 4a). These layers may show some grading and imbrication of the clasts and - if well developed - are interbedded with fine grained oolitic sands. In this well sorted oolitic sand one can commonly find thin micritic algal layers, however, some thicker algally formed beds occur independently of these sands (Fig. 4b). Fenestral pores are rarely developed in these grainstones.

In the Z5 well, well sorted oolitic sands grade upwards into coarse, badly to moderately sorted layers intercalated with fine grained layers near the top of the cored sections.

Description of the ooids

The centres of the ooids are often leached, so that moldic pores are formed. In non-leached ooids, however, the nuclei mostly seem to consist of some peloid particle or, more rarely, of a biogenic fragment (Fig. 3b, c).
The cortices are smoothly laminated and are often formed of an interlamination of darker, micritic laminae (2-4 µm thick) and light microsparitic laminae (4-8 µm thick), that sometimes may show incipient radial orientation of the crystallites. The cortices may be strongly affected by micritization and their thickness is correlated to the size of the ooids.

Description of the aggregate grains

Two types of aggregate grains may be recognized:

(1) Grains mainly consisting of ooids, which are cemented by a 20-80 µm thick, micropeloidal micritic cement, exhibiting meniscus characteristics. In these

compound grains the ooids are loosely packed and often fenestroid pores are developed between them (Fig. 4a).

The surface of these aggregate grains may be grapestone-like and is often coated with a relative thin cortex, making the grains to resemble the "botryoidal lumps" of Illing (1954). This cortex is dense and oolitic on the protruding subgrains, but may sometimes consist of more loosely packed fine peloids especially in the surficial depressions between the subgrains. The grains may be abraded before or after being coated and can be called intraclasts, then. The form of the particles vary from subrounded lumpy to chiplike.

Multiple-stage aggregate grains are made up of ooids, peloids and smaller irregular shaped and coated compound grains, that again may enclose ooids, peloids, and smaller aggregates (Fig. 3d). In these particles the subgrains also are cemented by micrite. The coatings of the compound subgrains sometimes seem to pass into the micritic cements.

Genesis of the aggregate grains

Clark (1980) interpreted the simple oolitic aggregates as eroded beachrock clasts. This interpretation corresponds to the meniscus features of the micritic cements and to the coatings of the aggregates. The often grapestone-like shape of the particles indicates only little erosion of the grains during transport. However, unreworked beachrock layers with their typical stratification and heavy vadose cementation could not be observed in the cores.

The shape of the aggregate grains may be the result of a grapestone-like formation of the grains, and the components may thus be bound by cements formed by algal activities. Later reworking produces the clasts. Such a grapestone-mode of formation does not require the genesis of the clasts in a separate environment (Bathurst, 1975).

A formation of the aggregate grains in the shoal environment is possible as indicated by the findings of Dravis (1979), who observed thin oolitic hardgrounds with algally induced lithification at Schooners Cay, Bahamas. The degree of lithification decreases from the sedimentary surface downwards.

In the multiple-stage grains several processes seemed to have been operating, but clear criteria for the genesis of these grains are not easily detected.

In these aggregate grains now and then micritic beardlike, but unoriented apophyses of the grains are developed, which may contain a smaller particle. In one aggregate grain a cortex lamina of a subgrain vanishes into the surroundings of the grain and is cut by the boundary of the aggregate.
These features might be interpreted as products of marine vadose diagenesis,

followed by reworking and redeposition. However, classical features of vadose diagenesis as brecciation and recoating of the grains or inverse grading are lacking. Moreover, these rare vadose features are restricted to the aggregate grains and no associated vadose cementation features are developed in the grainstones-rudstones. Thus clearly vadose features are rare, but multiple-stage aggregate grains form an important part of the sediments in the Z7 well. Other processes of grains formation may therefore be more important than vadose diagenesis.

Multiple reworking, coating and redeposition of the above mentioned algally cemented lumps are thought to have resulted in the formation of multiple-stage aggregate grains. As indicated by micritic algal mats occurring in close, even intercalating (Fig. 4b) association with them, a microbial origin of the grains might be the dominant process in their formation. The occurrence of thick finely peloidic agglutinated layers in the surficial depressions of the grains also suggest an algal origin of the grains. However, in a shoal environment the occurrence of some products of vadose diagenetic processes are to be expected.

The reworked coarse lumps and intraclasts may be swept together to form bars as described by Imbrie and Buchanan (1965), which may only migrate during high energy episodes.

Thus various microenvironments, each with their own characteristic energy level may have provided (1) oolitic sediments, (2) sediments with lumps and intraclasts, and (3) algal mats.

Intertidal facies

This facies type occurs mainly in the Z5 and Z6 wells with a thickness of about 29 m and subordinately at the top of the Z8 and Z3 wells. Two types of sediment are developed:

Fig. 4. (a) Rudstone from the Z7 well with coarse aggregate grains (lumps and intraclasts). The subgrains in these aggregates are cemented by a micritic cement with meniscus features. Note the thin coating of some grains. Grain in centre consists of several layers of agglutinated peloids (arrow). The interparticle porosity of the sediment has been partly filled with small ooids. **(b)** Bindstone from the Z7 well. Dark, irregular micritic algal layers in a poorly sorted sand with ooids, lumps and intraclasts. **(c)** Grainstone of the oncoid facies from the Z9. Loosely packed layers of coarse coated grains are interbedded with more densely packed layers with fine lumps and peloids. The coarse coated grains consist of a single or aggregate nucleus and a cortex with a fine, partly irregular lamination. Sediment partly recrystallized and strongly replaced by anhydrite. **(d)** Detail of the coarse sediment of the oncoid facies from the Z9, showing the irregular lamination of the cortices and the small lumps.

Length of black and white bar: 2 mm.

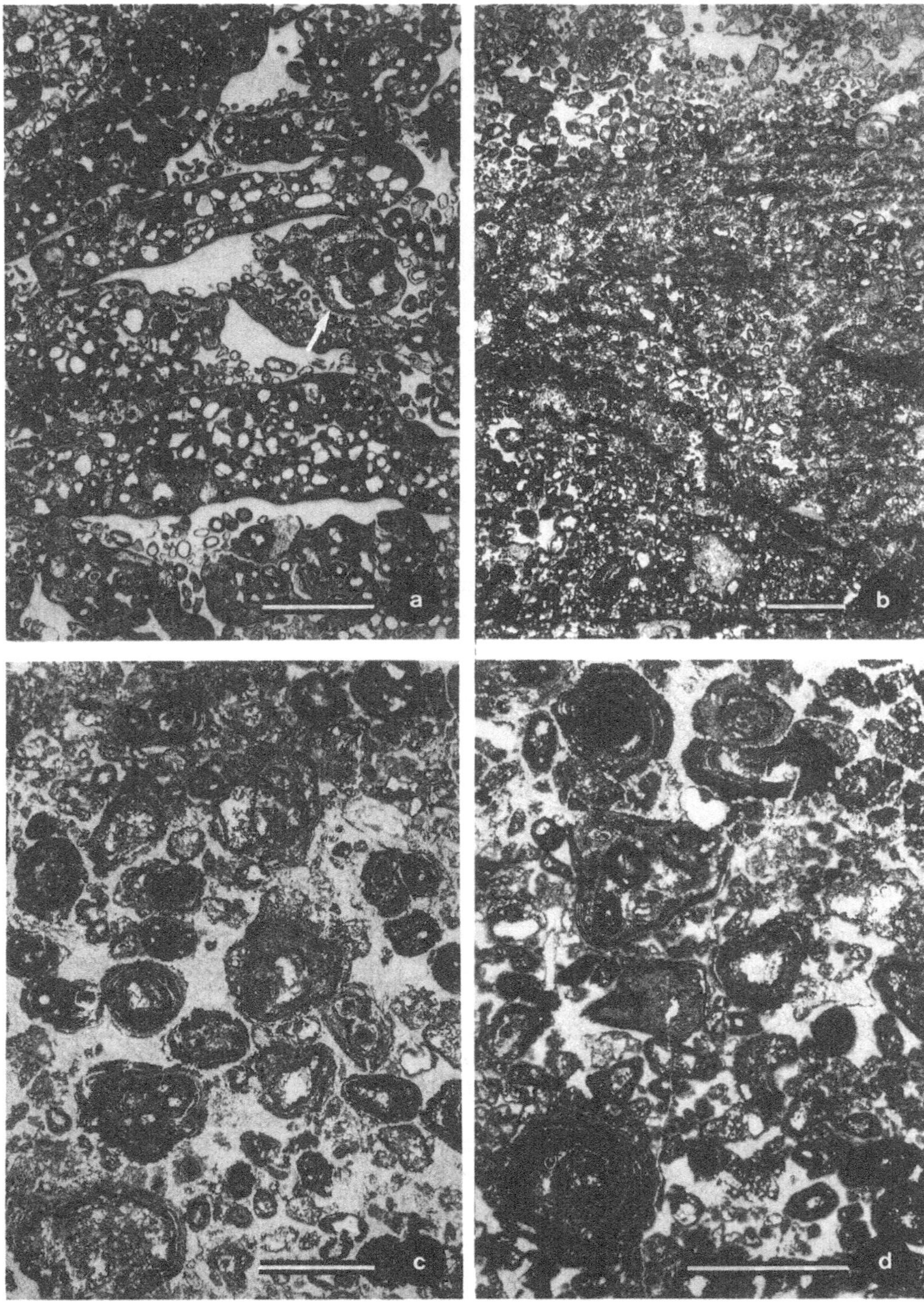
a
b
c
d

(1) The first type, representing the bulk of the intertidal sediments, consists of fine grained peloid packstones, with sizes of the peloid ranging between <50 and 150 µm, and some grainstones with ooids and/or peloids. In some areas dense packing of soft peloids has formed a mudstone-like fabric.

In these fine grained sediments a variety of sedimentary structures can be found. Small scale ripple lamination, with sets less than 1 cm in height, fine discontinuous laminations, and flaser bedding are commonly observed. Laminations with fine irregular boundaries of the laminae may be indicative for algal activities. Small channels with graded infills, sometimes showing imbrication of reworked early lithified intraclasts, are less common.
In higher parts of the Z6 well sequence irregular shrinking and dessication cracks are developed, especially near an anhydritic supratidal intercalation and a section rich in anhydritic nodules in the Z5 well.Bioturbation is generally not very common but can be observed in some cases.

(2) The second type of sediment is found only at the top of the Z6 well intertidal facies. Here, grainstones and rudstones are found in a coarsening upward sequence, about 2.5 m thick. The change from fine grained peloidal and oolitic packstones or grainstones is brought about by the upward increasing abundance of coarse clasts consisting of big ooids, lumps and intraclasts, which are either mixed with fine sediments or form discrete coarse layers with particle sizes up to 5 mm.
These coarser sediments are cross-bedded or are roughly horizontally bedded with fenestral pores.

Upwards the sediments pass into a 0.4 m thick dark horizon of algal mats, that mark the transition to supratidal "Basalanhydrit" facies.

An intertidal environment of deposition is indicated by a variety of observations:

(1) Channels, which are a typical feature of intertidal environments (Shinn, 1983); (2) shrinking and dessication cracks, which indicate temporary subaerial exposure of high intertidal sediments; (3) intercalation of a mosaic anhydrite layer, which must have occurred due to the proximity of supratidal environment; (4) the discontinuous and flaser bedding, which was caused by changing water energy; and (5) the possible algal structures.

The coarse sediments below the transition to the supratidal facies in the Z6 well might be interpreted as produced within a beach ridge. Such an interpretation is suggested by the association of rough horizontal bedding, cross-bedding and fenestral pores.

Supratidal facies of the "Basalanhydrit"

Anhydritic rocks and some intercalated dolomitic layers are the main features of this facies.

Several types of anhydrite are developed:

(1) A brownish anhydrite coloured by a host of dolomitic relics, that show a cloudy or streaky texture abd display an irregular bedding. Small spots of dolomite remain between the abundant small (<5 mm) nodules of anhydrite, which may be coarsely aligned according to the bedding.

(2) Less frequent white "chicken-wire" anhydrite occurs consisting of mm to cm sized nodules with the only relics of the former dolomitic sediment being the nearly opaque insoluble seams between the anhydritic bodies.

(3) Rare enterolithic layers of anhydrite may occur.

The different types of anhydrite tend to form separate layers, but may also be intergradational. The fabric of the anhydrite corresponds to the filthy fabric of lath-shaped crystals described by Shearman (1978), with crystal sizes up to 20 µm.

In these anhydritic rocks, a few cm to dm thick dense dolomitic layers may be intercalated, still showing sharp boundaries towards the anhydrite. However, in some instances anhydrite can replace parts of these layers. The microcrystalline dolomites are finely subhorizontally to wavily laminated. Occasionally some shrinking features and dessication cracks may be developed.

The evaporitic sediments may have been formed in a sabkha environment similar to that described from the coasts of the Persian Gulf (Bush, 1973; Kendall, 1984). The precipitation of displacive and replacive anhydrite in the capillary water fringe and of gypsum below the groundwater table may both have contributed to the formation of the A2 sulphate deposits. Gypsum may later be dehydrated to anhydrite during diagenesis.

The dense dolomitic layers are interpreted as being formed by algal mats (Kendall & Skipwith, 1968) and may represent episodes of intercalated high intertidal conditions in the generally supratidal environment.

History of deposition

First phase of deposition

At the end of the Werra Anhydrite a rise of sea level caused flooding of the thick (up to 300 m) sulphate platform that was built during A1 time at the southern margin of the basin (Sannemann *et al*, 1978).This flooding initiated carbonate production. In the deeper marine basin-slope environments in front of the sulphate platform the change from sulphate to carbonate deposition led to the formation of a carbonate basin-slope facies.

In more shallow marine conditions an open marine facies with low turbulence was formed, which prograded basinwards together with the basin slope facies. This resulted in the formation of a carbonate ramp, extending northward of the sulphate platform, while high-energy shoal and intertidal conditions were established on the platform (Sannemann *et al*, 1978; Clark, 1980).

This progradation produced a regressive facies sequence from basin-slope facies to high-energy oolitic shoal facies in the Z6 well. Basinwards of this ooid sand shoal a shallow marine "shelf" environment was established (comp. Fig. 2), which caused the scarcity of oolitic turbidites in the basin-slope facies.

During this phase of deposition a slightly enhanced subsidence in the area of the Z7 well is indicated by the higher sediment thickness of the basin-slope facies.

At the end of the first depositional phase a slight drop of the sea level caused the displacement of the high-energy oolite shoal environment from the Z6 area to a more basinward position in the Z7 area. Meteoric vadose and phreatic dissolution features in the former, more landward oolitic facies may mark a temporary subaerial exposure. However, no karst- or caliche-related features could be observed. Clark (1980) argued in favour of falls in sea level during Ca2 times and J. Paul (pers. comm., 1986) observed omission surfaces in the Ca2 of the Harz region.

Second phase of deposition

A slow rise in sea level led to the construction of an oolitic barrier in the Z7 area, while intertidal and lagoonal conditions were established landward of this barrier.

The new oolitic barrier was fairly stationary during the course of the deposition of oolitic sand. This compensated the slightly enhanced subsidence of the region, thus forming a thick sequence of oolitic sediments.

The intertidal-lagoonal complex landward of the barrier was slowly filled during a renewed regressive development in the area of the Z6 well, culminating in the formation of supratidal anhydrites.

In front of the barrier the open shallow marine facies and basin-slope facies again began to prograde basinwards and constructed a shallow muddy "shelf" area in front of the oolitic barrier.

During later Ca2 times the oolitic barrier could expand in southerly (Z5) and northeasterly directions (Z8) forming a broad oolitic belt with differentiated subenvironments.

In front of the barrier in late Ca2 times an oncoid facies was established in a moderately agitated environment, reflecting the changes to more restricted conditions with more elevated salinities in the basin before the transition to evaporitic deposition of the "Basalanhydrit".

Suggested facies model

The association of facies types indicates deposition on a "distally steepened ramp" (Read, 1985).

"Muddy shelf" sediments of the deep ramp are developed seaward of the ooid sand shoals and the break in slope occurs well basinward of the high-energy shoal environment. Thus, the onset of deformation features in the basin-slope facies demonstrates the increased depositional angle, but the sedimentary breccias of this facies lack clasts of the high-energy shoal sediments.

Conclusions

Genesis of the particles in the Ca2 of the region is probably more controlled by microbial activity than by physico-chemical processes.

Facies association corresponds to a prograding "distally steepened ramp", with an overall regressive facies sequence.

Two depositional phases can be distinguished, which are probably separated by a slight fall of the sea level. (1) The first phase of deposition resulted in the construction of a shallow marine "shelf" area in front of the Z6 oolite shoal. Changes in sea level resulted in a rather quick displacement of the oolite shoal conditions to a more basinward position. (2) Construction of a barrier-lagoon complex, with development of intertidal and lagoonal facies landward of an oolite barrier. This barrier was stationary during longer time intervals in spite of the overall progradation of the sedimentary wedge.

Acknowledgements

We want to thank BEB Ergas und Erdöl for the permission to publish this study. BEB and Mobil Oil provided the core material and BEB funded the work. Dr. J. Zimdars helped with many fruitful discussions and Dr. G. Multer reviewed the manuscript and made helpful comments.

References

Arakel, A.V., 1982. Genesis of calcretes in Quaternary soil profiles, Hutt and Leeman Lagoons, Western Australia. *Jour. Sed. Petrology* 52: 109-125.

Assereto, R. & Folk, R.L., 1976. Brick-like texture and radial rays in Triassic pisolites. *Sediment. Geology* 16: 205-22.

Bathurst, R.G.C., 1975. Carbonate sediments and their diagenesis. Sec. enl. ed., *Developments in Sedimentology* 12, 658 pp. Elsevier.

Bernoulli, P. & Wagner, C.W., 1971. Subaerial diagenesis and fossil caliche deposits in the Calcare Massiccio Formation (Lower Jurassic, Central Apennines, Italy). *N. Jb. Geol. Paläont. Abh.*, 138: 135-149.

Bush, P., 1973. Some aspects of the diagenetic history of the sabkha in Abu Dhabi, Persian Gulf. In: Purser, B.H. (Ed.), *The Persian Gulf*, 395-408. Springer, Berlin.

Chafetz, H.S. & Butler, J.C., 1980. Petrology of recent caliche pisolites, spherulites, and speleothem deposits from central Texas. *Sedimentology* 27: 497-518.

Clark, D.N., 1980. The sedimentology of the Zechstein 2 Carbonate Formation of eastern Drenthe, The Netherlands. *Contr. Sedimentology* 27: 131-165.

Cook, H.E. & Mullins, H.T., 1983. Basin margin environments. *Am. Assoc. Petroleum Geologists Memoir* 33: 539-617.

Dravis, J., 1979. Rapid and widespread generation of recent oolitic hardgrounds on a high energy Bahamian platform, Eleuthera Bank, Bahamas. *Jour. Sed. Petrology* 49: 195-208.

Dunham, R.J., 1969. Vadose pisolites in the Capitan Reef (Permian), New Mexico and Texas. *SEPM Spec. Publ.*, 14: 182-191.

Esteban, M. & Pray, L.C., 1983. Pisoids and pisolite facies (Permian), Guadalupe Mountains, New Mexico and West Texas. In: Peryt, T. M. (Ed.), *Coated Grains*, 503-537. Springer, Heidelberg.

Fabricius, F.H., 1977. Origin of marine ooids and grapestones. *Contr. Sedimentology* 7: 113 pp.

Füchtbauer, H., 1964. Fazies, Porosität und Gasinhalt der Karbonatgesteine des norddeutschen Zechsteins. *Z. dt. geol. Ges*, 114: 484-531.

Füchtbauer, H., 1968. Carbonate sedimentation and subsidence in the Zechstein Basin (North Germany). In: Müller, G. & Friedman, G. M. (Eds.), *Recent Developments in Carbonate Sedimentology in Central Europe*, 196-204. Springer, Berlin.

Füchtbauer, H., 1972. Influence of salinity on carbonate rocks in the Zechstein Formation, north-western Germany. *Earth Sciences* 7: 23-31.

Illing, L.V., 1954. Bahamian calcareous sands. *Am. Assoc. Petroleum Geologists Bull.*, 38: 1-95.

Imbrie, J. & Buchanan, H., 1965. Sedimentary structures in modern carbonate sands of the Bahamas. *SEPM Spec. Publ.*, 12: 149-172.

James, N.P., 1972. Holocene and Pleistocene calcareous crusts (caliche) profiles: criteria for subaerial exposure. *Jour. Sed. Petrology* 42: 817-836.

Kendall, A.C., 1984. Evaporites. In: Walker, R.G. (Ed.), *Facies Models*, 259-296.

Kendall, C.G.St.C. & Skipwith, P.A.d'E., 1968. Recent algal mats of a Persian Gulf lagoon. *Jour. Sed. Petrology* 38: 1040-1058.

Monty, Cl.V., 1967. Distribution and structure of recent stromatolitic algal mats, Eastern Andros Island, Bahamas. *Ann. Soc. Geol. Belgique* 90: 55-100.

Peryt, T.M., 1981a. Former aragonitic submarine hemispheroids associated with vadose deposits, Zechstein Limestone (Upper Permian), Fore-Sudetic area, Western Poland. *N. Jb. Geol. Paläont. Mh.*, 1981: 559-570.

Peryt, T.N., 1981b. Phanerozoic oncoids - an overview. *Facies* 4: 197-214.

Peryt, T.M., 1983a. Vadoids. In: Peryt, T.M. (Ed.), *Coated Grains*, 437-449. Springer, Heidelberg.

Peryt, T.M., 1983b. Coated grains from the Zechstein Limestone (Upper Permian) of Western Poland. In: Peryt, T.M. (Ed.), *Coated Grains*, 587-598. Springer, Heidelberg.

Read, J.F., 1976. Calcretes and their distinction from stromatolites. In: Walter, M.R. (Ed.), Stromatolites. *Developments in Sedimentology* 20: 55-71. Elsevier.

Read, J.F., 1985. Carbonate platform facies models. *Am. Assoc. Petroleum Geologists Bull.*, 69: 1-21.

Richter-Bernburg, G., 1955a. Übersalinare Sedimentation. *Z. dt. geol. Ges.*, 105: 593-645.

Richter-Bernburg, G., 1955b. Stratigraphische Gliederung des deutschen Zechsteins. *Z. dt. geol. Ges.*, 105: 843-854.

Sannemann, D., Zimdars, J. & Plein, E., 1978. Der basale Zechstein (A2-T1) zwischen Weser und Ems. *Z. dt. geol. Ges.*, 129: 33-69.

Schäfer, A. & Stapf, K.R.G., 1978. Permian Saar-Nahe Basin and Recent Lake Constance (Germany): Two environments of lacustrine algal carbonates. *IAS Spec. Publ.*, 2: 83-107.

Scholle, P.A. & Kinsman, D.J.J., 1974. Aragonitic and high Mg-calcite caliche from the Persian Gulf - A modern analog for the Permian of Texas and New Mexico. *Jour. Sed. Petrology* 44: 904-916.

Schmalz, R.F., 1969. Deep water evaporite deposition: a genetic model. *Am. Assoc. Petroleum Geologists Bull.*, 53: 798-823.

Shearman, D.J., 1978. Evaporites of coastal sabkhas. *SEPM Short Course Lecture Notes* 4: 6-42.

Shinn, E., 1983. Tidal flats. *Am. Assoc. Petroleum Geologists Memoir*, 33: 171-210.

THE ZECHSTEIN SULPHATES: THE STATE OF THE ART

Rolf Langbein
Section of Geological Sciences, University of Greifswald
Jahnstraße 17a, DDR-2200 Greifswald, GDR

Abstract: The origin and preservation of the features observed in the Zechstein sulphates exclusively depend on their diagenesis. The main factor of the diagenetic feature overprint is the cementation. It is controlled by the stability relations at the bottom of the basin. Different compensation-depth lines can be distinguished. The primary lithofeatures are preserved by the halite or anhydrite cementation. Thereby sedimentites may be recognised as inorganogenic or organogenic sedentites. By means of the gypsum cementation rocks are formed which are deformed during the diagenetic phase change from gypsum to anhydrite. The plastic deformation ranges from enterolites up to large folded blocks. Cementation which has been partial or which has not taken place favours total or differential compaction. During this process partial compactites (stratolites) and compactites (stratobolites) are produced. The breccia structures frequently occurring in the Zechstein anhydrites must be assigned to different processes. During the resedimentation they formed as intraclastites, during the progressive diagenesis as residual or collapse breccias. Completely new feature elements can be produced by retrograde gypsification. Compactites and cementites differ in their thickness in the ratio 1:4 to 1:5. The cementites are formed in shallow water at the margin of the basin and form there anhydrite platforms. Thin compactites are widely distributed in the basin facies. Because of the differential compaction between thick pletholites and thin stratolites in the Werra series, the facies distribution of the Stassfurt carbonate and the thickness of the halite are influenced as well. The giant breccias in the Upper Werra Anhydrite are interpreted as compaction or collapse breccias. Olisthostromes and, therefore, deep-water facies cannot be found in the Zechstein 1. Temporary drainage and differential compaction continue to take place in younger series and influence sedimentation there. Thus, the Sangerhausen Anhydrite is formed as an intercalation in the Stassfurt Halite. It is altered to a thick residual breccia by subrosion. The plastic deformation due to the drainage of gypsum cementites leads to "injections" into the overlying Leine halite in the form of high anhydrite pinnacles and to sedimentary anhydrite enrichment in the Leine halite.

Introduction

The Zechstein anhydrites have been a favourite matter of discussion in geology for more than 100 years. There are several reasons for this. First, the features

Lecture Notes in Earth Sciences, Vol. 10
T. M. Peryt (Ed.), The Zechstein Facies in Europe

and fabrics of gypsum and anhydrites are picturesque and so extremely variable that they represent a challenge to every geologist. Second, as far as stratification is concerned, the anhydrite facies rapidly change and because the calcium sulphates usually crop out in karst regions, bedding is commonly disturbed. Not only the facies changes, but also the big differences in the thickness between the sulphate ridges and the wide basin areas create problems during palaeogeographical interpretations. Of practical interest are the anhydrite breccias because they are closely related to the security of mines. Finally, the anhydrites play an important role in the discussion of general geological problems, e.g., the cyclicity of sedimentation, annual rhythmicity, etc. Accordingly, questions of the formation of anhydrite and gypsum have been discussed in the literature for decades. Typical examples are the origin of enterolitic sulphates or the question of the primary crystallization of gypsum or anhydrite. Some time ago, other problems led to serious discussions such as annual rings, the genesis of Sangerhausen Anhydrite, the anhydrite pinnacles, or nowadays the olisthostromes. Famous German geologists like G. Richter-Bernburg, H.R. v. Gaertner, E. Fulda, F. Lotze, and H. Borchert made important contributions to this debate. Despite this intensive discussion and the intensive investigation of drilling cores, some problems are still not solved. This fact becomes obvious in the last work of the master of Zechstein geology, G. Richter-Bernburg (1985).

What are the reasons that there are still so many open problems concerning the origin and alteration of the calcium sulphates after such a long and qualified investigation and discussion? For a long time scientists felt uncertain about the physicochemical fundamentals of the origin of anhydrite and gypsum, this partly being due to errors in the determination of the alteration point and wrong ideas about the significance of the pressure in this alteration. An important limitation was that the principle of actualism cannot be applied to most of the Zechstein anhydrites, because there are no recent examples of oceanic, thick evaporite deposits (saline giants). According to the author, a modern discussion of the Zechstein anhydrites became possible after the following results have been obtained. First, Shearman (1963) with his evidence and investigation of recent gypsum and anhydrite formation in the sabkha zone, second, the exact determination of the gypsum-anhydrite change point by Hardie (1967) and third, the sedimentological investigation of the Sicilian Solifera gypsum by Hardie and Eugster (1971).

The intention of the author is to give a comprehensive presentation of the anhydrites of the European Zechstein, basing on literature data and his own ideas and results concerning the facies and differential diagenesis of anhydrites. The examples were chosen from the southern and south-eastern foreland of the Harz Mountains.

The author considers it appropriate to discuss the theories at the beginning, since many sulphate structures are ambiguous in themselves or due to many discussions about them. Thus, from the beginning all possibilities of their interpretation should be clear.

The physicochemical principles of calcium sulphate origin

The formation of the calcium sulphates during the evaporation of aqueous solutions depends on the solubility products of the $CaSO_4^-$ modifications. The activity of the Ca^{++} and SO_4^{--} ions is important for the solubility of gypsum which is controlled by pressure and temperature as well as the activity of water, i.e. the decrease of the vapour pressure in connection with the increase of the concentration. The value of the solubility product of gypsum is about 0.3 mole% in pure water and depends only slightly on the temperature. The solubility product of anhydrite, however, shows a strong sensibility to temperature. At 20°C it is about 0.4 mole% and up to 80°C it decreases to about 0.1 mole%. Therefore, the question whether anhydrite or gypsum occurs as a primary product is mainly a question of temperature. Theoretically, at an increased temperature anhydrite could be precipitated from sea-water, since at higher temperatures with a concentration of 0.2 mole $CaSO_4$/l it is already oversaturated with anhydrite. In sea-water, however, the solubility is greatly increased by the influence of foreign ions. At 30°C the value for anhydrite increases up to about 0.6 mole% and that for gypsum up to 0.5 mole%, so that at 30°C the gypsum precipitation is only possible at an evaporation of the sea-water to a third of the starting volume, and the anhydrite precipitation is possible at a sevenfold increase of the concentration. This is already near the crystallization point of halite.

The state of equilibrium between gypsum and anhydrite does not only depend on the concentration of the brine. Fig. 1 shows the stability relations in sea water according to Kinsman (1966). In this diagram, only the stability of minerals already formed can be seen. The formation itself is also a kinetical problem.

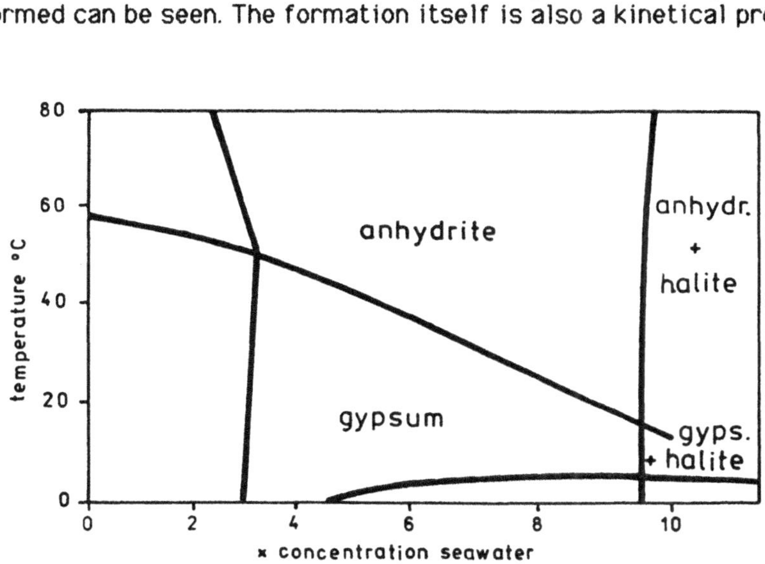

Fig. 1. Stability of gypsum and anhydrite in dependence on temperature and concentration (after Hardie, 1967).

In sea-water the Ca^{++} ions are not isolated but bound in Ca-aquo complexes. In comparison to the common Ca-ion, which is 0.1 nm in diameter, this complex is 0.27 nm. Since the SO_4-complex ion is also relatively big, the maximum lattice distance of 0.7 nm is relatively small. The nucleation and growth of the anhydrite are strongly inhibited under normal conditions. As far as gypsum is concerned the prerequisites are much better. In this case, the maximum lattice distance is about 1.5 nm and in addition, a part of anhydrite envelope of the Ca-ions can be built in as crystal water. Thus, it is no wonder that gypsum very easily forms nuclei, crystallizes and grows faster than anhydrite and at first developes metastably in the stability field of anhydrite. Therefore, an equilibrium is only reached when the gypsum has been removed from the sphere of influence of the relatively diluted sedimentary brine. A stable crystallization of primary anhydrite is only possible at a very small H_2O activity, i.e. at a high concentration (about that of halite field) and under diagenetic conditions as well as slow crystallization (slight oversaturation) in the presence of many anhydrite crystal nuclei.

So far, the influence of the pressure has not been taken into account, that means all considerations are applicable to normal pressure, sedimentation and weathering on the earth's surface. But regarding diagenetic processes, the pressure dependence of the phase transition of gypsum to anhydrite and *vice versa*, is of great importance. According to the principle of Le Chatelier, the phase with the smaller volume is favoured by the pressure. In the reaction:

$$\underset{\text{(gypsum)}}{CaSO_4 \cdot} \underset{\text{(water)}}{2H_2O} = \underset{\text{(anhydrite)}}{CaSO_4} + \underset{\text{(water)}}{2H_2O}$$

the sum of the specific volumes for anhydrite plus water is about 10% higher than that of gypsum. This results in a pressure dependence of the phase transition from gypsum to anhydrite of +0.012 K/kbar. So, gypsum is favoured with an increasing pressure, i.e. with an increasing subsidence of the sediment. This is only true for a closed sustem, when water is under lithistatic pressure, and for relatively low temperatures. Most diagenetic systems, however, are open systems. In the open system only solid phase is under lithostatic pressure and the liquid phase is under hydrostatic pressure. Under these conditions and with a sediment of a medium density of 2.4 g/cm^3 the changing point from gypsum to anhydrite alters to -0.025 K/kbar, which is a marked promotion of the anhydrite with increasing subsidence. If the geothermal gradient and the temperature of the surface are known, a diagram can be developed from the dependence of the alteration temperature with varying pressure, and it is possible to read off the dependence of depth of the change point at different concentrations. In Fig. 2, this is shown for NaCl saturation (i.e. saline subrosion) as well as for $CaSO_4$ saturation (i.e. humid subrosion). According to the temperature conditions in Central Europe, the conclusion can be drawn that in brines saturated with halite, gypsum is only stable at a depth above 125 m, but in brines saturated with sulphate it is already stable at a depth of 650 m onwards. In pure water it may also be found below 700 m. Therefore, anhydrite can only occur in deeper-lying formations, while near the surface is it again changed into gypsum. This does not mean that anhydrite could not reach the surface, because the re-change into gypsum is only incomplete due to a lack of residual porosity.

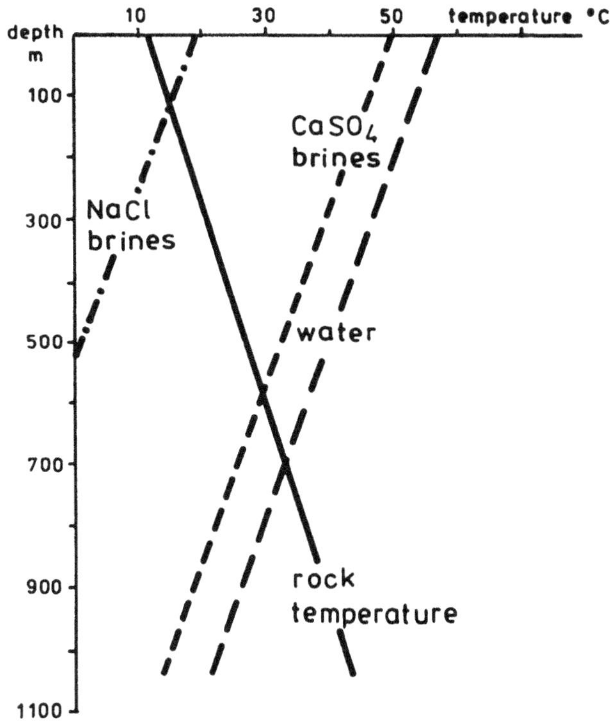

Fig. 2. The point of the changing of anhydrite and gypsum in dependence on temperature, depth, and the types of brines.

In a closed system, the gypsum first of all remains stable, but only until the alteration temperature is reached. Although the pressure of overburden may delay the formation of anhydrite, it is not able to prevent it. The temperature of the rocks increases with increasing sediment overload according to the thermal gradient. At present, the temperature of the rocks raises about 3 K/100 m of depth, whereas the alteration temperature from gypsum into anhydrite raises only about 1 K/350 m of sedimentary cover. If, in the course of subsidence, the alteration temperature is only 1 K higher, an excess pressure will develop in the rocks. According to Braitsch (1962) this pressure can be calculated by $dT/dP = +0.012$ K/bar or $dP = dT/(+0.012)$ bar/K at 83 bar/K. This pressure will be released if the sediment subsides 33 m. But it can only be compensated by a rock bed 350 m thick. Thus, this pressure is 10 times higher than the compaction pressure and must lead to intensive deformations. Heard and Rubey (1964) investigated these conditions by experiments and considered them to be the possible cause of movements of tectonic dimensions. It is difficult to state at which depth these processes take place, because of the strong dependence of the brines involved and their concentrations. Under temperature conditions of the Zechstein time and the influence of NaCl brines, these processes must have been possible already at a slight depth. At present, they can occur from 100 m downwards.

Sedimentological and diagenetic principles of $CaSO_4$ origin

A lot of processes are involved in the formation and alteration of sulphate rocks. These processes are integrated in the so-called sulphate cycle (*cf.* Murray, 1964), since on the surface usually gypsum crystallizes, which is diagenetically anhydritized and is rechanged into gypsum following uplifting and weathering.

As far as the Zechstein anhydrites are concerned, such a cycle can generally be presented as shown in Fig. 3. All parts of this cycle possess distinct features and textures but in practice, some processes produce nearly the same features, and the microtextures can be destroyed by later processes.

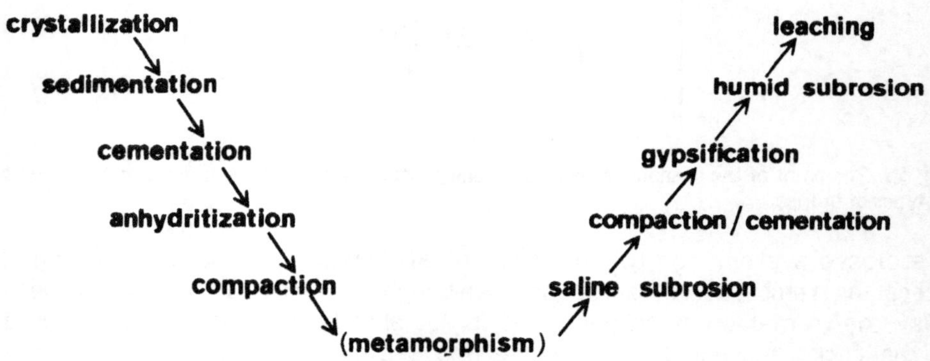

Fig. 3. Sulphate cycle of Zechstein anhydrites.

Essentially, there are 5 primary texture types, which are common in the Zechstein anhydrites and which build up various features (Figs. 4 &5) of which a lath-shaped type consists of tabular crystals. In a thin section usually lath-shaped sections with a length-width ratio of 10:1 can be seen. Very often these laths are contaminated with fine dust and therefore also show undulatory extinction. Just as frequent are the tables composed of a few minute crystals, i.e. they are pseudomorphic after anhydrite. A similar texture type occurring more rarely is the fibrous texture type in which thin needles with a length-width ratio of about 20:1 occur. They are also often contaminated with dust and optically undulatory. The isometric-crystallotopic type is equally important. It consists of nearly isometric, often also of subhedral crystals with a length-width ratio of about 3:1. This type is completely free of finest inclusions. Both variants are different in character concerning the main direction as well. The tables are extended to (100) and the isometric subhedral crystals to (010). Considering the intensity relations in X-ray diffractograms they can be differentiated as well (Langbein, 1986). Finally, there is still an irregularly amoeboid texture type which seldom forms rocks, and mainly replacements of carbonates.

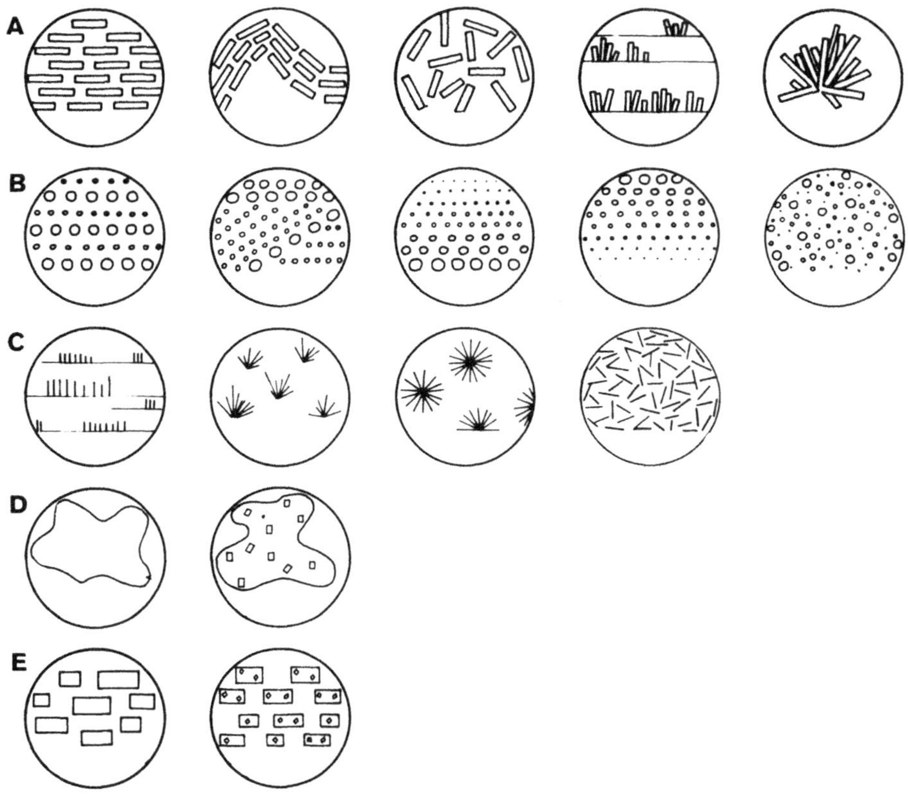

Fig. 4. Scheme of microscopic fabrics of anhydrite. **(A)** Lath-shaped table-like. **(B)** Isometric granular. **(C)** Fibrous. **(D)** Amoeboid. **(E)** Isometric crystalloblastic.

The isometric type (pile-of-brick texture) is always secondary compared with the tabular structure. Accordingly, considering the mineralogical properties mentioned above, it is clear that the lath-shaped type was formed during the alteration of gypsum to anhydrite which was topotactically conditioned. Thus, it informs about primary gypsum and is pseudomorphic to gypsum. It can also be concluded that the isometric type represents a stable crystal form of the anhydrite, i.e. it informs about a direct anhydrite crystallization. This conlusion is supported by the fact that the isometric type can overprint the tabular type called corrotopic anhydrite and may also occur subhedrally in lithophyses. The temporal succession can be very clearly seen in the cemented anhydrites in which the tabular to spherolitic type forms the present rock. The fibrous type forms the drusy cement A and the isometric type forms the blocky cement B.

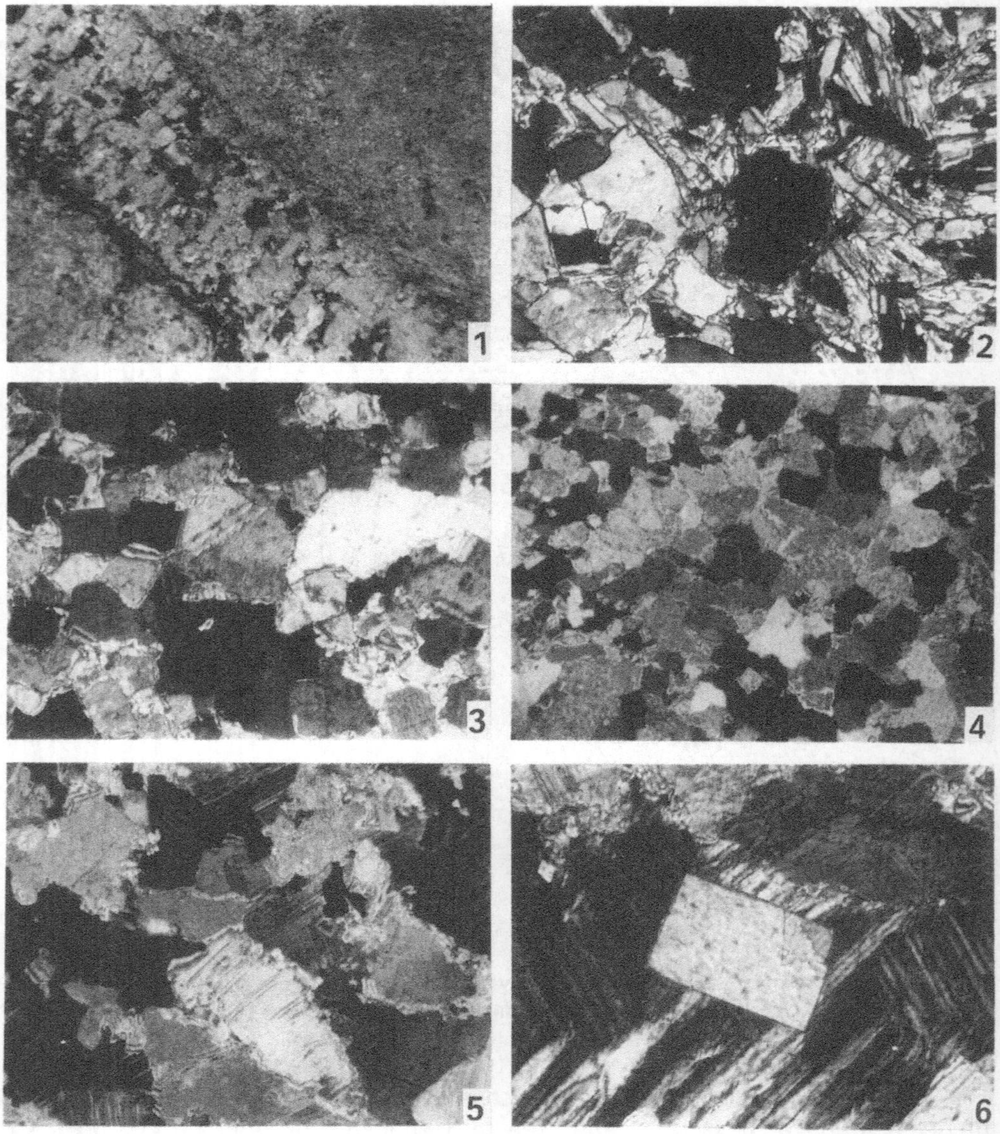

Fig. 5. Isometric primary anhydrite. **(1)** A layer of sparitic isometric anhydrite in a matrix with microspar. **(2)** Isometric sparitic anhydrite replacing sparry laths of pseudomorphs from anhydrite to gypsum. **(3)** Anhedral isometric sparitic anhydrite. **(4)** Subhedral isometric sparitic anhydrite, pavement fabric. **(5)** Anhedral isometric anhydrite with many pressure-twins. **(6)** Euhedral anhydrite pinacoidal crystal recrystallized in a metamorphic anhydrite rock; with pressure-twins.

According to the present knowledge the tabular variants always represent the primary formation in the Zechstein. Even in the direct neighbourhood of the halite crystallization, when halite fills the pores, gypsum was the primary phase. Recently, this fact was stressed by Richter-Bernburg (1985). In connection with the bedding, two main types of the tabular variants can be distinguished. One type is parallel oriented and the other one is normally oriented. The parallel oriented microtexture type clearly represents the real sedimentary type. Gypsum tables which are comparable to the micalith concerning the texture, settled on the bottom of the sea. Therefore, they are subparallel oriented. There is, however, no doubt that the type which is perpendicularly oriented to the bedding plane was not deposited but grew in place, in the bottom-water interface, and may be called sedentary. These two types were described (Langbein, 1964) as "Tonanhydrit" and "Hauptanhydrit" (Fig. 4). Replacement fabric types which are only seldom rock-forming, were described in detail by Clark and Shearman (1980) and Langbein (1973). They are also known from recent occurrences of the Trucial Coast investigated by Kinsman (1966) and others. Early, usually mechanical replacements of the soft sediments can be found as nodules and late chemical replacements of already consolidated carbonate rocks can be found as poikilotopic spots.

As a result of this discussion, from a genetic point of view, seven anhydrite microfacies types can be distinguished:
(1) settled features of parallel-to-bedding oriented pseudomorphoses to gypsum tables,
(2) sedentary features of normally-to-bedding oriented pseudomorphoses to gypsum tables,
(3) drusy cements of fibrous pseudomorphoses to gypsum,
(4) granotopic cements of pure primary anhydrite,
(5) overprinting, isometric-euhedral, crystallotopic primary pile-of-brick features,
(6) mechanical replacement features of table aggregate with snowball texture,
(7) chemical replacement fabrics of primary amoeboid poikilotopic anhydrite.

During the settlement and sedentary crystallization dislocations are also possible. These are turbations which can be partially followed up in the microscopic picture, and turbations due to sliding which may be recognized as microcrumpling in "Tonanhydritlagen" as intraformational pebbles or as felted tabular features. But it is impossible to differentiate these turbations from similar, later-diagenetic formations only by means of the microscopic picture (Figs. 6 & 7).

Although the diagenetic processes are not principally distinguishable from other sediments, they are very complicated as far as calcium sulphates are concerned since the final dehydration occurs nearly simultaneously with the processes of cementation and compaction. The mechanical compaction, the physicochemical compaction, the chemical cementation, and the sparitization or recrystallization indirectly induced that way are theoretically able to take place in a different order. If the cementation occurs very early, then the lithification is already finished near the surface. The primary features are preserved and the compaction has no more influence. **Cementites** are formed. If, however, the early cementation does not occur, i.e. the pores remain filled with brine, then a compaction can become active in the process of subsidence. Thus, the brine is mechanically pressed out and primary features are superimposed. **Compactites** are formed in this way. At some stage of that normal diagenetic process the dehydration, i.e. the phase change from gypsum to anhydrite must occur. Following the differences in the density of both minerals and the volume need difference between free water and crystal water, the phase change is still connected with a change of volume, which must have great influence on the features. Table 1 shows the theoretical possibilities of the sequences of processes between compaction, cementation, and phase change. The Roman numerals mean the temporal position of the respective processes. The respective loss of volume is given in brackets.

Table 1. Sequences of processes of the progressive diagenesis of calcium sulphates

Phase change	Compaction	Cementation/ cement mineral	Volume reduction	Type of features
I (0.38)	II (0.6)	III/anhydrite	0.25	(laminite) compactite
I (0.38)	III (0)	II/anhydrite	1.0	(selenite) cementite
II (0.38)	III (0.38)	I/gypsum	0.62	(ptygmalite) deformite
II (0.38)	I (0.6)	III/anhydrite	0.23	(laminite) compactite
III (0.38)	II (0)	I/gypsum	0.62	(enterolite) deformite
III (0.38)	I (0.6)	II/gypsum	0.25	(laminite) compactite

As shown in Table 1, a displacement in the sequence of the processes from the same sediment must lead to completely different feature variants. If the process of the alteration of gypsum to anhydrite is the first one followed by the compaction, a reduction of the sediment (**sedentite**) thickness will be the result.

Fig. 6. Anhydrite sedentites. **(1)** Anhydrite forming pseudomorphs to subhedral sedentary gypsum embedded in dolomite micrite. **(2)** Two layers with subhedral pseudomorphs from anhydrite to gypsum, of Hauptanhydrit texture, which are encrusted by mats of cyanos. **(3)** Anhydrite pseudomorphic to large gypsum crystals forms a typical chicken-wire texture. **(4)** Section through an anhydrite pseudomorph (the sparry Hauptanhydrit texture type can be seen). **(5)** A layer of pseudomorphs with sparry and spherolitic fabric. **(6)** Spherolitic pseudomorph from anhydrite to gypsum encrusted by filaments of cyanos (upper left). **(7)** Ghost crystal (gypsum) marked by dusty inclusions. **(8)** Dolomitic layer embedded in anhydrite; pseudo-ooids and hollow spheres are due to chlorophyceans.

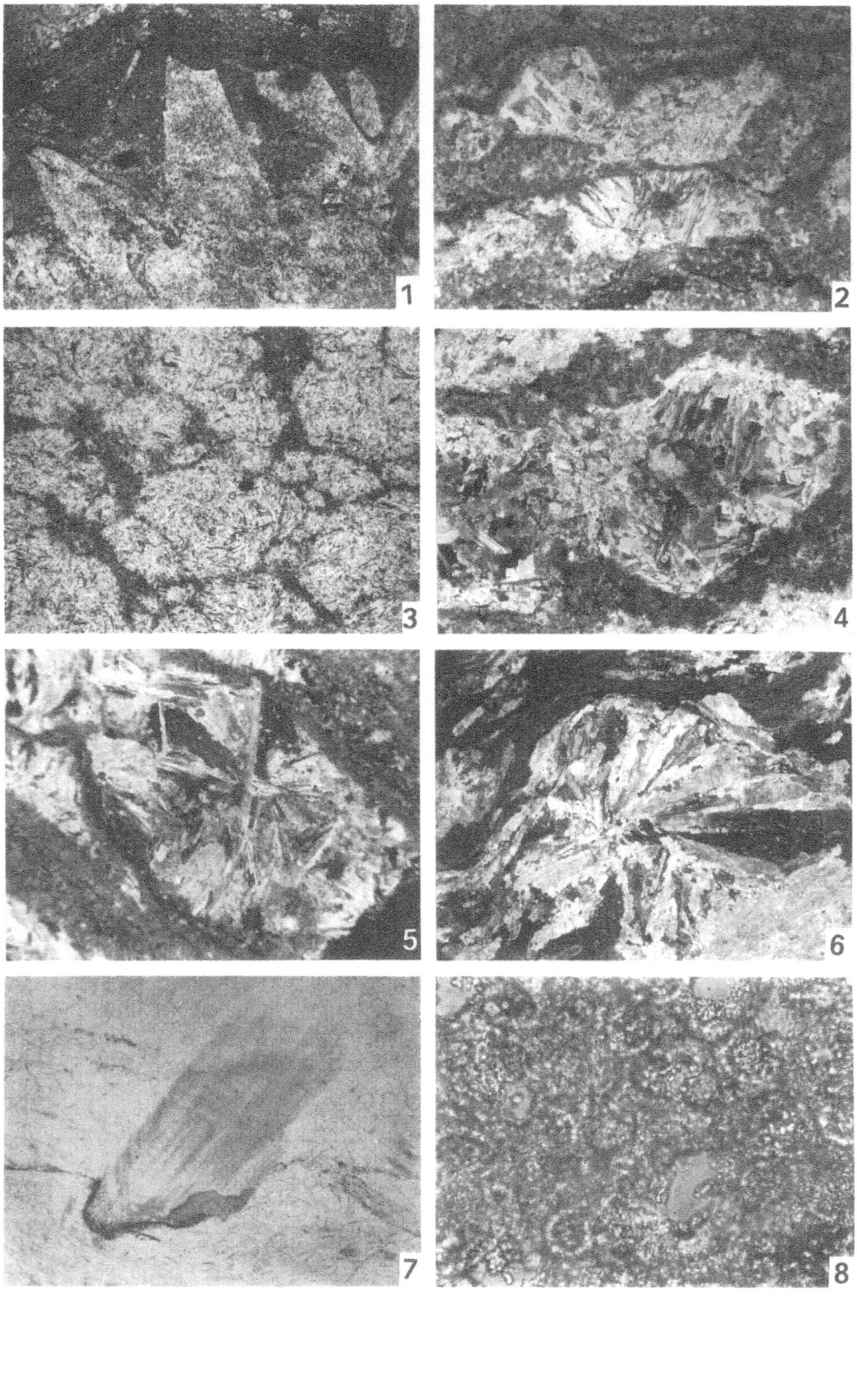
1
2
3
4
5
6
7
8

This reduction can be calculated by the specific difference in density between gypsum and anhydrite, the primary pore volume and by the crystal water released in the phase change. During the phase change a secondary pore volume of 21% (crystal water) plus 17% (density difference) is formed, that is 38% or 0.38 volume parts. There are only little data concerning the primary pore volume.

Generally, it is arranged between that of sand and that of clay. Blatt *et al.* (1972) estimate it to be of 60%. Some measurements of cementites (Langbein, 1986) resulted in average contents for the blocky cement B of 45% and for the drusy cement A of 15% and altogether of about 70 volume% of cement. But this refers to sedentites and not to actual deposits. Thus, in the case described, the thickness is reduced to less than a quarter of the primary value during the compaction. Primary sedimentary features are destroyed and the stripes with non-sulphate impurities move closer together. A very finely striped laminite compactite may develop. A later cementation is more or less without influence because no pore space is left. If, however, as shown in the second example, the phase change is immediately followed by cementation before an overburden or compaction can occur, the primary features are preserved in a cementite. During the phase change no general break down of the crystals occurs but a loose framework of $CaSO_4$ chains is formed which preserves the original form of the gypsum crystal until the secondary pore space fills with cement (Langbein, 1979). Since primary rocks mainly crystallize as gypsum, we can find pseudomorphoses to gypsum in such cementites, either by means of ghost crystals as in the "Hauptanhydrit" (Langbein, 1961) or as outlines in cumulus anhydrites (Gottesmann, 1964), as pseudomorphic anhydrite in varieties rich in dolomite (Langbein, 1968) or as selenite grass in the "Festungsanhydrit" and "Pegmatitanhydrit" (Stewart, 1953, Borchert & Baier, 1954, Richter-Bernburg, 1985).

Occurring after the phase change, the cementation of these anhydrites may also be caused by halite or even sylvite. Because the lithification of the features occurs near the surface, the compaction has no more influence. Generally speaking, this extensive cementation is only possible near the surface of the sediment. There are two reasons for it. First, these processes take place only if there is no overburden pressure, and second, the unhindered migration of large amounts of solutions must be possible, since the amount of cement is about three quarters of the total substance of the lithified sediment. The solubility of $CaSO_4$ in water, however, is relatively low. Depending on salinity and temperature it ranges between 1.5 and 7.5 g/l.

Fig. 7. Anhydrite fabrics. **(1)** Anhydrite compactite; the bitumen and siliceous impurites are enriched in layers. **(2)** Anhydrite sedimentite; the anhydrite (gypsum) laths are parallel arranged. **(3)** Finely laminated anhydrite with an interlayering of sedimentary and sedentary beds. **(4)** Detail of **3**; sedentary anhydrite laths are overlain by parallel arranged fine-grained anhydrite laths. Both laths are pseudomorphs from anhydrite to gypsum. **(5)** Palisade-like development of an anhydrite sedentite on a dolomitic layer. **(6)** Spherolitic sedentite weakly deformed by compaction. **(7)** Bulky deformed sedentite. **(8)** Anhydrite pseudomorph to a spherolite of gypsum.

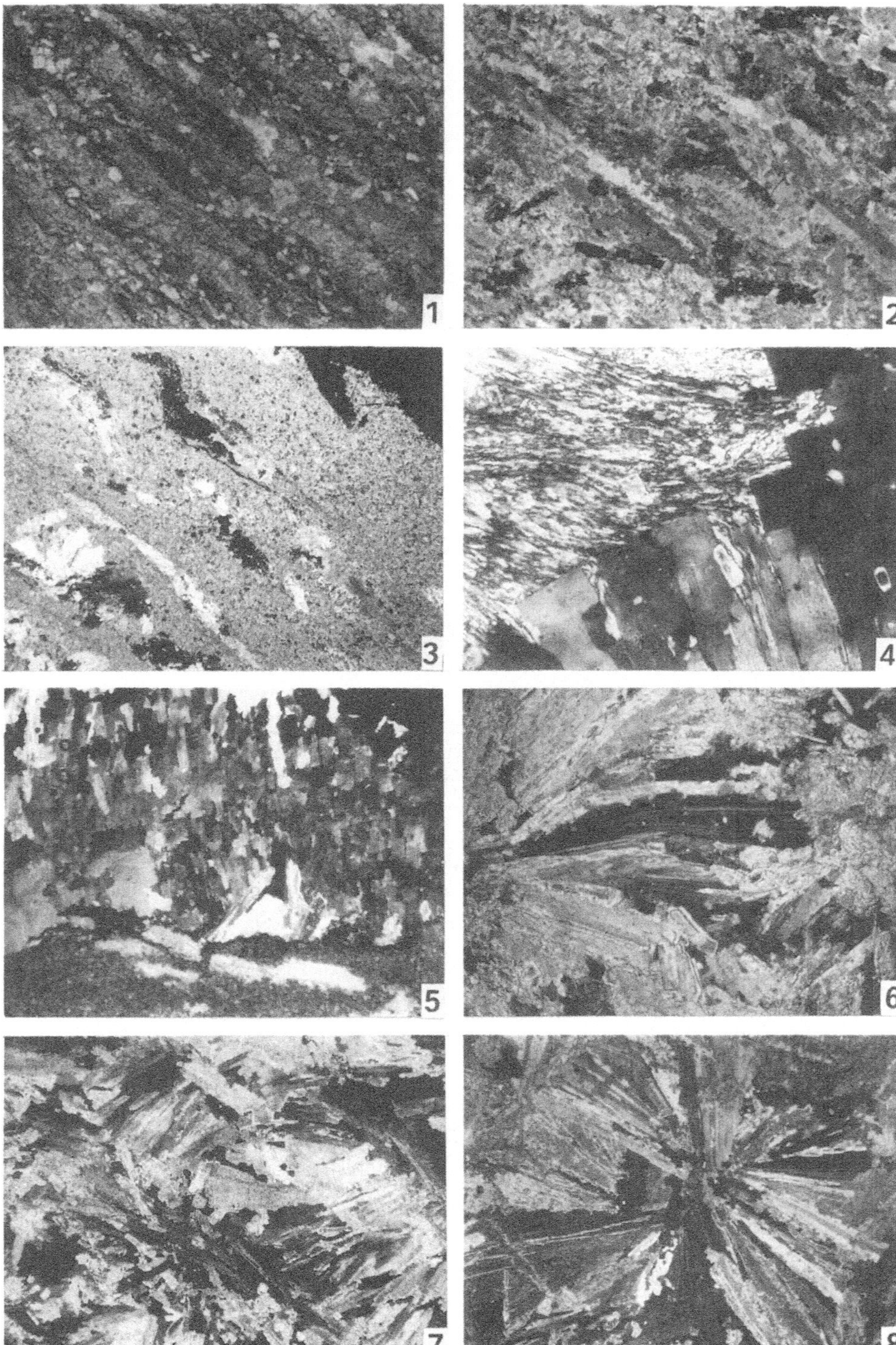
1
2
3
4
5
6
7
8

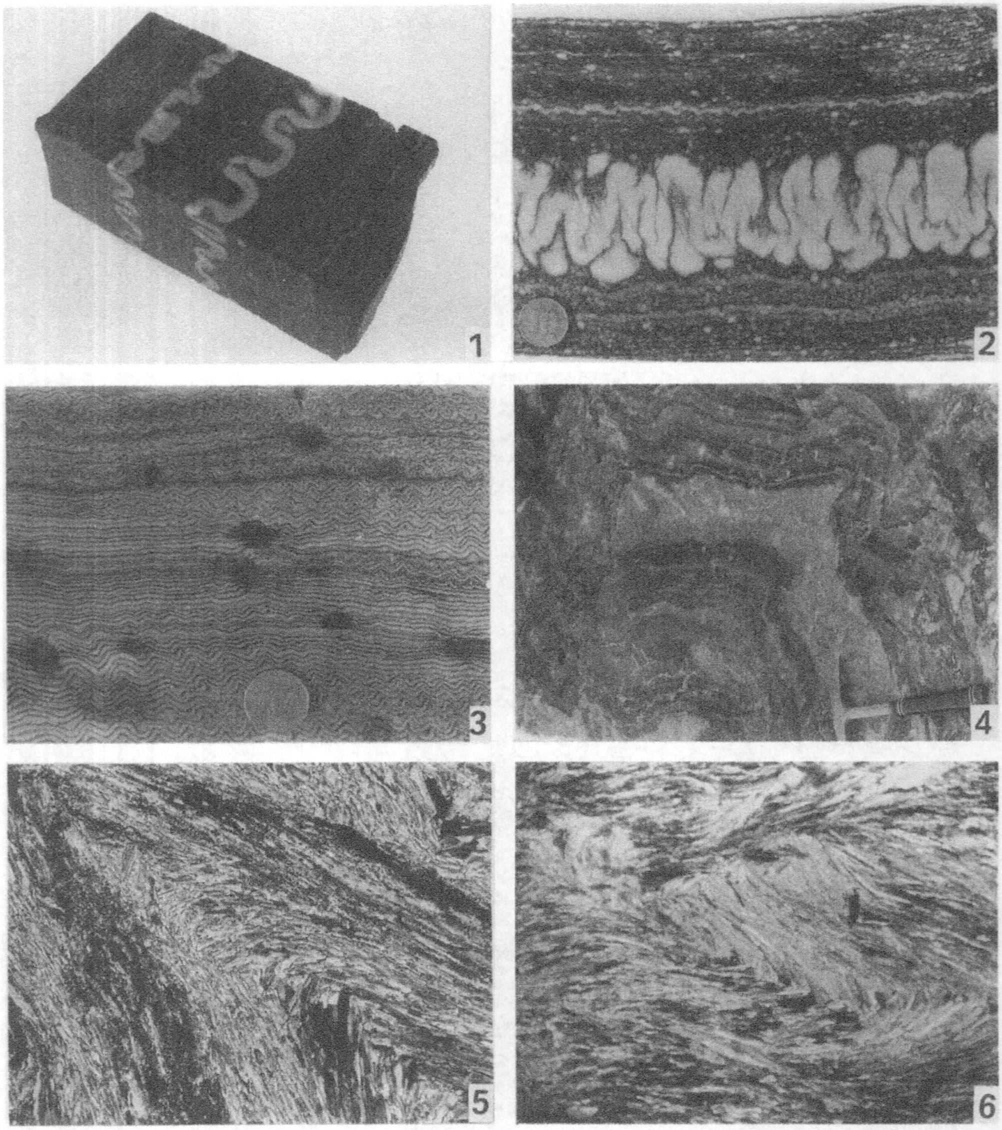

Fig. 8. Anhydrite deformites. **(1)** Layers of gypsum-cemented deformites in a matrix of compactites. **(2)** Intensively deformed layer of cementite sandwiched between compacted matrix with anhydritic micro-nodules. **(3)** Intensively deformed laminite, the so-called seismogram anhydrite. There are some dark spots of secondary gypsum porphyrotopics. **(4)** Anhydrite deformite with tectonic extension (the so-called diagenetic tectonics). **(5)** Structure of a deformed sedimentite; front of a fold. **(6)** Texture of a deformed sedimentite; S-like fold and sparry anhydrite laths.

However, 1 dm^3 of gypsum sediment contains only 0.9 kp of pore-free sulphate which increases to 2.9 kp of sulphate in the anhydrite. This requires at least 333 l/dm^3 of the solution, provided that all the sulphate could precipitate from the solution. In an anhydrite profile with a thickness of 10 m, this would be

33,000 l/dm^3. Hence it appears that the formation of cementites can only occur in an open system, i.e. at the surface, where direct contact with the inexhaustible sea-water reservoir is possible.

In the third example of the Table 1 the cementation is assumed to take place as the first diagenetic process. Since the phase change has not yet occurred, both sediment (40 vol.%) and cement (60 vol.%) consist of the mineral gypsum. For certain time compaction has either no influence or it stabilizes the gypsum. But as soon as the phase change begins, compaction increases. On the one hand escaping of the free water (formed from the 21% crystal water) is not yet possible due to the lack of porosity and permeability, and on the other hand an increase of the volume of about 10% is connected with the drainage. As a result of this increased internal pressure with the simultaneous release of water it can be assumed that the grain support in the rocks is lost and a fluid mass develops (*cf.* Blatt *et al.*, 1972, and Heard & Rubey, 1964). Thus an "explosive" phase is included in the diagenetic development, which leads to an internal deformation of the gypsum-cemented sulphate layers. Such rocks should have plastic deformation features and are called **deformites** (Fig. 8).

In contrast to sedimentary slidings and also to tectonic deformations, where linear anticlinal axes are required, such internal deformations should have irregularly or concentrically running axes. The recrystallization usually starts from nuclei and spreads concentrically outwards like a concentric water-wave. In contrast to the compactites, the deformites as well as the cementites are relatively thicker and have much fewer impurities, being usually light in colour. The other possible processes during the diagenesis of $CaSO_4$ do not reveal any new aspect . There is also a formation of compactites and deformites. Thus, the mode of the overprint of the texture mainly depends on the early cementation, i.e. on the eodiagenesis. If the cementation is caused by anhydrite, as it can be assumed at a high concentration near the halite saturation point at the bottom of the sea, then cementites are formed. If, however, gypsum is being deposited and the concentration at the bottom of the sea is very low or the rate of sedimentation is so high that a continuous cementation is impossible, then compactites are formed. When, because of the relatively low concentration, gypsum cementation occurs, deformites will be produced. Therefore, the diagenetic overprint of the texture essentially depends on the concentration and thus should lead to a palaeogeographic regional zonation ranging from highly saline cementites to deformites and to low saline compactites.

During the diagenesis not only a complete cementation or none is possible, but also a patchy or partial cementation occurs. The most famous variant of the sulphate rocks are the ball-shaped concretions, the alabastrine balls or ox eyes. Starting from a nucleus, such concretions are produced radially outwards by an eodiagenetic cementation of the pore space and partly in competition with the compaction (*cf.* Raiswell, 1971). Jung (1958) gave full details of such early diagenetic concretions from the Werra anhydrites and presented photographs of them which made clear a mechanism of their formation.

Anhydrite can occur as accessory component in shales and carbonates. In shales it produces small nodules which have pushed away the soft sediment by active concretionary growth. In this case the anhydrite is a displacement anhydrite characterized by a "concentric trend" of the shale bedding (or marlstones). As it is known from the Persian Gulf, the small sulphate nodules can develop into large nodules according to the evaporation pumping model. These nodules may coalesce and produce a thick layer of anhydrite with chicken-wire fabric (*cf.* Kinsman, 1966).

In limestones, the emplacement of diagenetic anhydrite occurs as the filling of voids or veins, and often as the chemical molecule-for-molecule replacement. This void-filling and replacement anhydrite was discussed in detail by Clark and Shearman (1980). The substitution of $CaCO_3$ by $CaSO_4$ is connected with an increase of the volume of about 25%. Therefore, both types occur jointly (Fig. 9).

Commonly, anhydrite postdates aragonite dissolution and first cementation, and the compaction due to overburden load. The microcrystalline aggregates of limestone are easily replaced, whereas coarse-crystalline material resist. Therefore, the anhydrite crystals tend to be irregular in shape and euhedral crystals are seldom.

In Fig. 10 the observed void-filling and replacement anhydrites are shown. The replacement anhydrite may also influence dolomite. In the Zechstein, the replacement anhydrite may occur in paragenesis with magnesite or calcite. Four reactions are possible:

$$\underset{\text{(dolomite)}}{MgCa(CO_3)_2} + SO_4^{--} = \underset{\text{(anhydrite)}}{CaSO_4} + Mg^{++} + 2CO_3$$

$$\underset{\text{(dolomite)}}{MgCa(CO_3)_2} + SO_4^{--} = \underset{\text{(anhydrite)}}{CaSO_4} + \underset{\text{(magnesite)}}{MgCO_3} + CO_3^{--}$$

$$\underset{\text{(dolomite)}}{MgCa(CO_3)_2} + Mg^{++} + SO_4^{--} = \underset{\text{(anhydrite)}}{CaSO_4} + \underset{\text{(magnesite)}}{2MgCO_3}$$

$$\underset{\text{(dolomite)}}{2[MgCa(CO_3)_2]} + SO_4^{--} = \underset{\text{(anhydrite)}}{CaSO_4} + \underset{\text{(calcite)}}{CaCO_3} + 2Mg^{++} + CO_3^{--}$$

Which of these reactions takes place depends on the partial pressure of the carbon dioxide and the magnesium content of the brine. The conditions of these replacement reactions are given by Usdowski (1967). If the sulphate brines are in the segment $CaSO_4$-a-f-e of the Fig. 11, then calcite is a paragenetic mineral. If they are, however, in a-b-g-f, then dolomite is paragenetic and in b-c-(d)-h-g magnesite occurs. Following these replacement reactions magnesite horizons were often produced in the Zechstein anhydrite horizons. In some cases, even magnesite-anhydrite mixed rocks were formed.

Apart from the replacement of carbonates, dolomite, and calcite can be found. These are processes of a regressive, retrograde diagenesis which, however, can seldom be observed, but influences of a retrograde diagenesis on the features and the minerals are of great importance. In this process which is connected with tectonic uplifting, denudation and leaching, and which started in the Central European Zechstein basin usually in the Upper Cretaceous, three stages can be distinguished: salinary subrosion, gypsification and humid subrosion by meteoric waters. The salinary subrosion was of particular influence on the origin of the texture. But it was only of importance if the calcium sulphates were interbedded with halite- and potassium-salt horizons or if they were underlain by them. The leaching of halite had only a limited influence on the anhydrite, since the anhydrite is the stable phase in brines saturated with halite. Thus, it did not change into gypsum. If, however, the subrosion was of greater extent, cavities of a halite karst were formed where residues of anhydrite and debris from the overburden were accumulated. So, monomictic breccias were developing which are divided into residual breccias and collapse breccias (Fig. 12).

The sedimentary breccias intercalated with horizontally bedded sulphates are produced by the sedimentary reworking of gypsum crusts or anhydrite crusts which have already been lithified in the intertidal region like the carbonate flat-pebble conglomerates. Such deposits were described in detail by Herrmann and Richter-Bernburg (1955). Other type of conglomerates consisting of abraded selenite crystals was described by Hardie and Eugster (1971).

For these intraformational clastites, intraclastites or cannibalistic clastics bedding textures, undisturbed over- or underlying, trough- or gully-shaped arrangements as well as a matrix of gypsum should be characteristic. The actual brecciation is already possible in the early diagenesis. Thus, it is possible to dissolve halite by undersaturated water and to brecciate anhydrite in an interlayering of anhydrite and halite, as shown by Richter-Bernburg (1979). A subrosion is also likely to occur in the early diagenetic phase if undersaturated brines are released by the compaction and transformation of gypsum into anhydrite. Breccia features may also be formed if there is an interlayering of cemented and non-cemented sulphate layers and if the cementation is not regular. Then, the compaction of such interbedded layers produces a nodular feature. The size of the fragments of such diagenetic breccias depends on the thickness of cemented layers and layers where compaction is possible. Interbedded layers some centimetres thick, produce breccias of the same size. Interbedded layer which have a thickness of metres, could cause giant breccias during the compaction, provided that there are greater inhomogenities in the basement. Such early-diagenetic and compaction breccias should occur in combination with plastically deformed matrix which diapirically fills in the interspaces between the fragments and fills them completely. These are matrix-supported features.

Theoretically, the brecciation in the stage of the retrograde diagenesis can be subdivided into two types.

During the subrosion of halite with a high percentage of anhydrite, the largest part of the anhydrite is left and a residual breccia is formed. Two pecularities of the feature can be assumed for such residual breccias. First, the halite subrosion in such rocks is usually parallel to the bedding plane. Therefore, the respective height of fall for the relic anhydrite is small and the breccia fragments should be hardly dislocated. The so-called breccia bands, i.e. fragments which are bedding-parallel and fit together, should be the rule. Second, for a long period of time the breccias have contact with brines saturated with halite or sulphate so that a comprehensive crystallization is possible, which may lead to a strong obscuration of the textures, unclear shapes, rounded edges and even to intensive recrystallization by microscopic corrotopic anhydrite. In such rocks only unclear lithofeatures and horn-like appearance, which occurs more frequently, can be supposed.

The second type of the subrosion breccia must be expected in the overburden or when horizons poor in sulphate have been subroded. In this case, large cavities are formed, into which the anhydritic roof falls, when load pressure becomes too high. Such collapse breccias are much more coarse and should be sharp-edged. The size of the fragments depends on the thickness of the anhydrite banks and increases from the bottom to the top. In the lower part of such breccias, the anhydrite can be assumed to be a cement mineral. Towards the top, gypsum may occur. In contrast to the residual breccia, the shape of the sharp-edged fragments should generally be distinct and sharp, and matrix should be missing.

Since the large boulders support each other and the loading due to overburden may be relatively small and a plastically reacting sediment is lacking, cavities may occur in the collapse breccias which are filled as lithophyses via the solution phase. In these lithophyses, elements such as boron and strontium, which may be released during the dissolution of halite and anhydrite, can be expected.

Fig. 9. Pore-filling and replacement anhydrite. **(1)** Pore-filling with fibrous anhydrite, cement A. **(2)** Pore-filling in a carbonate rock; the anhydrite forms fibrous cement (centre) and blocky cement B (upper right). **(3)** Anhydrite fills an ostracod shell. The anhydrite forms a large sparitic grain surrounded by a small fibrous seam. One part is filled with carbonate matrix (upper left). **(4)** Void-filling anhydrite in an algal micrite overlain by a stem-like crust. **(5)** Void-filling and replacement anhydrite in an algal micrite. **(6)** Isometric-euhedral anhydrite forming poikilotopic replacive grains in an algal dolomite micrite. **(7)** Dolomite forming pseudomorphs to a lath-shaped Ca-sulphate (gypsum?). **(8)** Anhydritic pore-fillings and needles of magnesite in a partially resorbed dolomite micrite (dark).

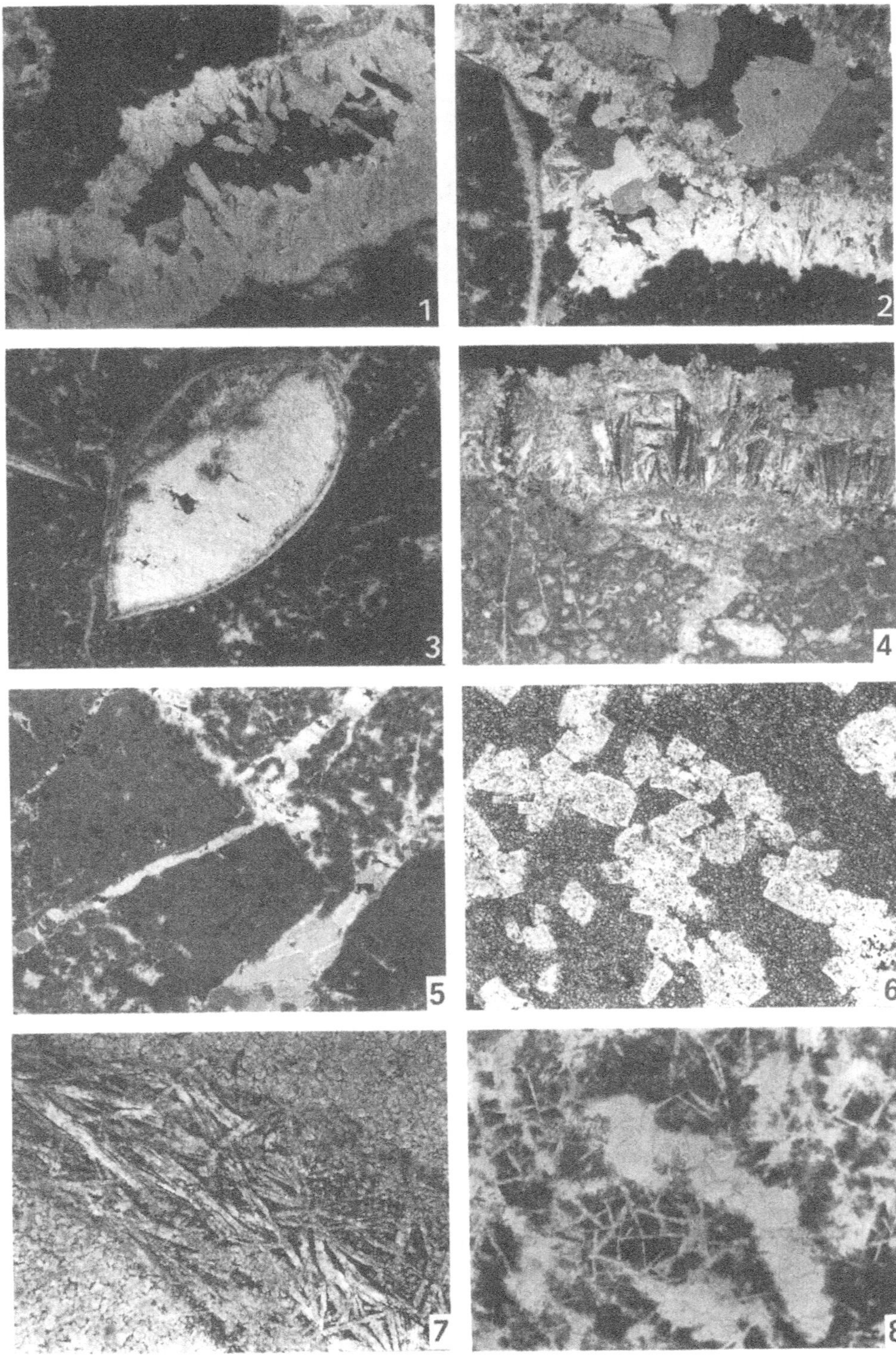

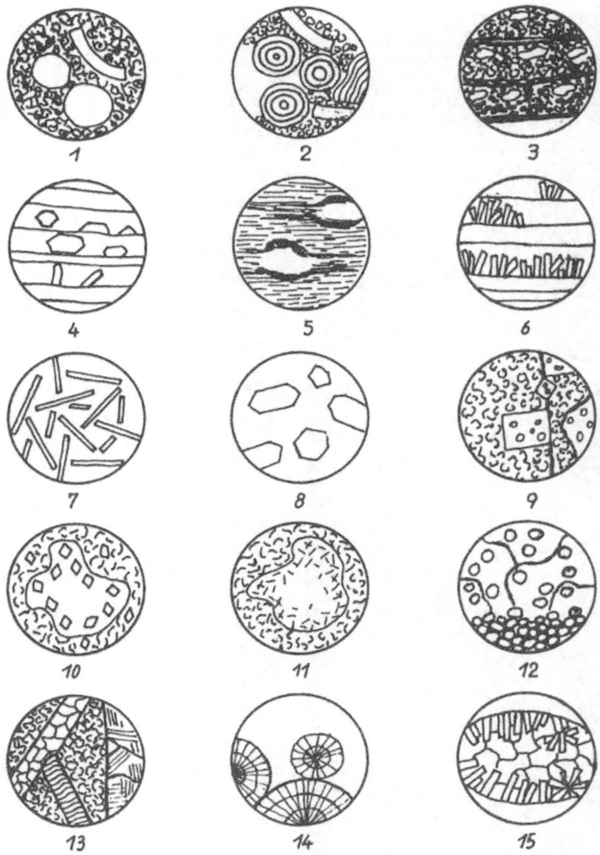

Fig. 10. Scheme of microscopic fabrics of void-filling and replacive anhydrites. **(1)** Fossil moulds filling. **(2)** Pore-fillings. **(3)** Spots in layered carbonates. **(4)** Idiotopic pseudomorphs after gypsum. **(5)** Concretions. **(6)** Crystal-grass pseudomorphs after gypsum. **(7)** Lath-shaped structure (intersertal). **(8)** Pseudomorphs after gypsum, not organized. **(9)** Porphyroblasts. **(10)** Poikilitic with dolomite. **(11)** Poikilitic with magnesite. **(12)** Coarse-sparitic replacement anhydrite. **(13)** Types of fracture-filling anhydrite. **(14)** Anhydrite spherules. **(15)** Void fillings.

Concerning the subrosion breccias it has been established that they can only be found in salt-bearing horizons and that they are frequently stratified. Ideally, a profile through such a breccia should by built up as follows:

- porous, loose sedimentites cemented by gypsum,
- breccias and plates of anhydrite with gypsum veinlets,
- monomictic giant breccias of pure anhydrite, anhydrite cement and lithophyses,
- residual breccia of anhydrite,
- undisturbed horizontally layered anhydrite.

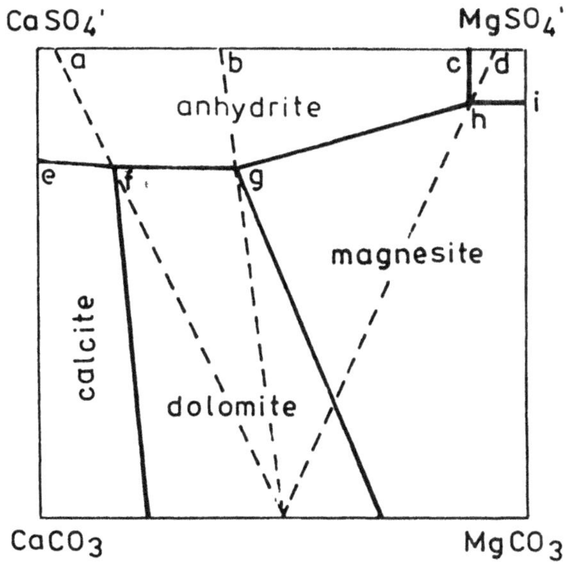

Field $CaSO_4$ - a - f - e = A + C + D
a - b - f - g = A + D + M
b - c - g - h = A + M + D

Fig. 11. The Ca-Mg-CO_3-SO_4 system after Usdowski (1967) to illustrate the replacement of dolomite by anhydrite. Segment $CaSO_4$-a-f-g = anhydrite+calcite+dolomite; a-b-f-g = anhydrite+dolomite+magnesite; b-c-g-h = anhydrite+magnesite+dolomite.

Lithological features of sulphate rocks

The manifold and polymorphic features of the gypsum and anhydrite rocks have been of great interest for a long time and have frequently been the matter of intensive discussion.

Gaertner (1932) described some typical features of outcrops of the southern part of the Harz Mountains and gave explanations to them. Richter-Bernburg (1941) presented a classification of the Werra anhydrites of the south-eastern part of the Harz Mountains, which is in principle still valid. He also started the discussion on the early-diagenetic feature overprint and on the varve stratigraphy (Richter-Bernburg, 1957, 1958, 1960). The research work of Hoyningen-Huene (1957) and Jung (1958) mainly aimed at documentation. The relatively comprehensive classification of the features according to Jung (1958) is still today frequently used, because it can simply and clearly be applied. Jung (1958) also dealt intensively with the origin of features and recognized the great importance of early-diagenetic concretionary processes for the formation of features. Meier (1975, 1977) and Schlager and Bolz (1977) discussed the resedimentation and deep-water sedimentation model for the Werra anhydrites.

Table 2. Genetic classification of features

Diagenetic category	Designation	Feature variants
halite cementites	sedentites	selenites, pseudomorphites, Festungsanhydrit, pegmatite anhydrite
anhydrite cementites	pletholites	banded anhydrite, mosaic anhydrite, massive anhydrite, linear anhydrite, chicken-wire anhydrite
gypsum cementites	deformites	ptygmalite, enterolite, ropy bedded anhydrite, seismogram anhydrite
differential compactites	stratobolites	flaser anhydrite, nodular anhydrite, cumulus anhydrite, cloudy anhydrite, ox-eye anhydrite
compactites	stratolites	varvite, lamellite, laminite, perlite, birdseye anhydrite
secondary features	conglomerates	intraclastites, flat-pebble conglomerates, crystal conglomerates
	brecciites	compaction breccia, residual breccia, collapse breccia
gypsification features	secondary gypsum	pseudoflaser gypsum, pseudolayered gypsum, porphyr gypsum, poikilotopic gypsum, gypsum spar

Recently, Richter-Bernburg (1985) developed a comprehensive model of the origin and the facies distribution of the Zechstein anhydrites combining features and genetic interpretations as well as the palaeogeographic position. He developed four categories for the classification of the features (and the palaeogeographic facies): the **pletholites** which are massive features free of matrix; the **stratolites** which are undisturbed, layered, lamellar features; the **stratobolites** which are quasi-stratolites with nodular intercalations; and the crystal **selenite grass**, i.e. the crystal pseudomorphs. Proceeding from this classification and from the feature-forming processes which are theoretically possible and were discussed in the preceding parts, the following genetic classification of feature observed can be developed (Table 2).

Fig. 12. Subrosional features. **(1)** Fine-grained stratiform residual breccia. **(2)** Coarse-grained platy collapse breccia from a cap rock. **(3)** Sharp-edged layered breccia from a saline subrosion horizon (Sangerhausen Anhydrite). **(4)** Detail from **3**; the fitting clasts can clearly be seen. **(5)** Thin-section photo of **4**; it can clearly be differentiated between the coarse sparitic clasts (lower left and upper right) and the microsparitic matrix. **(6)** The gypsification firstly influences the microsparitic matrix; the gypsum includes poikilitically the coarse sparitic anhydrite crystals therefore. **(7)** The secondary gypsum builds up megacrystals which include the impurities without any disturbance (width of the picture 20 cm). **(8)** Secondary gypsum crystals reaching large dimensions in the modification of selenite (Marienglas). Often the gypsum shows swallow-tail features (height of the picture 4 m).

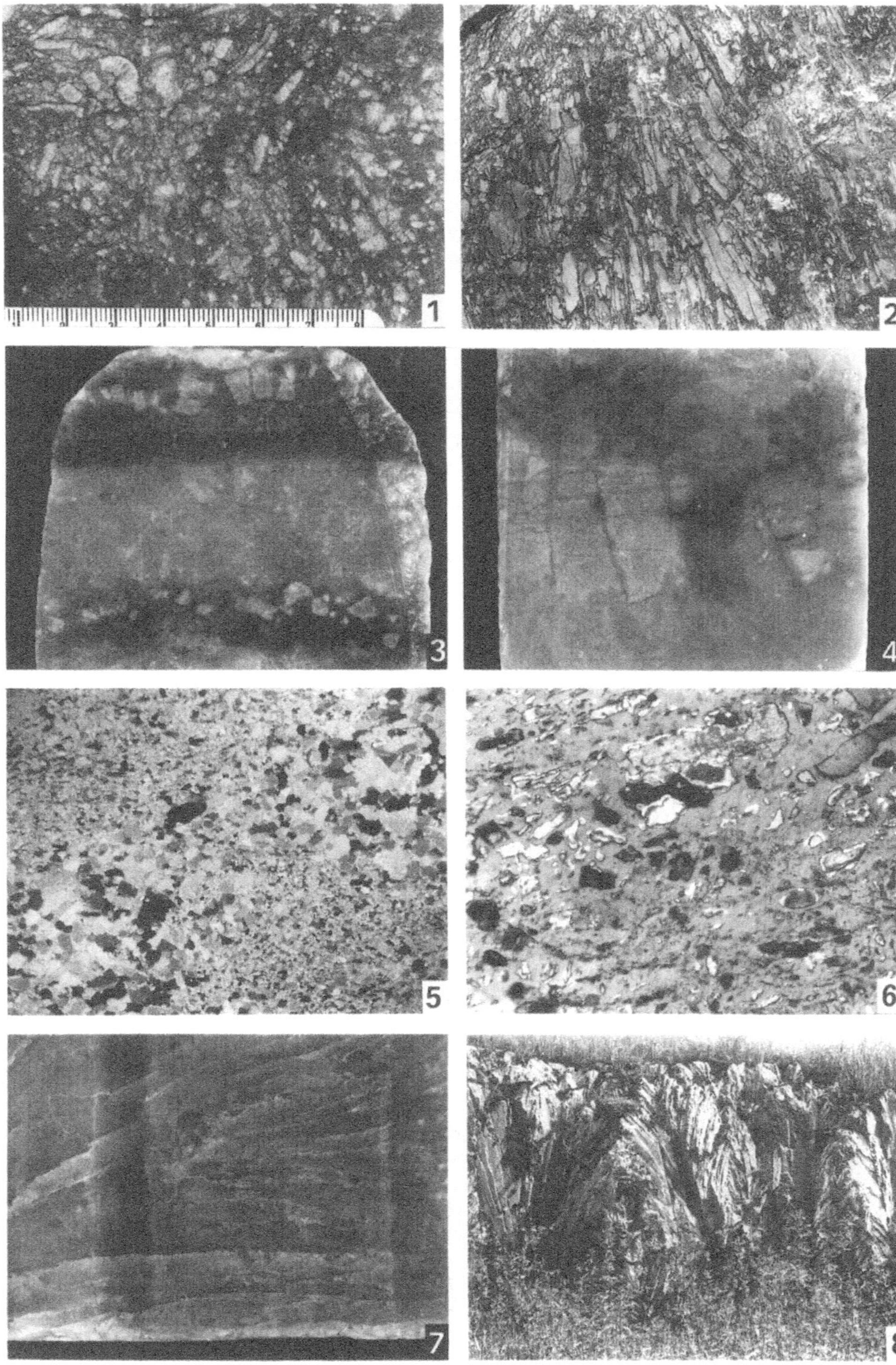
1
2
3
4
5
6
7
8

Gypsum crystals which originate on the air-water interface due to insolation and sink to the bottom of the basin, are either dissolved below a gypsum lysocline or preserved. In a seasonal or, more appropriately, in a climatic rhythm, they alternate with carbonate layers. A strong thickness reduction of the carbonate laminae (also referred to as compaction) is connected with the overburden by sediment and the phase change to anhydrite. Thin-bedded **sedimentites** such as **varvites, laminites, lamellites, pearl anhydrites** or **birdseye anhydrites** are produced. This genetic type of the compactites corresponds with the group of the stratolites.Thick sulphate interbeds may be due to favourable climatic conditions or to a lack of carbonate sedimentation.

With increasing aggregation and shallowing of the basin resulting from aggradation or due to a drier and hotter climate, the gypsum compensation depth is the first line to come into the region of the bottom of the basin (Fig. 13). In this case, the first cementation of the sediment occurs. This first cementation is usually very incomplete. There must be an equilibrium between the rate of sedimentation and the rate of cementation, since the cementation is only possible during the direct contact with the sea-water. The sediments produced in that way are only partially cemented and subsequently they have to be compacted partially and differentially (Fig. 14). These are **differential compactites** or **nodular anhydrites**.

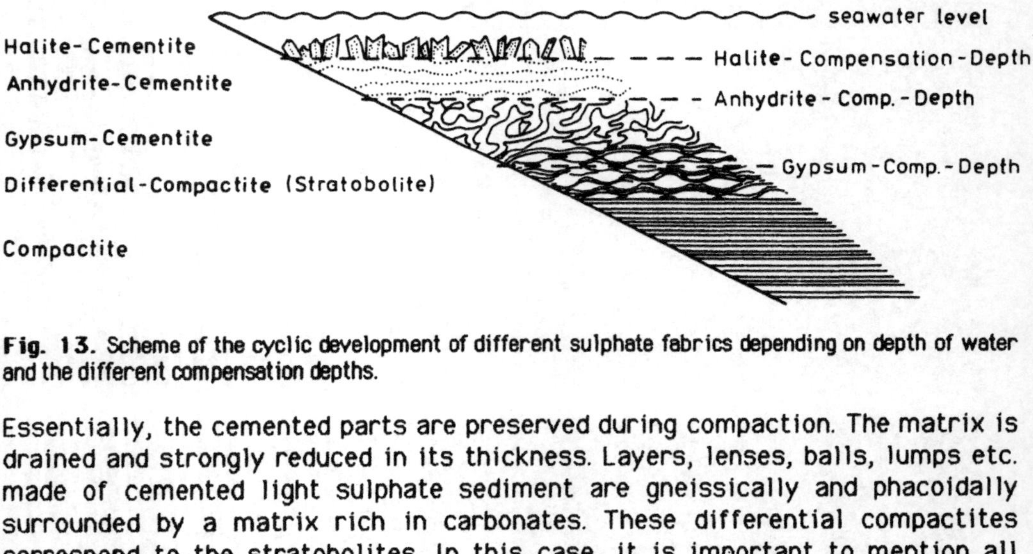

Fig. 13. Scheme of the cyclic development of different sulphate fabrics depending on depth of water and the different compensation depths.

Essentially, the cemented parts are preserved during compaction. The matrix is drained and strongly reduced in its thickness. Layers, lenses, balls, lumps etc. made of cemented light sulphate sediment are gneissically and phacoidally surrounded by a matrix rich in carbonates. These differential compactites correspond to the stratobolites. In this case, it is important to mention all flaser anhydrites, nodular anhydrites, knotted anhydrites, cumulus anhydrites, cloudy anhydrites, and so on. The classic representative of this group is the anhydrite with alabaster balls or ox eyes. These balls represent real concretions according ot Raiswell (1971) and were described by Jung (1958). By means of the laminae distance, which are large in the nucleus and small at the rim, cementation with increasing compaction in these balls, i.e. the simultanity of both processes can be shown (type 1 concretions according to

Raiswell, 1971). Other concretions have been formed earlier and do not show internal laminae. They are lenticularly surrounded by laminae (type 2 concretions).

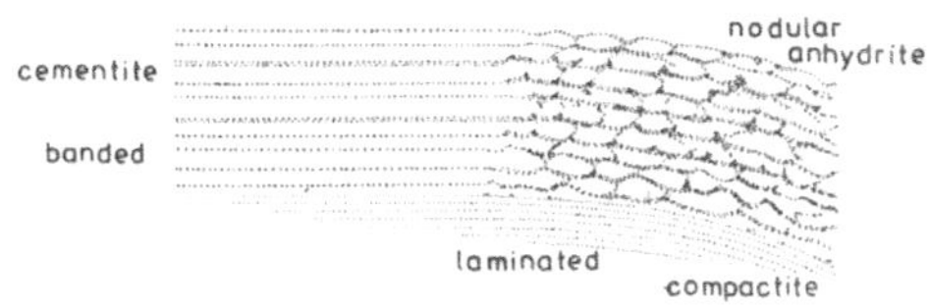

Fig. 14. Scheme of the origin of nodular anhydrite by compaction.

In differential compactites, the spot-like partial cementation is typical. The cementation of a relatively thick layer can be regarded as the next stage, whereas there is no cementation in the overlying or underlying sediment. The sedimentary result is an interlayering of a light cemented layer with dark compact sulphate. This feature is only then preserved in the sedimentite, if the cement mineral consists of anhydrite, i.e. if the basin bottom has been located above an anhydrite compensation depth (ACD) or if this line oscillated in this region. This can only be expected at relatively high temperatures or concentrations. If, however, the cementation occurs in the range of the GCD-line, the gypsum sediment is cemented by gypsum and gypsum cementites are formed. The diagenetic phase change to anhydrite leads to large dislocations and plastic deformations. **Deformites** are formed.

Lithotypes belonging to that group are the snakey anhydrites or ptygmalites and enterolites. Gypsum cementites are not restricted to particular layers or interlayerings with laminite, but they can also comprise thicker sediment masses. The diagenetic result of these gypsum cementites are **deformites** in the form of seismogram anhydrites, translational fold anhydrites and ropy bedded anhydrite.

A complete and total cementation of the sediment by anhydrite may occur if the bottom of the basin lies above the anhydrite compensation depth and if sedimentation and cementation are nearly in equilibrium. In this case, the primary sediment features are preserved. There is no compaction and because of the supply of a lot of cement sulphate the relative content of carbonate matter is low. Therefore, the rocks are pure and white as well as thick. These anhydrite cementites are pletholites. All anhydrite features with only slight impurities such as mosaic anhydrites, massive anhydrites, chicken-wire anhydrites and the broadly striped band anhydrites with undisturbed beds belong to this category. The difference in the cementation due to anhydrite or gypsum correspond to the different conditions in temperature and concentration. Since anhydrite becomes stable at higher temperatures and concentrations, the undisturbed anhydrite cementites or primary features can mainly be found in shallow water and in higher regions of evaporite cycle. The anhydrite cementites are clearly more saline than the deformites. The level of salinity during the development of the sulphate features can mainly be seen in the amount and type of the eodiagenetic cement. After anhydrite, the next more saline cement to be expected is halite. But under such highly saline conditions a sediment can hardly be expected. The whole solution is saturated to such a high degree and even oversaturated, that the crystallization mainly proceeds from the bottom of the basin, because the concentration there may be particularly high due to the brine layering. Under these conditions sedentites are produced. The selenite crystals being formed may become very large, following the rapid growth of the gypsum under such highly saline conditions. In this case, the most conspicuous feature type is the selenite grass. When it is formed, the halite compensation depth (HCD) lies in the sedentite and according to this a cementation is caused by halite. But there is also a great possibility that the first cement consists of anhydrite. The halite crystallization phase may then be reached a bit later, when the phase change of the selenite gypsum takes place. During this process, an intracement made of halite occurs in the nuclei of the crystals and the features of the pegmatite anhydrite are formed (Zimmermann, 1913) or if the gypsum crystals have been covered with algae (or cyanobacteria) the Festungsanhydrit is formed (Brochert & Baier, 1954). In addition, sedentites occur, which are covered with algal mats or clayey sediment and which have received their pseudomorph crystal grass feature which can be obliterated during a further gypsum sedimentation.

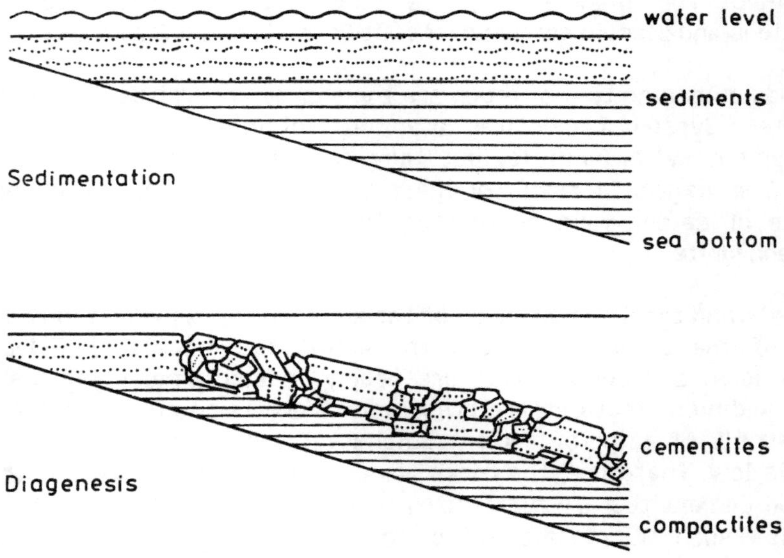

Fig. 15. Scheme of the origin of breccias by differential compaction.

To summarize, it can be said that the formation and/or preservation of most sulphate features depend on diagenetic processes and therefore the progressive diagenesis is an important feature-forming process. Thereby, the most important factor is the position of the bottom of the basin in relation to the equilibrium lines for gypsum, anhydrite and halite in the water body, in relation to compensation depths and thus to the level of salinity. These features mainly depend on the evaporitic concentration. Primary features as they are shown in sedentites or pletholites, have only been preserved by cementation. Apart from these features of the progressive diagenesis, there are also features which are post-lithificational. They are summarized as secondary features, as for instance, the resedimentation features. In the tidal shallow water region, early lithified rocks such as selenites and anhydrite cementites may be rearranged as brecciated conglomerates, flat-pebble conglomerates and crystal conglomerates. These formations resemble the carbonate tempestites and the tidal channel fillings of the marginal marine region and they seem to have been formed in a similar way.

The second group comprises the brecciated rocks, i.e. the **brecciites**. According to the criteria mentioned above, sharp-edged, matrix poor, cemented or also mortary breccias can be formed in different processes and at different times. The compaction breccias produced during the compaction of interbeds consisting of cementites and sedimentites (Fig. 15) belong to that group which have already been assumed to exist by Gaertner (1932).

Other brecciites are the products of retrograde diagenesis which is connected with a dissolution of rock salt horizons. In this way, either residual breccias or collapse breccias are formed. The moment of this subrosion cannot exactly be determined. Richter-Bernburg (1985, p. 45) assumes a dissolution of the halite in the Zechstein time. In many cases, the present author assumes an essentially later subrosion, i.e. in the Upper Cretaceous to Tertiary (Langbein, 1984).

Finally, the third group of secondary features to be defined are gypsification features. It is obvious that these are products of a retrograde, diaphtoritic diagenesis formed after the Upper Cretaceous-Tertiary uplift and subrosion of the evaporitic rocks. They are hardly connected with the features of the progressive diagenesis. The gypsification partially proceeds from veins and fissures. The anhydrite features can be recognized since the gypsification usually occurs mole per mole and thus the features are preserved. Therefore, neoformed gypsum rocks have at first the same features as anhydrites. Only during the actual weathering are new features formed.

First idiomorphic gypsum porphyroblasts appear which are constructed poikilitically and form porphyric gypsum as dark (because they are transparent) xenocrysts. In addition, finely granular-granotopic secondary gypsums may also occur. The microscopic features of these porphyrotopic and alabastrine secondary gypsum were described in detail by Holliday (1970).

The porphyroblasts are the first stages for the gypsum spar feature. The property of the gypsum crystals to enclose non-sulphate impurities poikilitically without dislocation leads to the formation of giant poikiloporphyroblasts or even to crystallized laminites. Gaertner (1932) described gypsum crystal poikiloporphyroblasts up to 20 cm, which have preserved the varvite layering. Concerning the giant crystals of the Sangerhausen anhydrite (gypsum) a length of 50 cm has been found (Langbein & Seidel, 1960). The giant gypsum crystals usually end on veins or impermeable layers and the more massive the anhydrite, the larger the gypsum crystals. Therefore, residual breccias are especially suitable for the formation of giant cystals.

Other secondary features are the pseudoflasering and pseudolayering according to Gaertner (1932). The pseudolayering is caused by dark gypsum layers occurring at a distance of 5-7 cm which gradually pass into anhydrite towards both sides. Thus they resemble a plane layering with slight waves and they are diagonal to the bedding plane. The dark stripes (pure gypsum) are parallel to the present surface and to the veins. They approximately correspond to the Liesegang rings. The pseudoflasering can mainly be found in massive structureless anhydrites. The gypsum shows an alabastrine structure. The impurities were mechanically replaced and enriched in fine stripes. These stripes form a dark network with mesh widths up to 7 cm.

Other specific appearances of the gypsification are the so-called intrusive domes or dwarfed holes which are bulges of the uppermost gypsum layers. On the veins and dolinas we can also find upwarpings of the particular gypsum layers. Both alabastrine gypsum and gypsum spar decay into gypsum earth during further weathering.

Organogenic remains

In the compacted anhydrite rock types no more organogenic structure relics can be found. In the anhydrite cementites, however, such remains are often found either in carbonate preservation or as relic features in the anhydrite. In addition to biogenic dolomite, round dolomite aggregates often occur in layers in basal parts of cycles. There are usually hollow spheres of a size up to 100 µm, which may be filled with clayey material, bitumen or anhydrite or large dolomite idioblasts with a spherical bitumen nucleus. These hollow spheres found in both the Hauptanhydrit (Langbein, 1961) and the Werra Anhydrite (Gottesmann, 1964) are interpreted as chlorophyceen remains. In addition, stromatolitic dolomite bands can frequently be found in anhydrite rich in carbonates. In the Festungsanhydrit type there is even a fine rhythmic change between gypsum and dolomite, whereby the dolomite stripes are interpreted as organic film. In this case, cyanobacteria must have formed the film. There is also a grass-like cover on the large sedentary gypsum crystals which is caused by algae (or bacteria). The bituminous remains of these algae are crystallized in the anhydrite without being deformed. Many internal sediment features in the anhydrite cementites look like parallel and deformed thin cryptalgal mats. Thus, the main part of the bitumen inclusions in the anhydrite must be of organic origin and organic activity during the gypsum sedimentation and sedentation had to play an important role.

Cyclic organization

Since the time when Lotze (1938) suggested investigating the cyclic development in the Zechstein as well as in smaller sedimentary units, there have been many attempts to show the cyclic development in the anhydrite horizons. Already in 1941 Richter-Bernburg presented a classification of the Werra Anhydrite in the south-eastern foreland of the Harz Mountains, which is based on a division into five small cycles. Seidel (1965) was able to prove this cyclic development of the Werra Anhydrite in the Thuringian Basin as well. As far as the Hauptanhydrit is concerned, a cyclic classification was proved by Langbein (1961) in Thuringia which in principle, may also be applied to the south-eastern foreland of the Harz Mountains (Jung *et al*, 1969) and to north-western Germany (Kosmahl, 1969). This cyclic approach has been shown to be of multiple use, because it can also be performed by means of drilling logs, especially electric logs (Seifert, 1967).

Table 3. Ideal cyclothem

Type of fabric	Textures	Chemistry	Typical minerals	Compensation depths
halite cementites	fine-grained spherolithic intersertal	very pure	halite, sylvite	higher than H C D
anhydrite-halite cementites	granotopic sparry	very pure	halite	higher than A C D
anhydrite cementites	coarse-spherolithic bedding-normal sparry lath-shaped	pure	talc, magnesite	higher than A C D
gypsum cementites	lath-shaped-fluidal deformed	impure	dolomite	higher than G C D
partial and differential gypsum cementites	fluidal porphyrotopic fine-grained granotopic	impure	dolomite	horizon of G C D
compactites	fluidal bedding-parallel	very impure	clay minerals quartz	below the G C D

One of the bases of the cyclic classification is the development of basin profiles of the Werra Anhydrite. Since there is a distinct boundary between the underlying varved anhydrite and the overlying flaser anhydrite, this boundary has partially been chosen as the lower boundary of the cycles. From the genetic point of view, however, it seems to be more valid to interpret the compactite horizons as least saline anhydrites or as cycle bases and to interpret the anhydrite types containing halite as cycle peaks. According to this approach, the small cycles are equidirectionally organized as are the large cycles of the Zechstein.

Table 3 and Fig. 13 present an ideal cycle, which would be the result of an increasing salinity or basin shallowing. Approximately, this cycle can be found in the Basalanhydrit of the Zechstein 2. The sedimentation starts at a low salinity with the uncemented compactites (Stinkschiefer, laminite or varvite). Impurities due to siliciclastics are as typical as calcitic carbonate. The anhydrite features show bedding-parallel compact "gypsum tables". Via partial cementites and deformites, anhydrites are formed as sedimentites or sedentites, which have been completely cemented. They show microscopically bulky lath features, spherolites, palisade aggregates or similar features, i.e. characteristic Hauptanhydrit features. As far as the English Zechstein (St. Bees Evaporites) is concerned, this cyclic feature succession has already been described by Arthurton and Hemingway (1972). According to Stewart (1949), in this case aphanic granular anhydrites of the varvites change via felted lath structures into fibroradiate structures. Dolomite and mainly magnesite, but also talc are typical impurities. Finally, halite-containing anhydrites occur as stromatolitic selenites with (dolomitic) zoning and intracrystalline cement (Festungsanhydrit) and as actual selenite grass with halite inter- and intracrystalline cement. The anhydrite features are often granotopic and finely fibrous-spherolitic (pegmatite anhydrite).

Generally speaking, the complete ideal cycle can seldom be found, i.e. the cycles comprise two or at most three such feature zones. In low saline anhydrites these are the stratolites and stratobolites, and in the higher saline ones these are the pletholites and sedentites. But, if the ideal cycle is known, it is also always possible to determine the cyclicity by two-texture-sequences and a stratigraphic comparison between the different profiles can be performed. Regarding the Werra Anhydrite up to 8 such cycles have been recognized (Langbein, 1968). In the Lower Werra Anhydrite, this is mainly an alternation between stratolites and stratobolites. In the Upper Werra Anhydrite, however, two extreme cases occur. In the basin profile the whole anhydrite complex may consist of one texture type, e.g. of 30 m of varvites, as in the Mansfeld facies according to Jung (1958). In the wall profiles, however, the pletholites predominate and the cyclicity is expressed by intercalations of pegmatite anhydrites. In the borehole Lubmin 1 (near Greifswald) for instance, three such horizons have been found in a profile 150 m thick (Helmuth *et al.*, 1968). In the Hauptanhydrit, there are generally four cycles with the first three mainly consisting of pletholites where there are only underdeveloped pegmatite anhydrite facies (Langbein, 1961, Kosmahl, 1969).

Selenite grass occurs frequently only in the uppermost cycle as for example in England (Richter-Bernburg, 1985). The anhydrite intercalations of the Zechstein

3 and the pegmatite anhydrite of the Zechstein 4 are often formed by only one texture type, i.e. by pegmatite anhydrite or selenite grass (Richter-Bernburg, 1985).

For a cyclic development only slight differences in the level of the brine column are necessary, which can be overcome by the sedimentation itself. In Fig. 1 it can be seen that the gypsum crystallization starts at an approximately threefold sea-water concentration. If the concentration goes beyond this level, then the sediment can be cemented with gypsum (deformite). At 30°C an approximately sevenfold sea-water concentration is necessary to make an anhydrite cementation possible (anhydrite cementite) and if the concentration is about ninefold, then cementation with halite is possible (selenite grass). If the original depth of the sediment surface were 10 m below the sea level, the ACD line could be reached after 5.8 m of sediment have been accumulated and the HCD line could be reached after 6.8 m of sediment have been deposited. This means that theoretically a small cycle 7 m thick could comprise all texture types from compactite to pseudomorphite.

Anhydrite ridges and changes in thickness

A particularly interesting problem, is the peculiar thickness distribution of the anhydrites, depending on the palaeogeographical situation.

The thickest Zechstein anhydrites, i.e. the Werra Anhydrite, Basalanhydrit, Sangerhausen Anhydrite, and Hauptanhydrit consist of thick coastal sulphate ridges and basin facies which is widespread but has only a fifth of the thickness of the ridges. The Werra Anhydrite is especially conspicuous and has been well investigated. If the Grauer Salzton which defines the end of the Stassfurt series is considered to be a horizontal time marker, the thickness of the sulphate increases (partly at the expense of the marginal carbonate) with the dipping sea bottom. At a certain distance from the coast, which is often between 50 and 100 km in the south, the thickness suddenly decreases. This does not actually result in a ridge but in a platform with a relatively steep slope at the edge as is known from carbonate and mainly from sandy sediments (progradation). Corresponding profiles in north Thuringia have been found by Seidel (1965), in the North Sea by Taylor (1980), in Western Germany by Richter-Bernburg (1985) and in the Isle of Ruegen by Muenzberger *et al.* (1966).

The platform accompanies festoon-like the south coast of the German Zechstein basin with a maximum thickness of 200 to 400 m. This platform of the south coast shows basement projections, islands as well as submarine swells. These relationships can be seen in Fig. 16.

This becomes obvious in the North Sea where there are, according to Taylor (1980), maximum thicknesses of 100 m for the Lower Werra Anhydrite and a markedly narrow junction of the isopachs. It also becomes clear on the north of the G.D.R. where according to Helmuth *et al.* (1968) the marginal ridge is only about 10 km wide and the thickness is 250 m for the total A1. This is probably connected with the dipping bottom of the Zechstein basin.

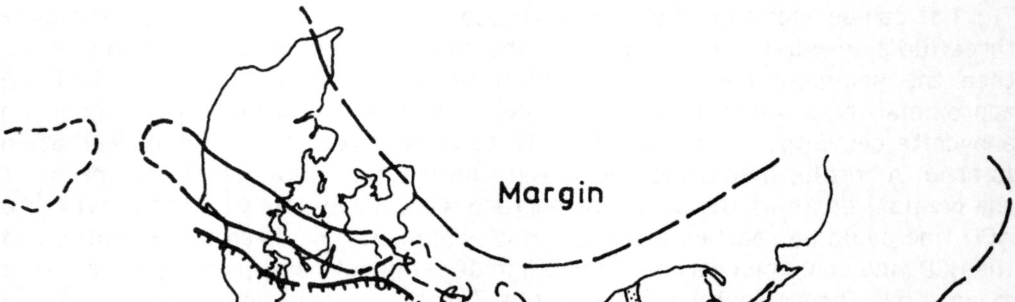

Fig. 16. Facies distribution of anhydrites in the German Zechstein basin (after Richter-Bernburg, 1985).

Beyond the platforms, i.e. in the actual basin area, a relatively constant thickness of about 30 to 50 m is maintained. This thin basin facies extends from Poland to the southern North Sea and comprises an area of more than 200,000 km^2. The basin facies is nearly uniformly developed in the whole area. As far as the features are concerned, the basin anhydrites consist of a rhythmic change of lamellites and flaser anhydrites, i.e. essentially of stratolites and stratobolites. The platform anhydrites are composed of pletholites with only thin stratolite intercalations. Sometimes, a belt of stratobolites is included between the two facies variants. Thus, Gaertner (1932) describes such zones of the southern edge of the Harz Mountains, which are about 5 km wide and show frequently occurring concretions (ox eyes) in fine-layered anhydrites surrounding the swells (Fig. 17).

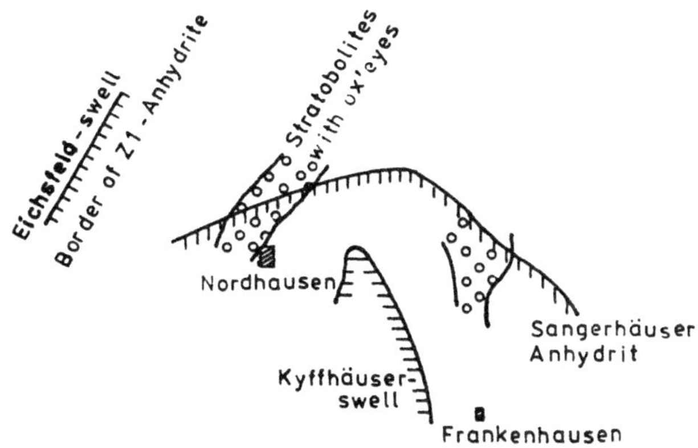

Fig. 17. The stratobolite facies of the Werra Anhydrite and the Sangerhausen Anhydrite facies around the Kyffhauser swell in northern Thuringia (after Gaertner, 1932).

The palaeogeographic succession: edge or swell - pletholite - stratobolite - stratolite or cementite - differential compactite - compactite with dipping sea bottom can be interpreted unambiguously by the sedimentary-diagenetic model. A dipping sea bottom successively crosses the different compensation depths and the different diagenetic anhydrite facies zones must be arranged parallel, what is demonstrated in Fig. 13.

Considering the thin stratolites and the thick pletholites as facies equivalents, it should be possible to correlate both stratigraphically. Despite multiple attempts it has remained problematic, since, firstly, there are only a few complete pletholite profiles and secondly, there are no comparable features. However, some examples are available where the stratigraphic correlation with some reservations was successfully performed. In this way, Richter-Bernburg (1985) correlated 213 m thick Werra Anhydrite of the borehole Wulfen 1 (near Osterode/Harz Mountains) with the stratolite facies of the borehole Hahausen which is about 40 m thick, Seidel (1965) correlated the 300 m thick Werra Anhydrite in borehole Niedersachswerfen 3 and the basin profile (Fig. 18) and Langbein (1968) the 180 m thick Werra Anhydrite in borehole Sundhausen (near Nordhausen) and the basin profile.

The most convincing attempt has been made by Helmuth *et al.* (1968), who succeeded in correlating a 50 m thick basinal profile, a 150 m thick transition profile and a 250 m thick pletholite profile (Fig. 19). It became obvious that the differences in thickness in the Lower Werra Anhydrite are relatively slight (40 to 100 m), whereas the greatest differences can be found in the Upper Werra Anhydrite (30 to 350 m). From the corresponding features it can be seen that there is an alternation between varvites (compactite) and flaser anhydrite (cementite) in the Lower Werra Anhydrite, whereas in the Upper Werra Anhydrite only lamellites occur. In the transition profile, the lamellar part is thicker and contains 2 horizons with greater distances between laminae (these are partial cementites). In the platform profile only pletholites with 2 horizons of pegmatite anhydrite are found. This cyclical stratification was showed by Jung (1958) in the Bottendorf facies to the south of the Harz Mountains.

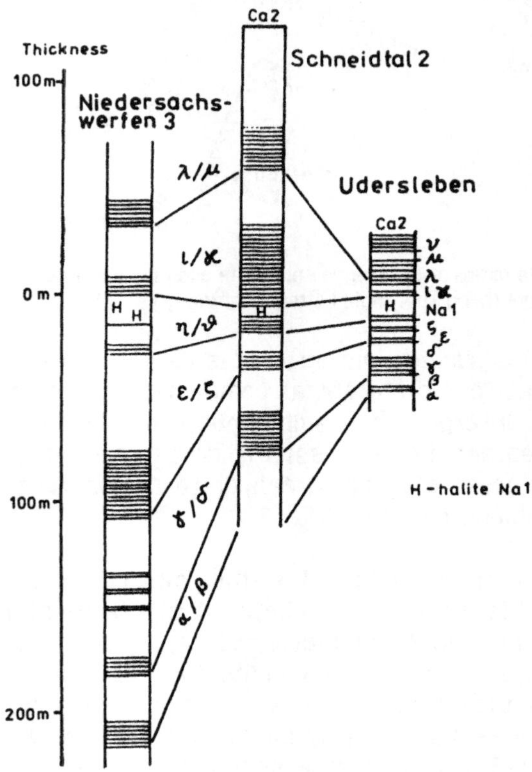

Fig. 18. Correlation of profiles from the Lower Werra Anhydrite between the basin and the Eichsfeld swell (after Seidel, 1965).

It may be assumed that the basin and the platform profiles can be correlated on a cyclical basis, provided that the upper boundary of the halite-bearing horizons is compared with the lower boundary of the stratolite horizons, which are the change points of cyclicity. The thickenss ratio of 1:5 corresponds approximately to the expected ratio between a compactite and an anhydrite cementite. In the upper anhydrite also further halite intercalations are possible as was shown by the borehole Sundhausen and generally speaking, thick halite is restricted to the platform areas.

If, however, platform anhydrites and basin anhydrites can be correlated and thus the only difference between both is that the cemented sedentites and sedimentites change via a partial transition zone into compacted sedimentites which have a fifth of a starting thickness, then there are no big original differences in the sedimentary conditions and the thickness of sediments. On the contrary, the only difference is that in one case, the concentration on the sea bottom for an anhydritic cementation was high enough and in the other case, it was not sufficient. The corresponding model can be compared with an alabastrine ball (ox eye') and its surroundings, where thick light cemented anhydrite changes into lamellar anhydrites, which are rich in carbonates and have been compressed to a fifth or sixth and which were not primarily cemented. These ideas are schematically shown in Fig. 20.

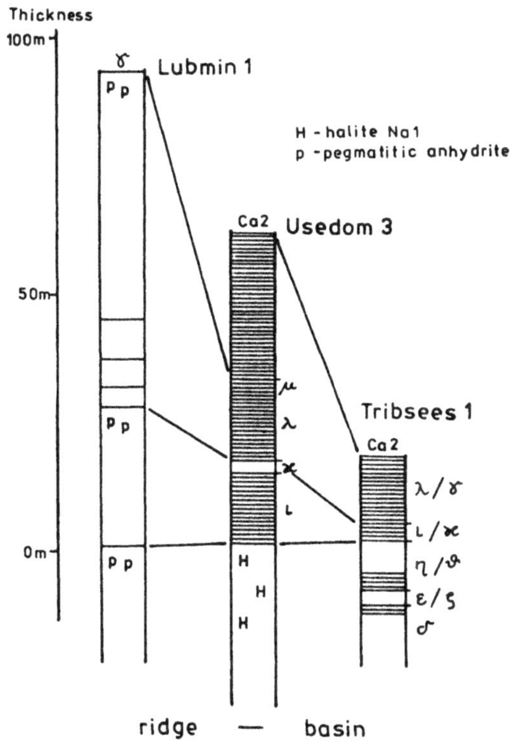

Fig. 19. Correlation of profiles from the Upper Werra Anhydrite in the north of Mecklenburg (after Helmuth *et al.*, 1968).

Other Zechstein anhydrite horizons which at a distance accompany the coasts and swells with relatively thick "wall facies", behave generally in the same way. However, regarding the Sangerhausen anhydrite primarily no massive anhydrite has been developed but an anhydrite-rich marginal facies of the Stassfurt halite (anhydrite region) can be found. In Fig. 17 the distribution of this anhydrite which may have a thickness up to 80 m, is inserted in the surroundings of the Kyffhauser swell.

Megabreccias and folded floes

In the Werra Anhydrite of the swell facies (pletholites) megabreccias can be found from the middle Fulda up to the region of Sangerhausen. There, they have often been described and variously interpreted. In north Thuringia these are, according to Gaertner (1932), sharp-edged angular flat fragments of a light grey anhydrite with mostly parallel wide bedding (distances of 3 to 5 cm), which float in a dark-grey matrix with carbonate impurities. According to Herrmann and Richter-Bernburg (1955) the matrix of the south-western edge of the Harz Mountains may be both finely layered and finely phacoidal. The fragments of the light anhydrites can form bedding-parallel blocks of considerable extent. Concerning Niedersachswerfen near Nordhausen, Gaertner (1932) described a block of a length of 30 m and a thickness of 1 m. These blocks may marginally be compressed and folded. Often some breccia horizons lie one upon another and are separated by thin layers of finely bedded anhydrite. Meier (1975, 1977) has comprehensively investigated and interpreted these breccias at the southern edge of the Harz Mountains.

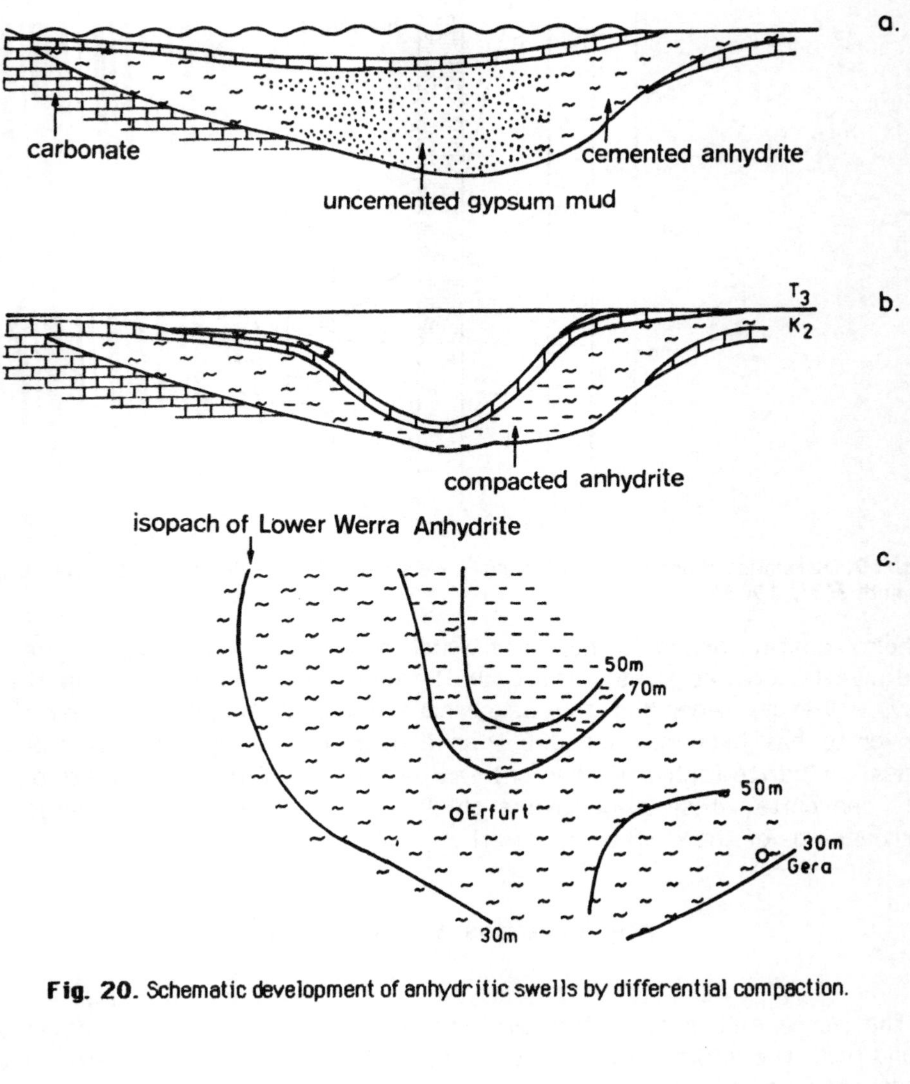

Fig. 20. Schematic development of anhydritic swells by differential compaction.

He interpreted the breccias as being sedimentary and olisthostromes and as being correlated with "turbidites" in the basin and with breccia banks especially occurring at the margins of the anhydrite occurrence. This interpretation, which is hardly compatible with the observations of the investigators previously dealing with this topic, is doubted today. Richter-Bernburg (1985) noted that nearly no bedding-parallel mass transport had taken place and that intensity and habit of the faults indicate rather vertical than horizontal movements. In addition, the breccias are restricted to the Upper Werra Anhydrite, i.e. to a period of time in which at least in the ridge profiles the greatest differences in thickness were already equalized and that they occur at the top of the platform and not at its foot as is shown from the sedimentary carbonate breccias.

Thus, the megabreccias must be interpreted as being of diagenetic origin (Herrmann & Richter-Bernburg, 1955, Gaertner, 1932). The fact that the blocks consist only of light banded anhydrite (cementite) and that the matrix is often composed of dark stratolite and stratobolite (compactite) supports this interpretation. Apart from that, the breccias also occur with features in which there is no matrix but lithophyses up to hollows. Here, subrosion breccias must be assumed. Unfortunately, these two diagenetic processes overlap in the Upper Werra Anhydrite, since cyclic interbedding between cementite and compactite occurs in the region of the platform near the basin and also some halite intercalations can be found in the profile.

Sangerhausen Anhydrite

In the region south of the Harz Mountains an anhydrite horizon occurs, which is a good example of a residual breccia. There have also been many discussions concerning this anhydrite. First it was supposed to be a thickness swelling of the Basal Anhydrite which accompanies the swells in a wall-like way (Gaertner, 1932). Richter-Bernburg (1942) interpreted it as a special development of the Deckanhydrit above the Stassfurt potash deposit. According to Jung (1958) it is a facies representative of the potash deposit and the uppermost Stassfurt halite since it only occurs if these horizons are leached. The intensive investigations of a borehole near Sondershausen (Langbein & Seidel, 1960) showed that it is the subrosion remainder of the Stassfurt halite. The great thickness, proved to be up to 60 m, are due to the leaching of a halite, particularly rich in anhydrite, which can be found around the Harz Mountains. This shows a palaeogeographic binding to the surroundings of swells. Jung (1966) presented appropriate distribution maps for the southern and eastern edge of the Harz Mountains and pointed out that also in other areas of the Zechstein basin corresponding anhydrite horizons between Basalanhydrit and Grauer Salzton occur. However, as far as they have been correctly recognized as subrosional remainders, they are not independent anhydrite horizons.

In the south Harz Mountains, a profile of about 60 m thick has been investigated (Langbein, 1984) and it has been shown that the whole horizon consists of typical flat, slightly dislocated breccias or breccia bands. The microscopic structures are granotopic and correspond to the anhydrites which otherwise occur as annual ring intercalations in halite (the so-called anhydrite middle structures, Langbein & Seidel, 1960, 1968, and Jung, 1966). The breccias are cemented by idiotopic anhydrite. Thus, the anhydrite contains only very few silty and carbonate impurities and is extremely pure (Langbein, 1978). Owing to a dense horn-like characteristic it is difficult to make the features visible and therefore misinterpretations can often result. Since the occurrence of the Sangerhausen Anhydrite as a subrosion remainder is restricted to the present tops of the Zechstein and according to Jung (1966) always a partial gypsification can be shown, the author assumes the subrosion occurred after Upper Cretaceous. But it is also possible that the leaching which occurred during the Zechstein time was of importance too.

Anhydrite enrichments in the Stassfurt halite which show clear palaeogeographic relations to such marginal areas characterized also by thick Werra Anhydrite, are the preconditions for the formation of thick Sangerhausen Anhydrite. It can be assumed that such anhydritic marginal developments, as for instance the Steinsalz anhydrite in Na2 (Seidel, 1965), have been formed under the influence of anhydrite compaction brines instead of the kieseritic halite mainly found in the basin centre, because the compaction of the Werra Anhydrite had certainly not finished at the beginning of the sedimentation of the Stassfurt halite.

Hauptanhydrit and anhydritic cliffs

The Hauptanhydrit differs from the Werra Anhydrites and from the Basal Anhydrite in that its features are mainly blurred. But according to Jung (1960), in addition to mainly marbled, veined and mosaic features also unclearly laminated features with pearls and particularly pegmatitic features can be found. Also, in this case all transitions form compactite to halite cementite are present. At the margins, however, the fine-grained sedentites, i.e. anhydrite cementite, predominate as can be seen by the numerous upright gypsum crystals, rosettes and palisade aggregates (Langbein, 1961, Jung, 1960). In the Hauptanhydrit, frequent replacements of the carbonates took place and that is why the features are more blurred. Among the carbonate alterations especially magnesitization of the dolomite (Langbein, 1961) can be found as well as the sulphatization with calcite precipitation (Jung & Knitzschke, 1961). According to Jung (1960), Seidel (1960) and Kosmahl (1969) a stratification is possible. This formation refers to the marginal region of the Zechstein Basin between Plattendolomit (Langbein, 1965) up to maximum primary thickness of about 40 to 55 m (ridge region). This formation is comparable to the pletholite- or platform facies of Werra Anhydrite.

In the basin region, the relationships are more complicated and more problematic and every detail has not yet been explained. The thickness is usually thinner, e.g., according to Hemmann (1968) it is on average 32 m for Subhercynian region. According to Richter-Bernburg (1985), corrosional and erosional areas, cross-beddings, and internal discordances may particularly be found in the lower part. Only in the upper part, banded anhydrites predominate. However, the anhydrite cliffs which have previously been described by Fulda (1929), remain especially problematic. Here and there, they may be tectonic slivers but in general these are steep-walled eminences of the Hauptanhydrit up to a maximum height of 50 m above its normal top. The halite bedding adjoins the flanks of the cliff discordantly. The halite is recrystallized into crystal salt and is coloured and it contains anhydrite fragments. Proceeding from the cliffs, stratiform anhydrite banks with a thickness up to 40 cm , can also be found in the surrounding halite (Hemmann, 1968).

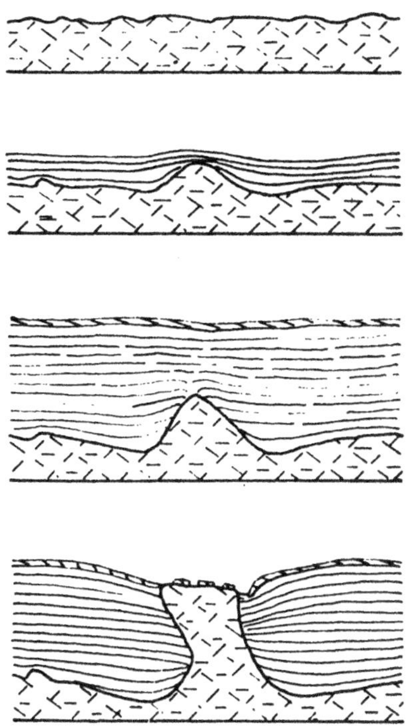

Fig. 21. The development of anhydritic cliffs by plastic intrusion of dehydrated gypsum cementites (after Hemmann, 1968). am - Anhydritmittel Zone.

Richter-Bernburg (1985) considers the cliffs as reef-like synsedimentary formations which were formed simultaneously with the halite. Hemmann (1968) links them to the diagenetic phase change from gypsum to anhydrite and describes them as "pasty intrusions" into the Leine halite connected with a dissolution of the salt and displacement and recrystallization up to synsedimentary outflow into the sedimentary basin (Fig. 21). According to this opinion, the cliffs would meet the requirements of my definition of gypsum cementites and deformites.

Conclusions concerning the sedimentary development in the Zechstein

One of the most interesting problems derived from the facies differences of the Zechstein anhydrites is that of the extremely different thicknesses between pletholite facies and stratolite facies and the regular facies change between stratolite facies and stratobolite facies in the basin profiles. If the stratolites are believed to represent a deep-water facies and the stratobolites and pletholites are believed to be a shallow-water one, differential subsiding movements should have taken place between margin and basin, and the basin area should have subsided or have been uplifted cyclically (or rhythmically) up to 200-300 m, respectively. If, however, the stratobolite facies is considered to be a variety of turbidite facies, then a horizontal transport of the pletholitic material towards the basin over more than 200 km should be assumed. This, and the hypothesis of vertical tectonic oscillations seem to be rather improbable (Richter-Bernburg, 1985). Thus the latter author holds slight changes of hydrographic parametres (salinity, temperature, currents) responsible for the differentiation. With the interpretation of the different anhydrite facies as cementites and compactites, there is a possibility to establish detailed models. Considering the simplest model in which the Zechstein sea bottom dips slightly, the concentration of the sea-water depends on the distance to the coast (assuming there are no currents). At a constant rate of evaporation controlled by the insolation, the concentration increases with decreasing water column.

At first, this has no influence on the amount of the deposited gypsum since due to the insolation, the sediment is formed at the surface of the water and that occurs in the whole basin at the same rate and the sediment sinks toward the bottom. Whether it is preserved there, compacted or even cemented, depends on the concentration at the sea bottom and thus on the depth of water. In the centre of the basin, the possibility of being preserved and cemented is smaller than at the margins because of the greater depth of water. Hence, cementites (and therefore pletholites) may be expected at the margins of the Zechstein basin and carbonate successions free of anhydrite and stratolites may be expected in the centre. Since stratolites are compactites, the present thickness of the sedimentites, in contrast to the pletholites, is not equal to the primary thickness of the sediment. The present thickness has to be multiplied by the factor of four in order to obtain the primary thickness. A sediment which is 200 m thick and also a cementite at the margin of the basin with the same thickness would correspond to a compactite 50 m thick. Thus the differences in thickness are largely corrected and easier to understand. Consequently, the top of the sediment would be essentially higher also in the centre of the basin than it has been expected so far. A transition of compactite into a region of partial cementation and therefore the formation of flaser anhydrites above the sulphate compensation depth can be better understood. This model is schematically shown in Fig. 20 which, however, cannot be applied to deep-water formations. At least in the centre of the basin the rate of the sedimentation is higher than the rate of subsidence or the latter occurs episodically. This results automatically in an elevation of the bottom of the basin and thus in a cyclic facies change when crossing the different compensation depths. Compaction and sedimentation are also inter-dependent. With increasing sediment load, compaction increases and the bottom of the basin subsides. At decreasing sedimentation rate this probably leads to a regressive cyclicity.

The compaction, however, is not yet finished with the formation of the last gypsum sediment. There must be influence also on the younger layers. Presumably, this becomes especially clear in the Stassfurt carbonate, where a shallow-water oolite facies above the noncompacted pletholites is opposed to deep-water oolite facies above subsiding compactites. But the whole compaction seems to have been finished only at the end of the Stassfurt time where potash and Grauer Salzton may be considered as compensation horizons. This means, however, that the differential subsidence during the Stassfurt time, and therefore the different thickness of the Stassfurt halite (perhaps even the facies in the potash deposit), are still due to the differential compaction in anhydrite. But this would mean that the facies distribution and sedimentation development in the Zechstein basin, at least up to the end of the Stassfurt series, depend to a high degree on the original morphology of the bottom of the basin and proceed according to inherent laws and need no differential tectonic movements for their explanation.

There are, of course, influences other than this compaction mechanism. If the surface of the anhydrite platform is supposed to have had a horizontal position at the moment of its formation, then it is possible to estimate the importance of further tectonical influences by comparing this surface and the compensation horizons of the Grauer Salzton, at least for the marginal region. This is illustrated in Fig. 22 by comparing three profiles. They show stagnation and parallel subsidence, respectively, tilting blocks and differential subsidence.

The anhydrites and their diagenesis do not only mechanically influence the sedimentary development in the Zechstein by compaction and differential compaction. Gypsum cementites and compactites release large amounts of brines saturated with sulphate during the diagenesis. This occurs to a great extent during the deposition of overlying halites and partly also potash deposits whose crystallization is influenced by these brines since the water of compaction should metamorphose the normal sea-water. Up to now this indirect influence has hardly been investigated. Therefore, only some possibilities should be pointed out here.

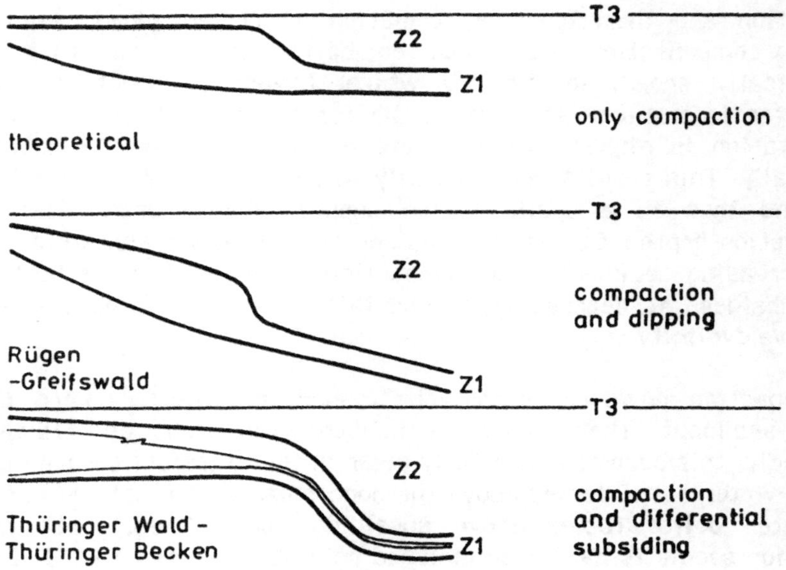

Fig. 22. The influence of postsedimentary compaction on the Werra Anhydrite platforms (profiles from Richter-Bernburg, 1985 and Münzberger *et al.*, 1966).

As far as the Subhercynian region is concerned, it is known that proceeding from the anhydrite pinnacles of the Hauptanhydrit sedimentary anhydrite layers occur in the Leine halite which have "corroded" the underlying halite and which have been overlain stratiformly by halite (Hemmann, 1968). That means that a favoured anhydrite sedimentation has taken place in the surroundings of the pinnacles due to the outflow of the gypsum cementite water. Thus, sedimentary influences on the facies differentiation of the potash seam Ronneburg are also possible. In the surroundings of the Kyffhauser swell in north Thuringia there are profiles of the halite of the Stassfurt series in which the so-called Steinsalzanhydrit occurs. In the drilling profile of Schneidtal 1 (Seidel, 1965), a zone with anhydrite banks was recorded which is about 50 cm thick and where the anhydrite predominates. This is the horizon which forms the Sangerhausen anhydrite after subrosion. Also, this abnormal anhydrite intercalation within a thick halite deposit can only be explained by the influence of waters of compaction. Finally, also the Stassfurt potash as an anhydritic sylvite halite may be well explained by means of a sulphate enrichment of the sedimentary brines by the water of compaction.

Summary

With reference to the literature and the author's investigations, this paper gives comprehensive picture of the origin and alteration of the Zechstein anhydrites. The most important conclusions are the following ones:

- As far as Zechstein anhydrites are concerned, the origin and preservation of the features exclusively depend on their diagenesis.
- The main factor of the diagenetic feature overprint is the cementation. It may be caused by halite, anhydrite or gypsum. It may be partial, total or missing.
- The cementation is controlled by the stability relations at the bottom of the basin. Similarly to the carbonate rocks, different compensation depths can be distinguished, i.e. the gypsum, anhydrite, and halite ones.
- The primary lithofeatures are preserved by the halite or anhydrite cementation. Thereby sedimentites may be recognised as inorganogenic or organogenic sedentites.
- By means of the gypsum cementation, rocks are formed which are deformed during the diagenetic phase change from gypsum to anhydrite. The plastic deformation ranges from enterolites up to large folded blocks.
- Cementation which has been partial or which has not taken place leads to total or differential compaction. During this process, partial compactites (stratolites) and compactites (stratobolites) are produced.
- The breccia structures frequently occurring in the Zechstein anhydrites must be assigned to different processes. During the resedimentation they are formed as intraclastites, during the progressive diagenesis as residual or collapse breccias.
- Completely new feature elements can be produced by retrograde gypsification.
- Compactites and cementites differ in their thickness in the ratio 1:4 to 1:5. The cementites are formed in shallow water at the margin of the basin and form there anhydrite platforms. Thin compactites are widely distributed in the basin facies.
- Because of the differential compaction between thick pletholites and thin stratolites in the Werra series, the facies distribution in the Stassfurt carbonate and the thickness of the halite are influenced as well.
- The giant breccias in the Upper Werra Anhydrite are interpreted as compaction or collapse breccias. Olisthostromes and therefore deep-water facies cannot be found in the Zechstein 1.
- Temporary drainage and differential compaction continue to take place in younger series and influence sedimentation there. Thus, the Sangerhausen Anhydrite is formed as an intercalation in the Stassfurt halite. It is altered to a thick residual breccia by subrosion.
- The plastic deformation due to the drainage of gypsum cementites leads to "injections" into the overlying Leine halite in the form of high anhydrite pinnacles and to sedimentary anhydrite enrichment in the Leine halite.

References

Arthurton, R.S. & Hemingway, J.E., 1972. The St. Bees Evaporites - a carbonate -evaporite formation of Upper Permian age in West Cumbrland, England. *Yorks. Geol. Soc. Proc.*, 38: 565-592.

Blatt, H., Middleton, G. & Murray, R., 1972. *Origin of Sedimentary Rocks.* Prentice Hall, Englewood Cliffs, pp. 653.

Borchert, H. & Baier, E., 1954. Zur Metamorphose ozeaner Gipsablagerungen. *N. Jb. Miner., Abh.*, 86: 103-154.

Braitsch, O., 1962. Entstehung und Stoffbestand der Salzlagerstätten. Springer, Berlin.

Clark, D.N. & Shearman, D.J., 1980. Replacement anhydrite in limestones and the recognition of moulds and pseudomorphs: a review. *Rev. Inst. Invest. Geol.*, 34: 161-186.

Fulda, E., 1929. Über "Anhydritklippen". *Kali*, 23: 129-133.

Gaertner, H.-R. v., 1932. Petrographie und paläogeographische Stellung der Gipse vom Südrand des Harzes. *Jb. preuß. geol. L. A.*, 53: 656-694.

Gottesmann, W., 1964. Petrogenese und Fazies des Werraanhydrits aus einigen Bohrungen in Südbrandenburg. *Geologie* 13: 1163-1190.

Hardie, L.A., 1967. The gypsum-anhydrite equilibrium at one atmosphere pressure. *Amer. Mineral.*, 52: 171-200.

Hardie, L.A. & Eugster, H.P., 1971. The depositional environment of marine evaporites: a case for shallow, clastic accumulation. *Sedimentology* 16: 187-220.

Heard, H.C. & Rubey, W.W., 1964. Possible tectonic significance of transformation of gypsum to anhydrite plus water. *Geol. Soc. Amer. Spec. Pap.*, 76: 77-78.

Helmuth, H.-J., Wiegratz, A. & Zagora, K., 1968. Zum Werra-Anhydrit (Z1) NE-Meclenburgs. *Geologie* 17: 378-387.

Hemmann, M., 1968. Zechsteinzeitliche Gips/Anhydrit-Umwandlung, Anhydritklippenbildung und zugehörige Erscheinungen in der subherzynen Leine-Serie. *Mber. dt. Akad. Wiss.*, 10: 454-462.

Herrmann, A. & Richter-Bernburg, G., 1955. Frühdiagenetische Störungen der Schichtung und Lagerung im Werraanhydrit (Z1) am Südharz. *Z. dt. geol. Ges.*, 105: 689-702.

Holliday, D.W., 1970. The petrology of secondary gypsum rocks: a review. *Jour. Sed. Petrology* 40: 734-744.

Hoyningen-Huene, E.v., 1957. Die Texturen der subsalinaren Anhydrite im Harzvorland und ihre stratigraphische und fazielle Bedeutung. *Geologie* Beiheft 18.

Jung, W., 1960. Zur Feingliederung des Basalanhydrits (Z2) und des Hauptanhydrits (Z3) im SE-Harzvorland. *Geologie* 9: 526-555.

Jung, W., 1966. Nochmals zum Sangerhäuser Anhydrit (Z2). *Geologie* 15: 443-460.

Jung, W., Gerlach, R. & Knitzschke, G., 1969. Zur Feingliederung des Hauptanhydrits (A3) im zentralen Zechsteinbecken. *Geologie* 18: 1164-1172.

Jung, W. & Knitzschke, G., 1961. Kombiniert feinstratigraphisch-geochemische Untersuchungen des Basalanhydrits (Z2) und des Hauptanhydrits (Z3) im SE-Harzvorland. *Geologie* 10: 288-301.

Kinsman, D.J.J., 1966. Gypsum and anhydrite of recent age, Trucial Coast, Persian Gulf. In: Rau, J.L. (Ed.), *2. Symp. on Salt* 1: 302-326. Cleveland.

Kosmahl, W., 1969. Zur Stratigraphie, Petrographie, Paläogeographie, Genese und Sedimentation des gebänderten Anhydrits (Z2), Grauen Salztons und Hauptanhydrits (Z3) in Nordwestdeutschland. *Beih. Geol. Jb.*, 71: 1-129.

Langbein, R., 1961. Zur Petrographie des Hauptanhydrits (Z3) im Südharz. *Chemie Erde* 21: 248-264.

Langbein, R., 1965. Zur Petrographie eines Hauptanhydrit-Plattendolomit Übergangsprofils. *Geologie* 14: 47-57.

Langbein, R., 1968. Zur Petrologie des Anhydrits. *Chemie Erde* 27: 1-38.

Langbein, R., 1973. Über die petrographische Strukturen aczessorischen Anhydrits, sowie Geochemie und Mechanismen seiner Bildung. *Chemie Erde* 32: 45-79.

Langbein, R., 1978. Petrologisch bedingte Veränderungen der Zusammensetzung von Anhydritgesteinen. *Z. geol. Wiss.*, 6: 889-896.

Langbein, R., 1979. Petrologische Aspekte der Anhydritsbildung. *Z. geol. Wiss.*, 7: 913-926.

Langbein, R., 1984. Über subrosionsbedingte Gefüge in Anhydrit- und Gipsgesteinen. *Z. geol. Wiss.*, 12: 349-362.

Langbein, R., 1986. The porespace of anhydrite and its lithological significance. *7. Regional Meeting IAS, Abstr.*, 103-104.

Langbein, R. & Seidel, G., 1960. Zur Frage des "Sangerhäuser Anhydrits". *Geologie* 9: 778-787.

Langbein, R. & Seidel, G., 1968. Zur Auslagung am Südrand des Harzes. *Geologie* 17: 529-542.

Lotze, F., 1938. Steinsalz und Kalisalze. Bornträger, Berlin.

Meier, R., 1975. Zu einigen Sedimentgefügen der Werra-Sulfate (Z1) am Osthang der Eichsfeld-Schwelle. *Z. geol. Wiss.*, 3: 1333-1248.

Meier, R., 1977. Turbidite und Olisthostrome, Sedimentationsphänomene des Werra-Sulfats am Osthang der Eichsfeld-Schwelle. *Mitt. ZIPE Akad. Wiss.*, 50.

Münzberger, E., Rost, U. & Wirth, J., 1966. Vergleichende Darstellung der Sedimentationsverhältnisse des Zechsteins von Thüringen mit denen des Nordostdeutschen Flachlandes. *Ber. dt. Ges. Geol. Wiss., A, Geol. Paläont.*, 11: 161-174.

Murray, R.C., 1964. Origin and diagenesis of gypsum and anhydrite. *Jour. Sed. Petrology* 34: 512-523.

Raiswell, R., 1971. The growth of Cambrian and Liassic concretions. *Sedimentology* 17: 147-171.

Richter-Bernburg, G., 1941,1942. Zur vergleichenden Stratigraphie des Zechsteins in Mitteldeutschland. *Z. Kali, Salze u. Erdöl* 35: 193-197, 36: 4-12.

Richter-Bernburg, G., 1955. Über salinare Sedimentation. *Z. dt. geol. Ges.*, 105: 593-645.

Richter-Bernburg, G., 1957. Isochrone Warven im Anhydrit des Zechstein 2. *Geol. Jb.*, 74: 601-610.

Richter-Bernburg, G., 1958. Die Korrelation isochroner Warven im Anhydrit des Zechstein 2. *Geol. Jb.*, 75: 629-646.

Richter-Bernburg, G., 1960. Zeitmessung geologischer Vorgänge nach Warven-Korrelation im Zechstein. *Geol. Rdsch.*, 49: 132-148.

Richter-Bernburg, G., 1979. Eodiagenetische Vorgänge nach Warven-Korrelation im Zechstein. *Geol. Rdsch.*, 68: 1055-1065.

Richter-Bernburg, G., 1985. Zechstein-Anhydrite -- Fazies und Genese. *Geol. Jb.*, A 85: 3-82.

Schlager, W. & Bolz, H., 1977. Clastic accumulation of sulphate evaporites in deep water. *Jour. Sed. Petrology* 47: 600-609.

Seidel, G., 1960. Zur Gliederung des Hauptanhydrits im Thüringer Becken. *Z. angew. Geol.*, 6: 383-385.

Seidel, G., 1965. Zur geologischen Entwicklungsgeschichte des Thüringer Beckens. *Geologie* Beiheft 50: 1-115.

Seifert, J., 1967. Das Perm am SE-Rand des Thüringer Beckens. Unpublished Dissertation Hoch. Archit. Bauwesen, Weimar.

Shearman, D.J., 1963. Origin of marine evaporites by diagenesis. *Trans. Inst. Min. Metall.*, B 75: 215-230.

Stewart, F.H., 1949. The petrology of the evaporites of the Eskdale no. 2 boring, east Yorkshire. Part I. The lower evaporite bed. *Miner. Mag.*, 28: 621-675.

Stewart, F.H., 1953. Early gypsum in the Permian evaporites of north-eastern England. *Proc. Geol. Assoc.*, 64: 33-39.

Taylor, J.C.M., 1980. Origin of the Werraanhydrit in the U.K. Southern North Sea - a reappraisal. *Contr. Sedimentology* 9: 91-113.

Usdowski, H.E., 1967. Die Genese von Dolomit in Sedimenten. Springer, 95 pp., Berlin.

Zimmermann, E., 1913. Der thüringische Plattendolomit und seine Vertreter im Stassfurter Zechsteinprofil. *Z. dt. geol. Ges.*, 65.

PALAEOGEOGRAPHY AND SEDIMENTARY MODEL OF THE KUPFERSCHIEFER IN POLAND

Slawomir Oszczepalski and Andrzej Rydzewski
Instytut Geologiczny
ul. Rakowiecka 4
00-975 Warszawa
Poland

Abstract: The Kupferschiefer in Poland consists of clayshales and shaly, laminated and argillaceous dolomites/limestones, with minor terrigenous and non-laminated carbonate interbeds. In the Kupferschiefer two main microlithofacies types were identified: (1) clayshales and (2) laminated argillaceous dolo- and calcilutites. These microlithofacies are interbedded throughout the Kupferschiefer sequences. Regional microlithofacies distribution and their mutual proportions in the profiles allowed two main palaeogeographical zones of the Kupferschiefer to be distinguished: (A) deep shelf, which comprises the center of sedimentary basin, characterized by lithofacies of low and fairly consistent thickness (20-60 cm), and alternately occurring microlithofacies 1 and 2; and (B) shallow shelf, which comprises marginal areas of the basin, characterized by lithofacies of highly variable thickness (0-170 cm), predominance of microlithofacies 2 and local occurrence of terrigenous and carbonate tempestites.

The spatial distribution of the lithofacies indicates that the Kupferschiefer was deposited in a relatively shallow, mud-dominated stratified shelf sea. Deposition of the Kupferschiefer generally took place below fair-weather wave base, in low-energy environment; the deep shelf lithofacies was formed in anaerobic-to-dysaerbic waters mainly below storm wave base, whilst the shallow shelf lithofacies was deposited within storm wave base in dysaerobic-to-aerobic waters. Conditions of oxygenated and agitated waters predominated around basin margins and on intrabasinal elevations, where the Kupferschiefer is lacking.

Variability of the Kupferschiefer sequences is attributed to fluctuating redoxcline, which intersected submarine topography. Vertical movements of the redoxcline, related to sea-level fluctuations could account for large lateral shifts of depositional environments along a gently dipping shelf.

Introduction

The Kupferschiefer (T1) is a shale formation lying in the basal part of the first cyclothem (PZ1) between the Rotliegendes/Zechstein siliciclastics and Zechstein carbonate-evaporite deposits (Fig. 1). The lithology and petrology of the Kupferschiefer have been described by Rydzewski and Wazny (1962), Haranczyk (1964), Rydzewski (1964, 1969, 1978), Jarosz (1968), Salski (1968), Oszczepalski (1978, 1980), Tomaszewski (1978), Niskiewicz (1980),

Lecture Notes in Earth Sciences, Vol. 10
T.M. Peryt (Ed.), The Zechstein Facies in Europe

Preidl and Metzler (1984). However, most previous studies have been limited to small parts of the Fore-Sudetic area and Pomerania. Too little attention has been paid to palaeoenvironmental interpretation of the Kupferschiefer. This study attempts to summarize the palaeogeographical aspects of Kupferschiefer sedimentation in the Polish part of the Zechstein basin. Depositional conditions of these sediments have also been considered to provide a general framework for a sedimentary model of the Polish Kupferschiefer.

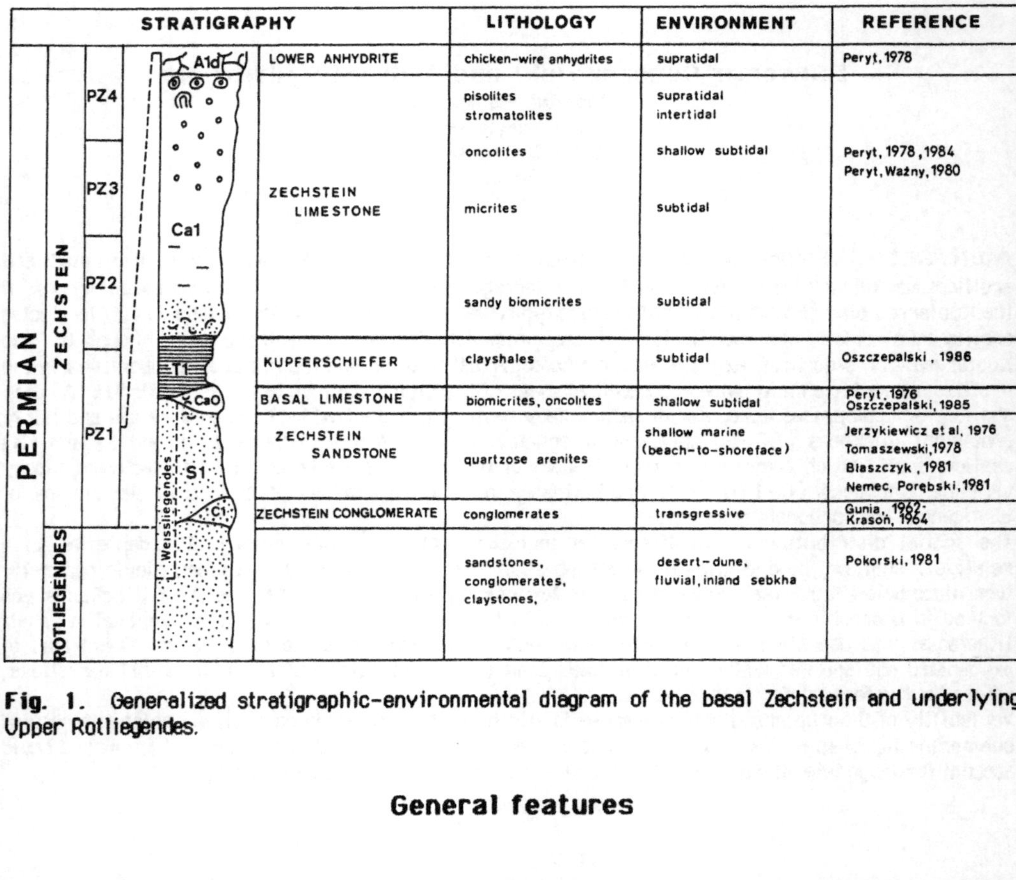

Fig. 1. Generalized stratigraphic-environmental diagram of the basal Zechstein and underlying Upper Rotliegendes.

General features

In this paper, the Kupferschiefer includes laminated shales (sensu Potter *et al.*, 1980) and laminated shaly dolomites/limestones together with associated non-laminated sediments (Oszczepalski, 1986). The lower boundary is the contact surface of the Kupferschiefer with the Zechstein Sandstone, Zechstein Conglomerate or with the Basal Limestone, or, rarely with the older bedrock. The upper boundary is the bottom surface of the Zechstein Limestone.

The Kupferschiefer covers a major part of the Polish Zechstein area, covering ca. 170,000 sq. km. Its extent is illustrated in Fig. 2. Except for isolated outcrops in the North Sudetic Trough, the Kupferschiefer is known from the subsurface. The thickness of this formation is approximately stable, varying generally from 30 to 60 cm, rarely exceeding 1 m (Fig. 2). Evidence of considerably higher thickness of the Kupferschiefer to about 10 m in the marginal part of the basin (e.g., North Sudetic Trough - Gunia, 1962; Krason, 1964) should be treated with great caution because of different criteria of distinguishing the Kupferschiefer and uncertainty of the correlations. Only some parts of the Spotted and Copper-bearing Marls from the North Sudetic Trough may be comparable - in view of the definition of the formation used here - with the typical Kupferschiefer sediments (cf. Figs. 3 & 4).

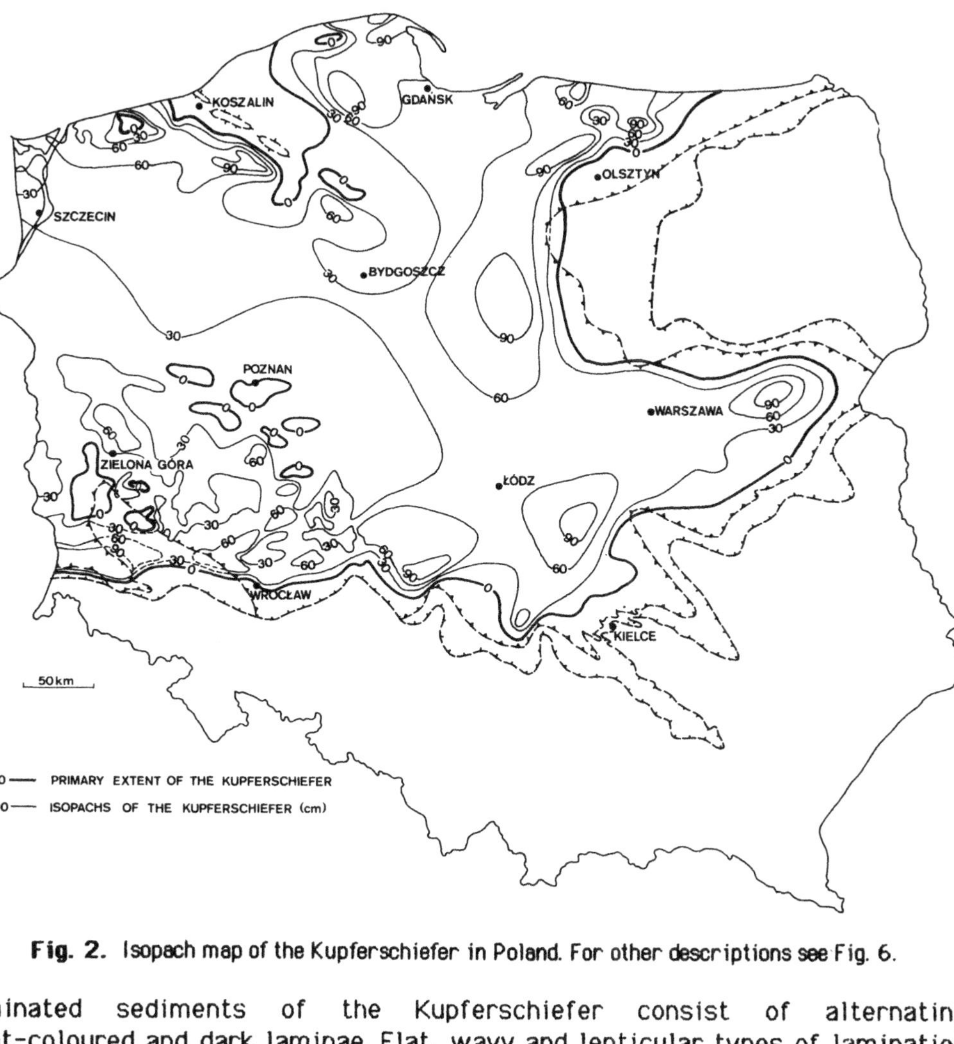

Fig. 2. Isopach map of the Kupferschiefer in Poland. For other descriptions see Fig. 6.

Laminated sediments of the Kupferschiefer consist of alternating light-coloured and dark laminae. Flat, wavy and lenticular types of lamination are dominant (Figs. 3 & 4). The light laminae are built of clay or carbonate material, whereas dark laminae combine a blend of organic, clay and carbonates. Illite dominates the clay material although traces of montmorillonite, kaolinite, chlorite and glauconite also occur.

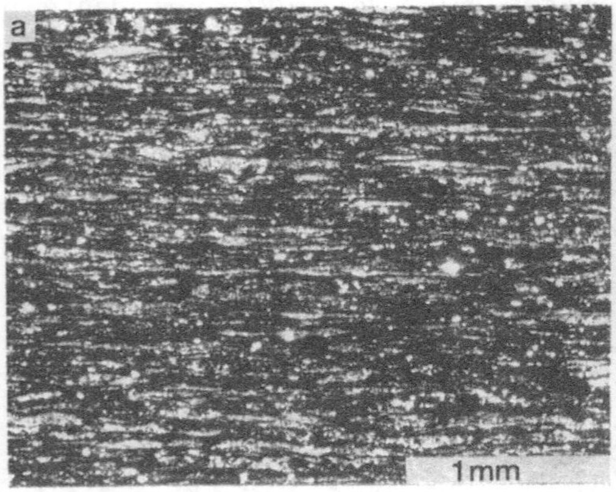

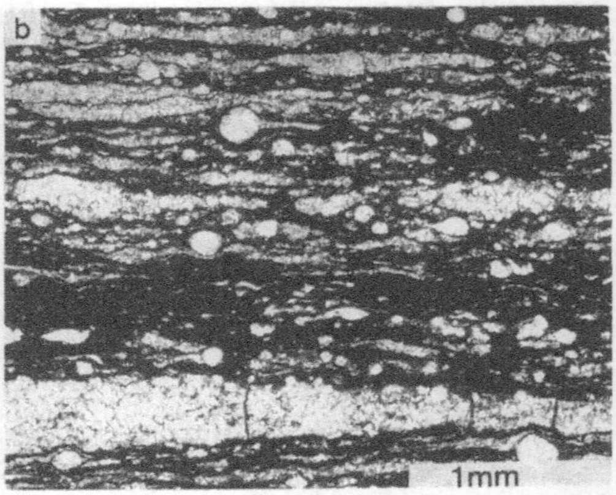

Fig. 3. Kupferschiefer microlithofacies from the deep shelf. **(a)** Microlithofacies 1 - organic-rich clayshale with fine planar laminae composed of clay. S-41 well. **(b)** Microlithofacies 2 - dolomitic-calcarous clayshale with planar and wavy lamination, rich in non-skeletal carbonate grains. M-1 Lipowiec well.

Table 1. Characteristic features of the Kupferschiefer microlithofacies.

General features	Microlithofacies 1 (Figs. 3a & 4a, b)	Microlithofacies 2 Figs. 3b, 4c, d, e,f & 5)
lamination	flat	wavy, lenticular
main component of light laminae	clay	carbonate
thickness of light laminae	<0.03 mm	>0.03 mm
share of dark matrix	>70%	<70%
skeletal grains	very rare	rare
non-skeletal grains	no	present (locally)
sand admixture	low	high (in places)

Carbonates - dolomite and calcite (mainly as micrite or microspar) are locally abundant. In places, a cosiderable admixture of detrital grains (quartz and rare muscovite, feldspar and lithic grains) of silt to very fine sand grade, dispersed or arranged in laminae is common (Fig. 4f). Metal sulphides are abundant (framboidal pyrite is ubiquitous), locally iron oxides occur. Fauna is very rare and little diversified. *Lingula credneri* Geinitz and fish remains are most common. Occasionally foraminifers, brachiopods, molluscs, bryozoans and ostracods are present. In places bioturbation occurs (Fig. 5). Locally non-skeletal carbonate grains, mainly peloids of 0.05 to 0.15 mm in diameter abound (Figs. 3b & 4e). Plant fragments are common too.

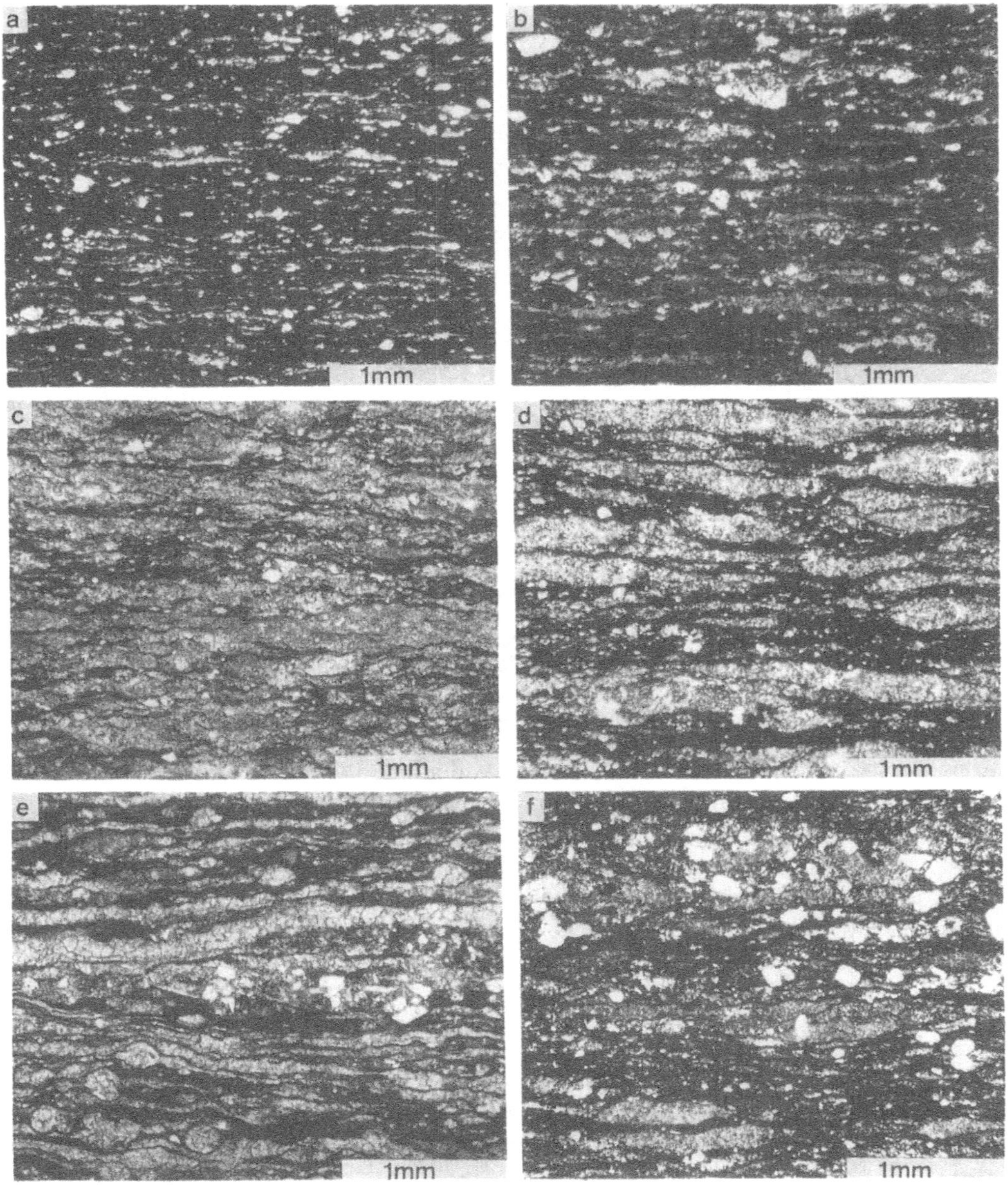

Fig. 4. Kupferschiefer microlithofacies from the shallow shelf. **(a)** Microlithofacies 1 - organic-rich clayshale with fine, even and discontinous laminae composed of clay, abundant dolomite matrix. Bartoszyce IG1 well. **(b)** Microlithofacies 1 - organic-rich silty and finely laminated clayshale. Zebrak IG1 well. **(c)** Microlithofacies 2 - organic-poor dolomitic clayshale with wavy lamination. Bartoszyce IG1 well. **(d)** Microlithofacies 2 - dolomitic clayshale with lenticular lamination. Ostrzeszow 1 well. **(e)** Microlithofacies 2 - calcareous clayshale with wavy lamination, rich in non-skeletal carbonate grains. Czarny Mlyn IG1 well. **(f)** Microlithofacies 2 - sandy dolomitic clayshale with lenticular lamination. Szklarka 3 well.

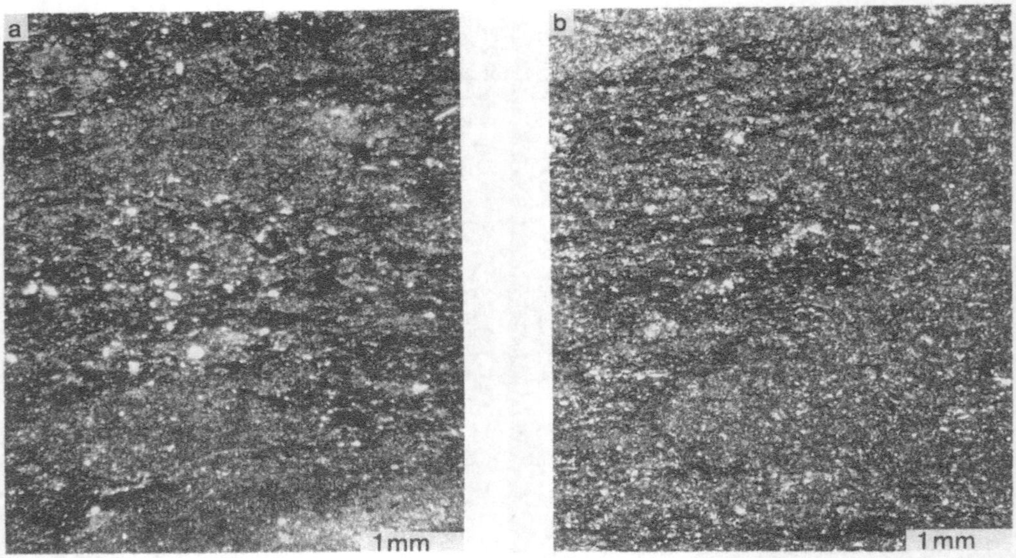

Fig. 5. Bioturbated Kupferschiefer microlithofacies. (**a**) Microlithofacies 2 - dolomitic unfossiliferous bioturbated clayshale, some irregular laminations are preserved. Lenartowice IG1 well. (**b**) Microlithofacies 2 - dolomitic unfossiliferous bioturbated clayshale (Kupferschiefer) passing upward to homogenous (totally bioturbated) dolomicrite of the Zechstein Limestone. Grundy Gorne IG1 well.

Considering sediment fabric and compositional variations of the Kupferschiefer laminites, two main microlithofacies may be distinguished: (1) clayshales, and (2) laminated argillaceous dolo- and calcilutites (Table 1). A thorough study of their sequence and mutual proportions served as the basis for broader palaeogeographic interpretations.

Principles of palaeogeographic zonation

The Zechstein transgression in the area of the Rotliegendes depositional basin, to which the Kupferschiefer sedimentation is related, was caused by the eustatic rise in sea level resulting from the recession of the polar Permian glaciation (Smith, 1980). The water inflow came from the NW direction, from the region of early Permian Protoatlantic rift and the system of the North Sea troughs connected with it (Ziegler, 1978).

A considerable role in the formation of the Polish-German fault-controlled Permian basin was played by tectonic process (Wagner *et al.*, 1980) resulting in a further subsidence of the Polish Zechstein basin (Pokorski, 1981). Denudation processes and synsedimentary block movements in Rotliegendes times resulted

in a fairly diversified palaeorelief of the Zechstein basement. At the early stage of the Zechstein sea stabilization, there were two large embayments in the east, divided by the Mazury Peninsula, and many bays along the southern sea-margin (Fig. 6). In the western part of Poland elevations were abundant and some of them could have built islands (Czajor & Wagner, 1973; Wagner, 1976; Tomaszewski, 1978).

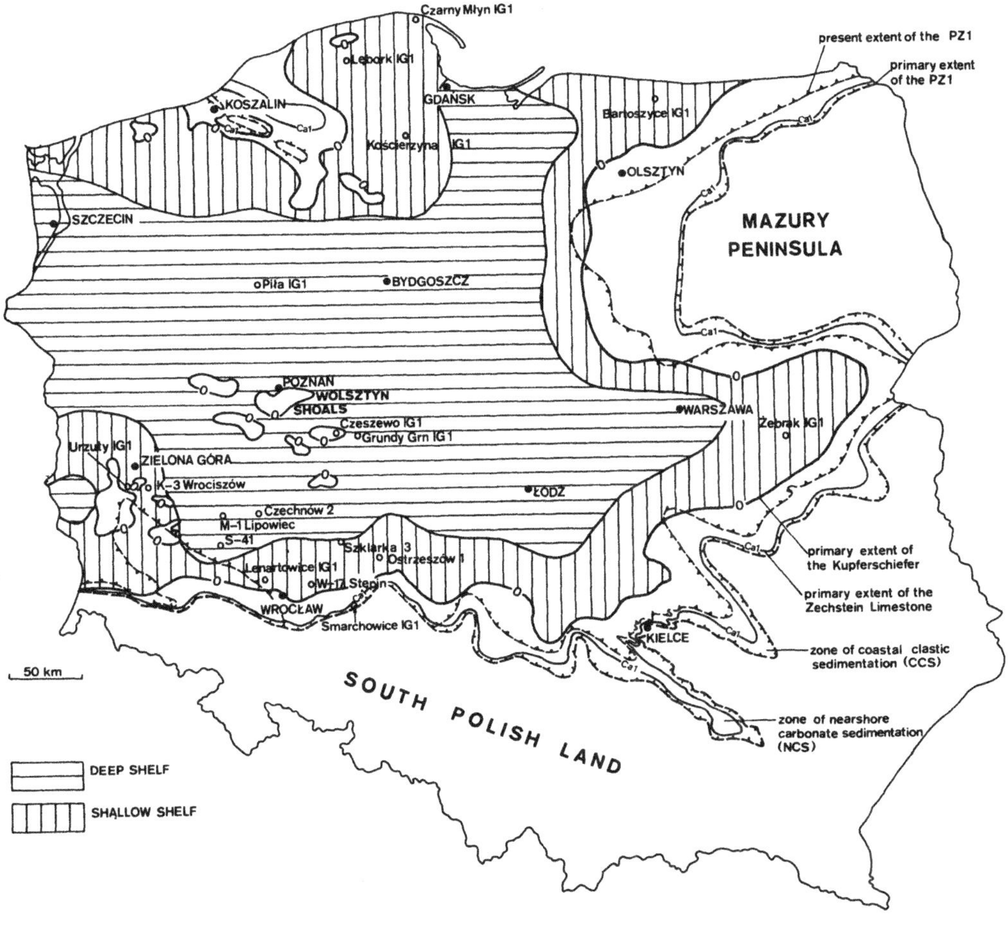

Fig. 6. Palaeogeography of the Kupferschiefer in Poland. Original extent of the PZ1 and Zechstein Limestone after Peryt and Piatkowski (unpublished). Of over 250 studied wells, only those mentioned in the text are shown.

The transgressing sea caused the resedimentation of upper part of the fluvial-aeolian Rotliegendes complex. As a result, the Zechstein Sandstone (Figs. 1 & 7; see also Glennie & Buller, 1983) or its facies equivalent - the Zechstein Conglomerate (e.g., Gunia, 1962; Jordan, 1969; Seifert, 1972; Paul, 1982) developed throughout almost the entire basin. These sediments at their top are highly enriched in carbonate or anhydrite cement and locally pass into a thin bed of sandy limestones and dolomites (< 10 cm thick), indicating low energy and proper oxidation of the bottom waters (Peryt, 1976; Oszczepalski, 1985). In the marginal parts of the basin, in aerobic environments of moderate-to-high energy, the sedimentation of fossiliferous micrites, biomicrites and oncolites took place leading to the formation of the Basal Limestone (*sensu* Oszczepalski, 1985) from several dozens of centimetres to over 4 m thick.

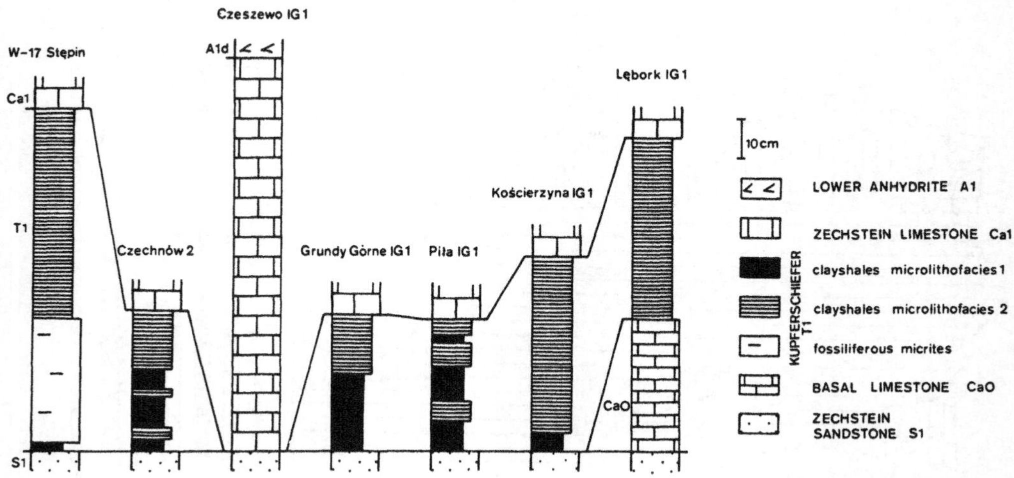

Fig. 7. Comparison of thickness and lithologic sequences for the Kupferschiefer. Measured sections of the W-17 Stepin, Koscierzyna IG1 and Lebork IG1 wells are representative for the shallow shelf, whilst sequences of the Czechnow 2, Grundy Gorne IG1 and Pila IG1 wells are characteristic of deep shelf. Condensed sequence of the Czeszewo IG1 well is representative for intrabasinal shoals (*cf.* Peryt & Wazny, 1980). For location see Fig. 6.

The sediments of the Zechstein Sandstone or Zechstein Conglomerate and of the Basal Limestone are covered by the Kupferschiefer over most of the Zechstein basin. Such a distinct change in lithology was probably due to a considerable rise of the sea level but only to such a degree that it allowed the basinal eminences to have influenced the microlithofacies differentation of the Kupferschiefer.

The Kupferschiefer basin was an epicontinental sea of a weak connection with the ocean. As for modern shelf seas (Heckel, 1972; Swift, 1976; Byers, 1977; Johnson, 1978; Potter *et al.*, 1980) the following main factors controlled the sedimentation in the Kupferschiefer basin: bottom energy and relief, supply of the sediment and oxidation of bottom waters, as discussed elsewhere (Oszczepalski, 1980, 1986, in press).

Taking into consideration the regional distribution of the Kupferschiefer microlithofacies and their mutual relations in particular profiles two main palaeogeographical zones may be distinguished: (A) deep shelf, and (B) shallow shelf (Fig. 6). Sediments of the Kupferschiefer vainsh 10 to 60 km from the inferred shoreline, passing shoreward into the zone of nearshore carbonate sedimentation (NCS) and farther into the zone of coastal clastic sedimentation (CCS). Also in the open-sea area there were regions where the Kupferschiefer was not deposited.

Deep shelf

The deep shelf is characterized by the Kupferschiefer profiles of a fairly small but consistent thickness (usually 20-60 cm), occurring between the Zechstein Sandstone and Zechstein Limestone formations (Fig. 7; Czechnow 2, Grundy Gorne IG1 and Pila IG1 wells). The Kupferschiefer consists of alternately occurring microlithofacies 1 (Fig. 3a) and 2 (Fig. 3b). In extensive depressions of the basin floor the microlithofacies 1 prevails and in extreme cases it forms the entire Kupferschiefer sequence. Sequences composed of repeated microlithofacies 1 and 2 predominate towards the shore.

These microlithofacies form a characteristic association of the following diagnostic features: regular, flat or wavy lamination, dominance of clay and a considerable admixture of the organic matter. Tempestites connected with local shoals and bioturbation limited to the topmost parts of the Kupferschiefer sequences at the contact with the Zechstein Limestone (Fig. 5b) are extremely rare.

Taking the above features into account one may assume that the sedimentation in the discussed part of the depositional setting was taking place within stagnant bottom waters, directly from suspension. At the deepest part of this zone, chiefly the sediments of the microlithofacies 1 formed, indicating the deep subtidal conditions of sedimentation, below the storm wave base. At slightly shallower depths of the deep shelf, a seasonal influence of the neighbouring shallow shelf was marked by interlayers of the microlithofacies 2. The deep shelf bottom predominantly stayed within the anaerobic zone or at its transition into the dysaerobic zone; the fact demonstrated by the lack of bioturbation, the paucity of fauna and also by the high content of the sapropelic organic matter. The Wolsztyn Shoals protruded over the zone of oxygen deficiency, thus they could not have been covered by the Kupferschiefer sediments (Fig. 7, Czeszewo, IG1 well).

Shallow shelf

Sequences of the Kupferschiefer of different thickness, from several to over 150 cm, lying above the Zechstein Sandstone/Zechstein Conglomerate or Basal Limestone and below the Zechstein Limestone, are characteristic of the shallow shelf (Fig. 7, W-17 Stepin, Koscierzyna IG1 and Lebork IG1 wells). The microlithofacies 2 predominates here (Fig. 4c, d , e , f), whilst the microlithofacies 1 occurs only at the base of the profiles situated at the border with the deep shelf (Fig. 4a, b). Because of dominance of microlithofacies 2, the Kupferschiefer is characterized mostly by lenticular and wavy lamination, abundance of the carbonate material and a small content of organic matter. Cryptobioturbational disturbances of the lamination (Fig. 5a), especially close to the nearshore zone of the carbonate sedimentation are frequent. Sometimes thin interlayers of terrigenous tempestites represented by mudshales with horizontal and cross-lamination (Fig. 8a) or hummocky cross-lamination as well as by massive siltstones and sandstones are present. The packstone lenses occur as well (Fig. 8b). At the bottom parts of numerous profiles from the Fore-Sudetic area, a dolomite interset has been recorded.

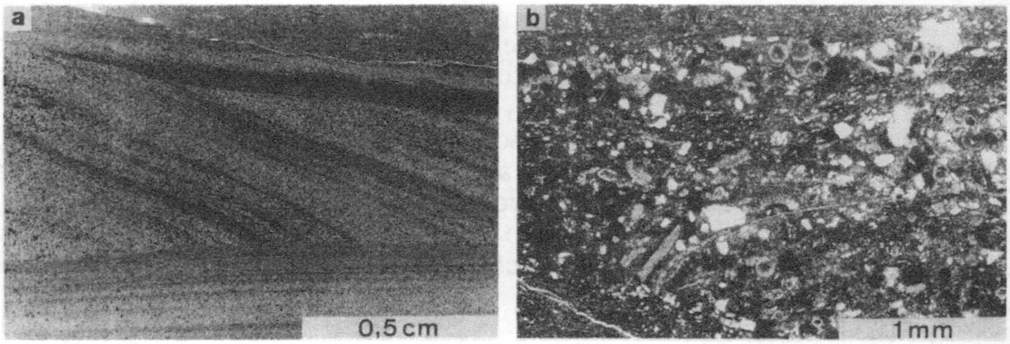

Fig. 8. Storm beds in the Kupferschiefer. **(a)** Tempestitic mudshale bed with plane parallel lamination at the base, followed by cross-lamination and overlain by organic poor hematitic clayshale. K-3 Wrociszow well. **(b)** Tempestitic calcarenite bed composed of foraminifers and silt to sand quartz, within organic-poor hematitic clayshale. Urzuty IG1 well.

The sedimentation within the shallow shelf was taking place in the environment of moderate and fairly differentiated energy. The features of the sediments as well as frequent cases of the sedimentary pinching out of the Kupferschiefer indicate the deposition in the transition area from deep to shallow subtidal zone. Carbonates, which predominate in the shallow shelf, seem to be a result of lateral sediment transport from the shoals and nearshore high-energy environments. Similarly, fossils (e.g., *Lingula*) underwent the post-mortem redeposition to a quiet environment of the Kupferschiefer deposition. Bioturbation together with paucity of skeletal grains indicate the sedimentation within the dysaerobic zone and seasonally - as for the terrigenous and carbonate interlayers - at the transition from dysaerobic into aerobic zone, probably near the fair-weather wave base. Such conditions were favourable for synsedimentary, oxic degradation of the organic matter.

Mechanism of sedimentation

The described palaeogeographic zones give a general picture of the Zechstein shelf during the Kupferschiefer sedimentation. A slow gravitational settling out of mud from suspension in the low-energy environment prevailed. The presence of lamination within the Kupferschiefer deposits testifies to an alternating and probably seasonal supply of the organic and inorganic substances to the bottom of the sedimentary basin.

The fact of deposition generally taking place in quiet waters does not however indicate a negligible role of the hydrodynamic factors in the processes of sedimentation. On the contrary, their activity caused that the deposited muddy sediment did not form a static system but, like in modern shelf seas (e.g., Swift, 1976; Johnson, 1978), underwent repeated resuspension and transport resulting most likely from the changes of the wave base. Obviously the influence of the hydrodynamic factors on the character of the Kupferschiefer sedimentation could only occur in case of a shallow sea. Such a relatively shallow nature of this sea is indicated by the frequently observed impact of the palaeorelief on the differentiation of the basal Zechstein sedimentation (Czajor & Wagner, 1973; Jung & Knitzschke, 1976; Wagner, 1976; Oszczepalski, 1978, 1980; Peryt & Wazny, 1980; Smith, 1980; Blaszczyk, 1981; Paul, 1982).

The spatial distribution of the Kupferschiefer lithofacies indicates that the accumulation was highly influenced by the activity of waves and currents generated by meteorological factors causing the redistribution of the sediment from the regions of strong turbulence to calm waters (Fig. 9). Such process usually leads to the levelling of the bottom topography, thus, the isopach map of the Kupferschiefer (Fig. 2) may be generally regarded as the reflection of palaeorelief. The map clearly shows that the zones of higher thickness extend along the nearshore carbonate sedimentation zone, the fact which should be regarded as the result of sedimentation on the foreslope running around basin flank areas, which occurred within the fair-weather wave base. In regions of high thickness of the Kupferschiefer carbonate material predominates over clays and most plausibly it may be explained by the resedimentation from the agitated environments of nearshore carbonate sedimentation (Figs. 6 & 9). The terrigenous material was supplied to the basin from the land and from the zone of coastal clastic sedimentation as well as from the intrabasinal shoals. Due to these phenomena the shallow shelf Kupferschiefer deposits locally contain substantial amounts of silt and sand quartz admixture and terrigenous or carbonate tempestite interbeds (Fig. 8). Only fine-grained terrigenous material (mainly clay) reached the deepest centre of the basin.

Organic matter, mainly autochthonous of phytoplanktonic and bacterial origin gathered mostly in the deepest parts of the basin. Within the shallow shelf it was partially degraded and diluted in the bulk of carbonates intensively supplied there. The lack of organics with the simultaneous enrichment in hematite (Fig. 8) in the oxidized non-tempestitic Kupferschiefer sediments

("Rote Fäule") is attributed either to (1) the syndepositional and syndiagenetic oxidation of the sediments due to their sedimentation on the submarine elevations and in the nearshore zones of the basin (e.g., Jung & Knitzschke, 1976; Paul, 1982; Preidl & Metzler, 1984; Tomaszewski, 1985) or to (2) the post-depositional hematitization due to the flow of oxidizing fluids through the sediment (Rydzewski, 1978; Oszczepalski, 1980, in press; Jowett *et al.*, in press).

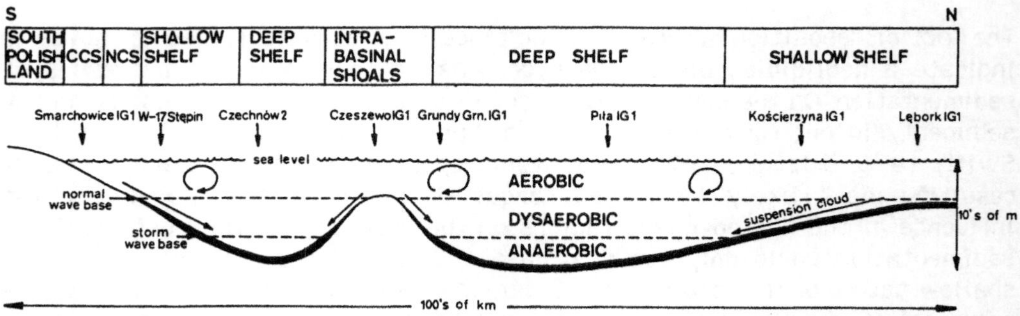

Fig. 9. Depositional model of the Kupferschiefer for Polish part of the Zechstein basin.

Sedimentary model

In conditions of generally low energy of bottom waters in the entire Kupferschiefer depositional system and substantial organic productivity, the basin waters were vertically stratified (Fig. 9). The following observations indicate that the Kupferschiefer formed almost exclusively on the bottom situated below the normal wave base wihin waters of dysaerobic and anaerobic zones: (1) an extensive distributions and relative continuity of the Kupferschiefer (beside its small and differentiated thickness), (2) very fine-grained sediments, (3) dominance of clay and carbonate mud, (4) regular lamination, (5) abundance of organic matter. (6) lack of current and erosional structures (except those related to tempestites), (7) almost complete lack of skeletal fauna and burrows, (8) presence of bioturbation at the margins of sedimentary basin, and (9) absence of features indicative of subaerial exposition. In light of these facts it seems that suggestions of an extremely shallow environment of the Kupferschiefer deposition (e.g., Alexandrowicz, 1973; Jerzykiewicz *et al.*, 1976; Blaszczyk, 1981; Preidl & Metzler, 1984) may be plausible for local bottom elevations only - but should not be generalized on the entire Kupferschiefer basin.

Reconstruction of the bathymetry is speculative. It seems that the Kupferschiefer generally formed below fair-weather wave base. i.e., at the depth beyond 10-40 m (Oszczepalski, 1980, 1986; Peryt, 1984). Within the shallow shelf the deposition might have taken place below the fair-weather wave base, but on the bottom lying shallower than 50-60 m (see also Bell *et al.*, 1979). Yet, for the deep shelf it could have been mainly within the zone beneath the storm wave base, i.e., in water depths over 50-60 m. For certain regions of the open sea, the water depth might have exceeded 100-200 m (Smith, 1980; Paul, 1982).

The application of the shallow stratified basin model allowed to estimate the depositional slope at the transition from the zone of nearshore sedimentation into the deep shelf, which should have been with a gradient of 0.4 to 1.2 m/km. These values agree with average bottom slopes of modern epicontinental seas (e.g., Heckel, 1972).

Summing-up, the extent of the Kupferschiefer and its facies differentiation are the result of sea water stratification and the intersection of the morphologically varied bottom by redoxcline. Beside the environmental factors related to the bottom topography, a considerable part of the sedimentation was played by pulsative shifts in the position of the redoxcline. Its rise caused the transgression of the anoxic waters and the widening of the accumulation area of the microlithofacies 1, whilst its lowering favoured transport of the carbonate material towards the open sea and deposition of the microlithofacies 2, as well as progradation of the nearshore carbonate flats and outwashing of shoals. It results in cyclic character of sedimentation (Fig. 7). Thus, the microlithofacies 1 indicates the retrogradational phases and the microlithofacies 2 the progradational phases. A special attention is attracted to the form of the upper part of the lowest cycle which can be considered as a marker horizone; in the open sea it usually is represented by the microlithofacies 2 enriched in peloids (see also Rentzsch *et al.*, 1965). This horizon can be correlated with the dolomite or limestone bed occurring within the Kupferschiefer of outer shallow shelf (*cf.* W-17 Stepin well , Fig. 8), and farther shorewards with the Basal Limestone. The intertonguing relationships of the Kupferschiefer and the Basal Limestone has been already suggested by Paul (1982) and Oszczepalski (1985).

The mechanism responsible for the fluctuations of the redoxcline is not clear yet. Gerlach and Knitzsche (1978) relate these changes to the epeirogenic movements and Paul (1982) to the log-term fluctuations in the productivity of phytoplankton. It seems that the pulsative changes of the sedimentary conditions (caused by the redoxcline oscillations) feasible to correlate in a considerable area of the Zechstein basin were due to eustatic shifts of the sea level, like during Zechstein Limestone sedimentation (Peryt, 1984, and in press). Because the eustatic.fluctuations take place very fast (Guidish *et al.*, 1984) the isochronism of the boundaries of particular sedimentational cycles may be assumed.

The cause why the Zechstein Limestone deposition substituted the Kupferschifer sedimentation has not been clearly explained. It is likely that, as

in successive phases of the Kupferschiefer sedimentation, a further shallowing of the sedimentary basin followed.
A more plasusible cause of this fact seems to be at the better circulation of the sea water (Smith, 1980; Peryt, 1984) or the lowering of the redoxcline below the storm wave base due to, for instance, a decrease of the surface water productivity. Despite the cause an intense expansion of the skeletal fauna and the homogenization of laminated deposits (Fig. 5b) related with it, terminated the deposition of the Kupferschiefer.

Conclusions

The Kupferschiefer together with the other basal Zechstein lithostratigraphic units represent a marine transgression to form a pre-evaporite semi-enclosed and gently sloping epicontinental sea. There is a gradual transition across the shelf from land through shoreline siliciclastics and nearshore carbonate belts to open marine deposition within the center of sedimentary basin. Distribution of the Kupferschiefer lithofacies suggests a deposition in an oxygen-starved stratified basin. It is suggested that the redoxcline existed between the oxygenated surface waters and anoxic bottom waters. It is reasonably to envisage the redoxcline as a transition zone separating agitated waters existing above fair-weather wave base from stagnant bottom waters below storm wave base. The deep shelf lithofacies which predominates in the central sections of the basin appears consistent with deposition in anaerobic-to-dysaerobic zone. This is supported by the fine regular lamination, lack of bioturbation and high clay and organic-carbon content. The shallow shelf lithofacies which occurs around margins of the Kupferschiefer depositional setting can be explained by sedimentation in dysaerobic-to-aerobic conditions. This is evidenced by irregular lamination, occasional bioturbation, low organic-carbon content, high amount of carbonates and presence of minor tempestites.

In summary, the general pattern of the Kupferschiefer deposition reflects decreasing oxygenation, bottom energy and terrigenous/carbonate supply as a function of increasing water depth and distance from the shore. Interfaces between energy and oxygenation zones intersected morphologically diversified sea bottom controlling the Kupferschiefer extent and lithofacies distribution. Pulsative changes of sedimentary conditions caused by vertical shifts of the redoxcline position have been of primary importance. The rise of the redoxcline caused transgression of anoxic waters and extension of the area of anoxic sediment accumulation, whereas its lowering caused the area of anoxic sediments to be restricted and carbonates to extend down into deeper parts of the basin. Such fluctuations resulted in multicycle sequences of the Kupferschiefer and its lateral interfingering with nearshore sediments.

Acknowledgements

We are grateful to K. Jaworowski for much helpful advice. We also wish to thank Z. Deczkowski, K. Dyjaczynski, J. Kulick, L. Lenartowicz, M. Rup, D.B. Smith, M. Solak and R. Wagner for offering valuable information and suggestions. We acknowledge M. Kado for his field assistance; H. Chojeta, drafting; J. Modrzejewska, photos. T.M. Peryt kindly criticized the manuscript and offered useful comments.

References

Alexandrowicz, S.W., 1973. Fauna lingulowa jako wskaznik warunkow tworzenia sie lupkow miedzionosnych. *Rudy Met. Niezel.*, 18: 525-526.

Bell, J., Holden, J., Pettigrew, T.H. & Sedman, K.W., 1979. The Marl Slate and Basal Permian Breccia at Middridge, Co. Durham. *Proc. York. Geol. Soc.*, 42: 439-460.

Blaszczyk, J.K., 1981. Wplyw paleoreliefu stropu bialego spagowca na zmiennosc facjalna serii zlozowej w zaglebiu lubinskim. *Geol. Sudetica* 16: 195-217.

Byers, Ch.W., 1977. Biofacies patterns in euxinic basins: A general model. *SEPM Spec. Publ.*, 25: 5-17.

Czajor, E. & Wagner, R., 1973. Typy genetyczne skal oraz mikrofacje i paleogeografia wapienia cechsztynskiego (Ca1) w strefie Koszalina - Chojnic. *Kwart. Geol.*, 17: 471-486.

Gerlach, R. & Knitzschke, G., 1978. Sedimentationszyklen aus der Zechstein Basis im SE-Harzvorland und ihre Beziehungen zu einigen bergtechnischen Problemen. *Z. angew. Geol.*, 24: 214-221.

Glennie, K.W. & Buller, A.T., 1983. The Permian Weissliegend of NW Europe: the partial deformation of aeolian dune sands caused by the Zechstein tresngression. *Sediment. Geol.*, 35: 43-81.

Guidish, T.M., Lerche, I., Kendall, C.G.St.C. & O'Brien, J.J., 1984. Relationship between eustatic sea level changes and basement subsidence. *Am. Assoc. Petroleum Geologists Bull.*, 68: 164-177.

Gunia, T. 1962. Cechsztyn synkliny leszczynskiej. *Biul. Inst. Geol.*, 173: 57-114.

Haranczyk, C., 1964. Petrographic classification of the Zechstein copper-bearing rocks from Lower Silesia. Bull. *Acad. Pol. Sc. Ser. geol. geogr.*, 12: 19-25.

Heckel, P.H., 1972. Recognition of ancient shallow marine environments. *SEPM Spec Publ.*, 16: 226-286.

Jarosz, J., 1968. Charakterystyka mineralogiczno-petrograficzna zloza Lubin. *Rudy Met. Niezel.*, 13: 625-634.

Jerzykiewicz, T., Kijewski, P., Mroczkowski, J. & Teisseyre, A.K., 1976. Geneza osadow bialego spagowca monokliny przedsudeckiej. *Geol. Sudetica* 11: 57-97.

Johnson, H.D., 1978. Shallow silicilastic seas. In: Reading, H.G. (Ed.), *Sedimentary Environments and Facies*, 207-258. Blackwell, Oxford.

Jordan, H., 1969. Zur Biostratigraphie und Fazies des Zechsteins in Germanischen Becken, unter besonderer Berucksichtigung des Thuringer Beckens. *Freiberger Forschungsh.*, 245: 27-45.

Jowett, E.C., Pearce, G.W. & Rydzewski, A., in press. A mid-Triassic paleomagnetic age of the Kupferschiefer mineralization in Poland, based on a revised apparent polar wander path for Europe and Russia. *Jour. Geoph. Research*

Jung, W. & Knitzschke, H., 1976. Kupferschiefer in the German Democratic Republic (GDR) with special references to the Kupferschiefer deposits in the southeastern Harz Foreland. In: Wolf, K.H. (Ed.), *Handbook of Strata-Bound and Stratiform Ore Deposits*, 6: 353-406. Elsevier, Amsterdam.

Krason, J., 1964. Podzial stratygraficzny cechsztynu polnocnosudeckiego w swietle badan facjalnych. *Geol. Sudetica*, 1: 209-262.

Nemec, W & Porebski, S.J., 1981. Sedimentary environment of the Weissliegendes sandstones in Fotre-Sudetic Monocline. *Proc. Intern. Symp. Central Europ. Permian*, 273-293. Wyd. Geol., Warszawa.

Niskiewicz, J., 1980. Zjawiska metasomatozy w cechsztynskich zlozach rud miedzi Dolnego Slaska. *Geol. Sudetica* 15: 7-75.

Oszczepalski, S., 1978. Utwory cyklotemu werra w zachodniej czesci niecki polnocnosudeckiej i poludniowej czesci perykliny Zar. *Przegl. geol.*, 26: 413-418.

Oszczepalski, S., 1980. Paleogeography, sedimentation and mineralization of the Z1 carbonate series (Zechstein) in the western part of the Fore-Sudetic Monocline (western Poland). *Contr. Sedimentology* 9: 307-323.

Oszczepalski, S., 1985. Sedymentacja utworow cechsztynskich wapienia podstawowego w rejonie Wroclawia. *Przegl. geol.*, 33: 192-199.

Oszczepalski, S., 1986. On the Zechstein Copper Shale lithofacies and palaeoenvironments in SW Poland. Geol. Soc. Spec. Publ., 22: 171-182.

Oszczepalski, S., in press. Kupferschiefer in southwestern Poland: sedimentary environments, metal zoning, and ore controls. *Geol. Assoc. Canada Spec. Paper.*

Oszczepalski, S. & Rydzewski, A., 1983. Miedzionosnosc utworow permu na obszarze przylegajacym do zloza Lubin-Sieroszowice. *Przegl. geol.*, 31: 437-444.

Paul, J., 1982. Zur Rand- und Schwellen-Fazies des Kupferschiefers. *Z. dt. geol. Ges.*, 133: 571-605.

Peryt, T.M., 1976. Ingresja morza turynskiego (gorny perm) na obszarze monokliny przedsudeckiej. *Rocz. Pol. Tow. Geol.*, 46: 455-465.

Peryt., T.M., 1978. Charakterystyka mikrofacjalna cechsztynskich osadow weglanowych cyklotemu pierwszego i drugiego na obszarze monokliny przedsudeckiej. *Studia Geol. Pol.*, 54, 88 pp.

Peryt, T.M., 1984. Sedymentacja i wczesna diageneza utworow wapienia cechsztynskiego w Polsce zachodniej. *Prace Inst. Geol.*, 109, 80 pp.

Peryt, T.M., in press. Basal Zechstein in SW Poland: sedimentation, diagenesis and gas accumulations. *Geol. Assoc. Canada Spec. Paper.*

Peryt, T.M. & Wazny, H., 1980. Microfacies and geochemical development of the basin facies of the Zechstein Limestone (Ca1) in western Poland. *Contr. Sedimentology*, 9: 279-306.

Pokorski, J., 1981. Paleogeography of the Upper Rotliegendes in the Polish Lowland. *Proc. Intern. Symp. Central Europ. Permian*: 56-68. Wyd. Geol, Warszawa.

Potter, P.E., Maynard, J.B & Pryor, W.A., 1980. *Sedimentology of Shale*, 306 pp, Springer.

Preidl, M. & Metzler, M., 1984. The sedimentation of Copper-Bearing Shales (Kupferschiefer) in the Sudetic Foreland. *Mineralium Deposita*, 19: 243-248.

Rentzsch, J., Ludwig, H. & Meyer, E., 1965. Der Kupferschiefer im Bereich der Scholle von Calvörde. *Jb. Geol.*, 1: 177-205.

Rydzewski, A., 1964. Petrografia i mineralogia osadow gornego permu na monoklinie przedsudeckiej i peryklinie Zar. *Przegl. geol.*, 12: 476-480.

Rydzewski, A., 1969. Petrografia lupkow miedzionosnych cechsztynu na monoklinie przedsudeckiej. *Biul. Inst. Geol.*, 217: 113-167.

Rydzewski, A., 1978. Facja utleniona cechsztynskiego lupku miedzionosnego na obszarze monokliny przedsudeckiej. *Przegl. geol.*, 26: 102-108.

Rydzewski, A. & Wazny, H., 1962. Badania petrograficzno-geochemiczne utworow dolnego cechsztynu wiercenia w Leborku. *Kwart. Geol.*, 6: 583-603.

Salski, W., 1968. Charakterystyka litologiczna i drobne struktury lupkow miedzionosnych monokliny przedsudeckiej. *Kwart. Geol.*, 12: 855-873.

Seifert, J., 1972. Das Perm am Südostrand des Thuringer Beckens. *Jb. Geol.*, 4: 97-179.

Smith, D.B., 1980. The evolution of the English Zechstein basin. *Contr. Sedimentology* 9: 7-34.

Swift, D.J., 1976. Continental shelf sedimentation. In: Stanley , D.J. & Swift , D.J. (Eds.), *Marine Sediment Transport and Environmental Management* 311-350. Wiley Intersc. Publ.

Tomaszewski, J.B., 1978. Budowa geologiczna okolic Lubina i Sieroszowic (Dolny Slask). *Geol. Sudetica* 13: 85-132.

Tomaszewski, J.B., 1985. Zloze rud miedziowo-polimetalicznych monokliny przedsudeckiej i jego zwiazki z osadami cechsztynu. *Przegl. geol.*, 33: 375-385.

Wagner, R., 1976. Cechsztyn. *Prace Inst. Geol.*, 79: 18-39.

Wagner, R., Pokorski, J. & Dadlez, R., 1980. Paleotektonika basenu permu na Nizu Polskim. *Kwart. Geol.*, 24: 553-569.

Wyzykowski, J., 1971. Cechsztynska formacja miedzionosna w Polsce. *Przegl. geol.*, 19: 117-122.

Ziegler, P.A., 1978. North-Western Europe: Tectonics and basin development. *Geol. Mijnbouw* 57: 589-626.

SEDIMENTARY FACIES IN THE OLDEST ROCK SALT (Na1) OF THE LEBA ELEVATION (NORTHERN POLAND)

Grzegorz Czapowski
Instytut Geologiczny
ul. Rakowiecka 4
00-975 Warszawa, Poland

Abstract: The evaporitic basin of the Oldest Rock Salt Formation in the Leba elevation area, was differentiated into deeper and shallow parts from the very begining. This division was inherited after Lower Werra Anhydrite basin. The analysis of dynamic conditions of salt deposition and distribution of the distinguished facies indicates that medium and high dynamic facies prevailed in the basin. The salt deposits have formed in two megacycles. During the first megacycle, the "clear" salt series originated, without traces of land influence. The partial infilling of bottom depressions at the end of the first megacycle caused the gradual shallowing and predominance of more dynamic facies during the second megacycle. In the nearshore zone the "clayey salts" series accumulated and in centres of larger basins the "clear" salts were deposited.

General characteristic

The salt deposits (0-210 m thick) in the area of the Leba elevation occur at the depths of 700-1100 m and are known from numerous boreholes (Figs. 1 & 2). A characteristic feature of the salt thickness distribution is reciprocally correlated with the Lower Werra Anhydrite thickness (Fig. 2). The thickest rock salt series are generally connected with the areas of the very thin Lower Werra Anhydrite of deep-water origin. Thin salt series overlie thick anhydrite platform deposits of mostly shallow-water origin (Peryt *et al.*, 1985). The major part of the thickest salt series have originated in deeper basin areas, while thin salt series have formed on the basin bottom elevations (Figs. 2 & 3).

The main lithological components of the Oldest Rock Salt Formation are *rock salts*. They consist mainly of structurally differentiated halites. The colour and transparence of halite depend on the kind and volume of admixtures, such as

Lecture Notes in Earth Sciences, Vol. 10
T. M. Peryt (Ed.), The Zechstein Facies in Europe

sulphates (anhydrite, polyhalite) and terrigenous material (clay minerals, quartz pelite). A large amount of the latter (up to several percent) allowed one to distinguish macroscopically the type of the so-called "clayey salts", previously regarded as descendent salts (Poborski, 1980). These are grey-brown salts, 0 to 70 m thick, occurring in the upper part of the Oldest Rock Salt. They have similar structures as interlayering and surrounding "clear halites" and were interpreted as shallow-water deposits formed in salt pans and nearshore zones of emerged shoals within evaporitic basin (Czapowski, 1983).

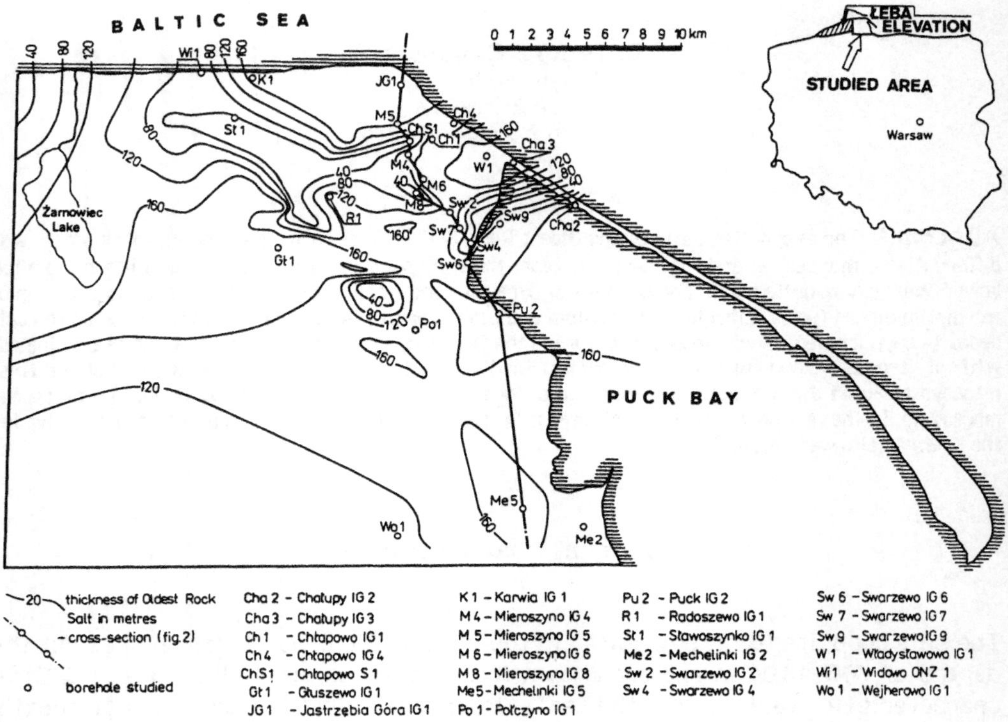

Fig. 1. Thickness of the Oldest Rock Salt (Na1) Formation in the Leba elevation.

Sedimentary structures in halites and their environmental significance

The halites are strongly structurally differentiated (Figs. 4 A-D & 5 A). These differences supposingly reflect the primary variability of salt deposits. There are distinguished three main *structural types* of halites, based on the size relations of halite crystals:

(1) A-halites with crystals of the same size, well selected (Figs. 4 A-B, 5 A-B & 6 B),

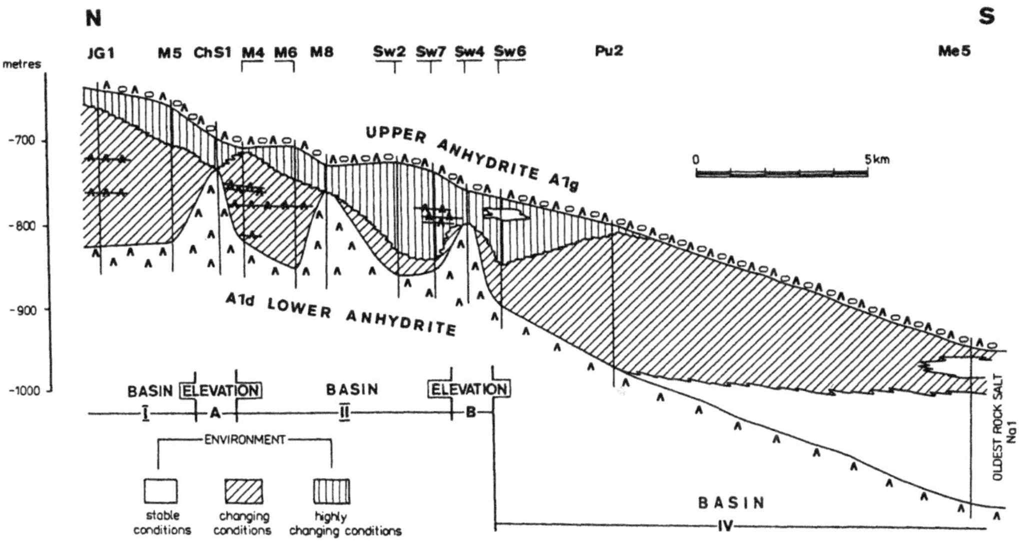

Fig. 2. Cross-section of the Oldest Rock Salt Formation with environmental reconstruction.

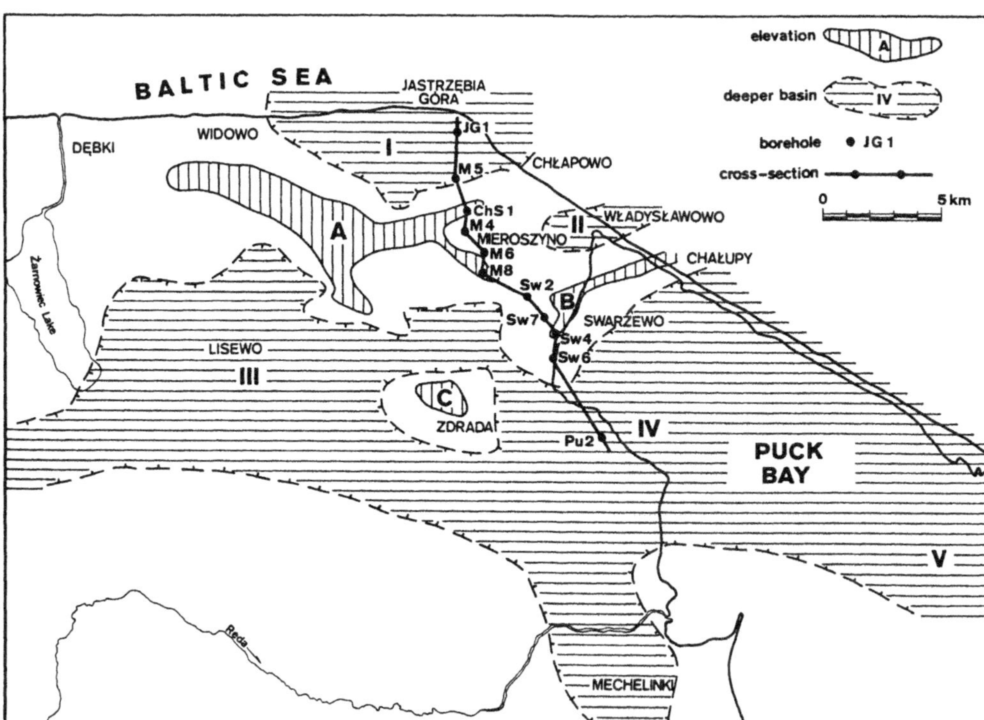

Fig. 3. Paleobathymetry of the Oldest Rock Salt basin in the Leba elevation.

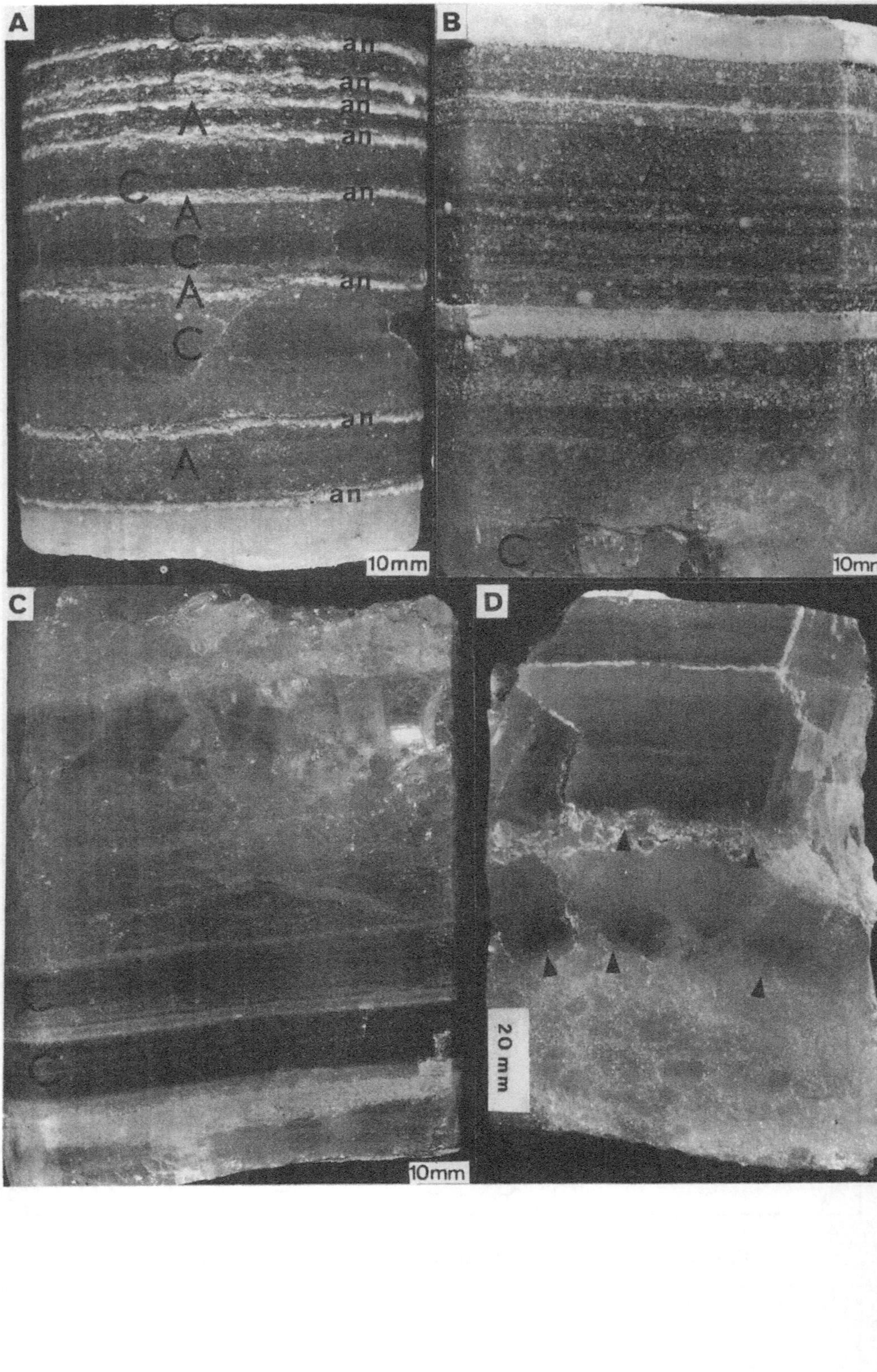
A
C
an
an
an
A
an
C
an
A
C
A
an
A
C
an
A
an
10mm
B
A
C
10mm
C
C
C
10mm
D
20 mm

Fig. 4. Deep basin facies. Explanation of symbols in the text. (**A**) A,C halite rhythmites with thin anhydrite laminae (an). Mechelinki IG 5 well, depth 1103.8 m. (**B**) White anhydrite layers, laminae and nodules within fine A-halite. Wladyslawowo IG 1 well, depth 860.8 m. (**C**) "Internal lamination" (arrows) within C-halite bands. Wladyslawowo IG 1 well, depth 805.1 m. (**D**) Slightly undulated "internal lamination" and lower recrystallized boundaries (arrows) of C-halite layers. Chalupy IG 2 well, depth 781.8 m.

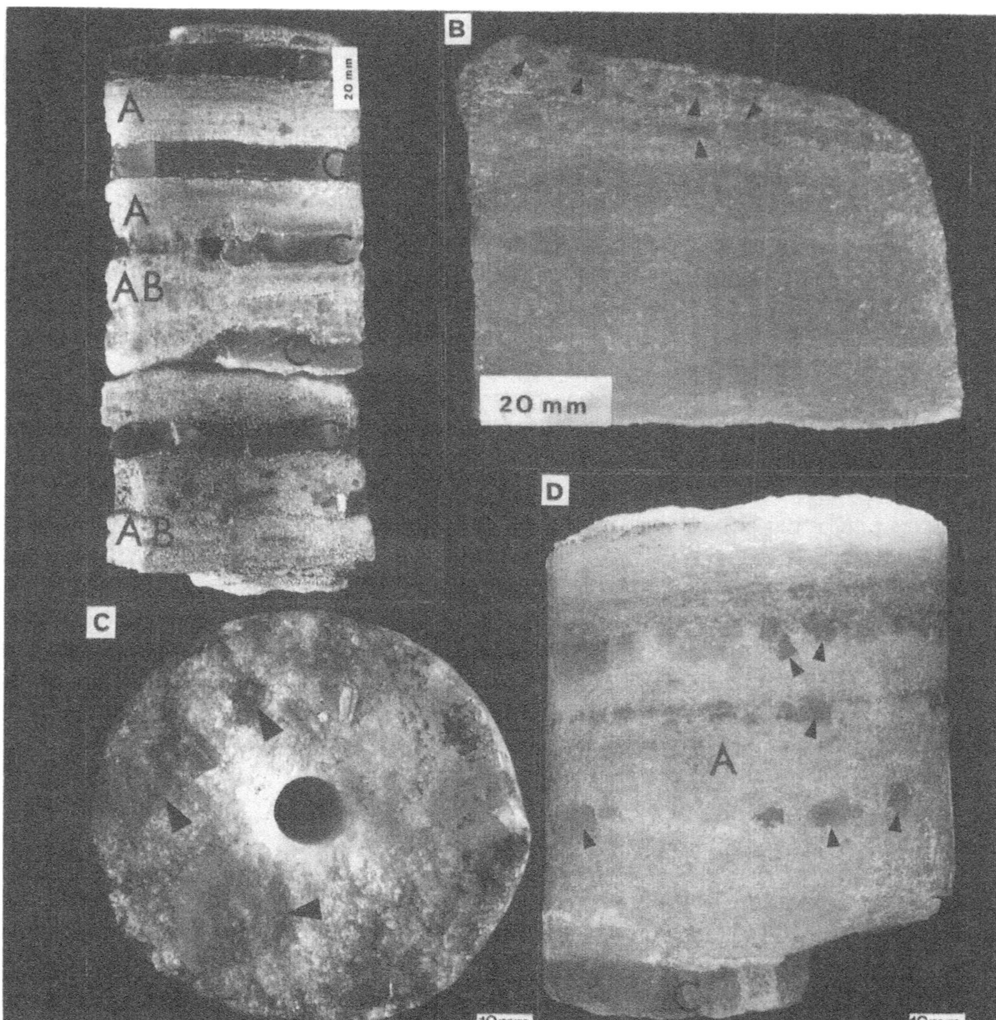

Fig. 5. Shoal facies.(**A**) A,AB,C structural halite sequence, finely laminated A-halite layer at top. Głuszewo IG 1 well, depth 931.6 m. (**B**) Fine A-halite, small raft fragments (arrowed) at top. Chalupy IG 3 well, depth 724 .0 m. (**C**) Large halite intraclasts (arrows) within fine A-halite. Chlapowo IG 1 well, depth 844.1 m. (**D**) Halite intraclasts (arrows) within A,C halite sequence. Polczyno IG 1 well, depth 943.7 m.

(2) B-halites with great variation of crystal size (Figs. 5 A, 6 A-B & 7 A-C),
(3) C-halites, composed of large platy halite crystals, forming the bands of various thickness (Figs. 4 A,C-D, 5 A,D & 6A).

The A- and B-halites are developed as fine (crystal size below 1 mm), medium (1-5 mm) and coarse (over 5 mm) varieties depending on the medium crystal diameter. Compound layers: AB, BA occur as well (Fig. 8).

The layers of the distinguished halite types repeat vertically many times in various order, giving characteristic structural sequence (Figs. 4 A & 5 A), termed "*structural unit*".

The structure, called "*internal lamination*", was recorded in C-halites (Fig. 4 C,D). It is composed of fine sulphate crystals and brine/gas inclusions, arranged in parallel laminae within platy halite. This lamination was formed (Czapowski, 1986) during the first stages of salt precipitation, when the rain of sulphate crystals formed in more diluted surface waters of evaporitic pan, was falling into the bottom brines where slowly growing large halite crystals were forming (Fig. 8). The lower bromine content in halite bands (Fig. 9) compared to the other halite layers in the same structural unit, confirms their origin from first condensed brines. Continuous formation of such halite crystals could persist in the stable chemical conditions of deeper basin where surface water mixing (see Raup, 1970; Sonnenfeld, 1984) did not disturb the precipitation from bottom brines.

The fine horizontal anhydrite laminae with dispersed halite crystals, visible in A-halite layers (Figs. 4 A-B, 5 A &10), mark the dilution events of brines when halite formation was partially stopped and sulphate precipitation increased. The lack of corrosion/dissolution features in surrounding halites suggests that these chemical changes have only influenced the brine concentration and the character of precipitation but they have not affected the already deposited salt. Such laminae, similarly as "internal lamination" in C-halites, can reflect the disturbances of brine stratification.

The mm-cm thick laminae and layers of anhydrite, occasionally with pseudomorphs after gypsum (Figs. 4 B & 6 B), have recorded the longer stages of brine dilution and sulphate precipitation (Fig. 10).

The A- and B-halite layers contain often among the mass of inclusion-free halite crystals many "cloudy" crystals with traces of hopper structure (Figs. 6 C,D, 7 C &10). These crystals, more or less changed by diagenetic processes,

Fig. 6. Shoal facies (A, B) and salt lagoon facies (C, D). (**A**) Dissolution/erosion scour (arrow) in C-halite band. Mechelinki IG 2 well, depth 1060.4 m. (**B**) Fine anhydrite layer (an) with pseudomorphoses after gypsum (arrows) within fine A-halite. Wejherowo IG 1 well, depth 1028.8 m. (**C**) A, B halite sequence with fine anhydrite (white) laminae and microaggregates. Swarzewo IG 6 well, depth 895.5 m. (**D**) A, B halite sequence with cloudy/hopper crystals. Radoszewo IG 1 well, depth 821.9 m.

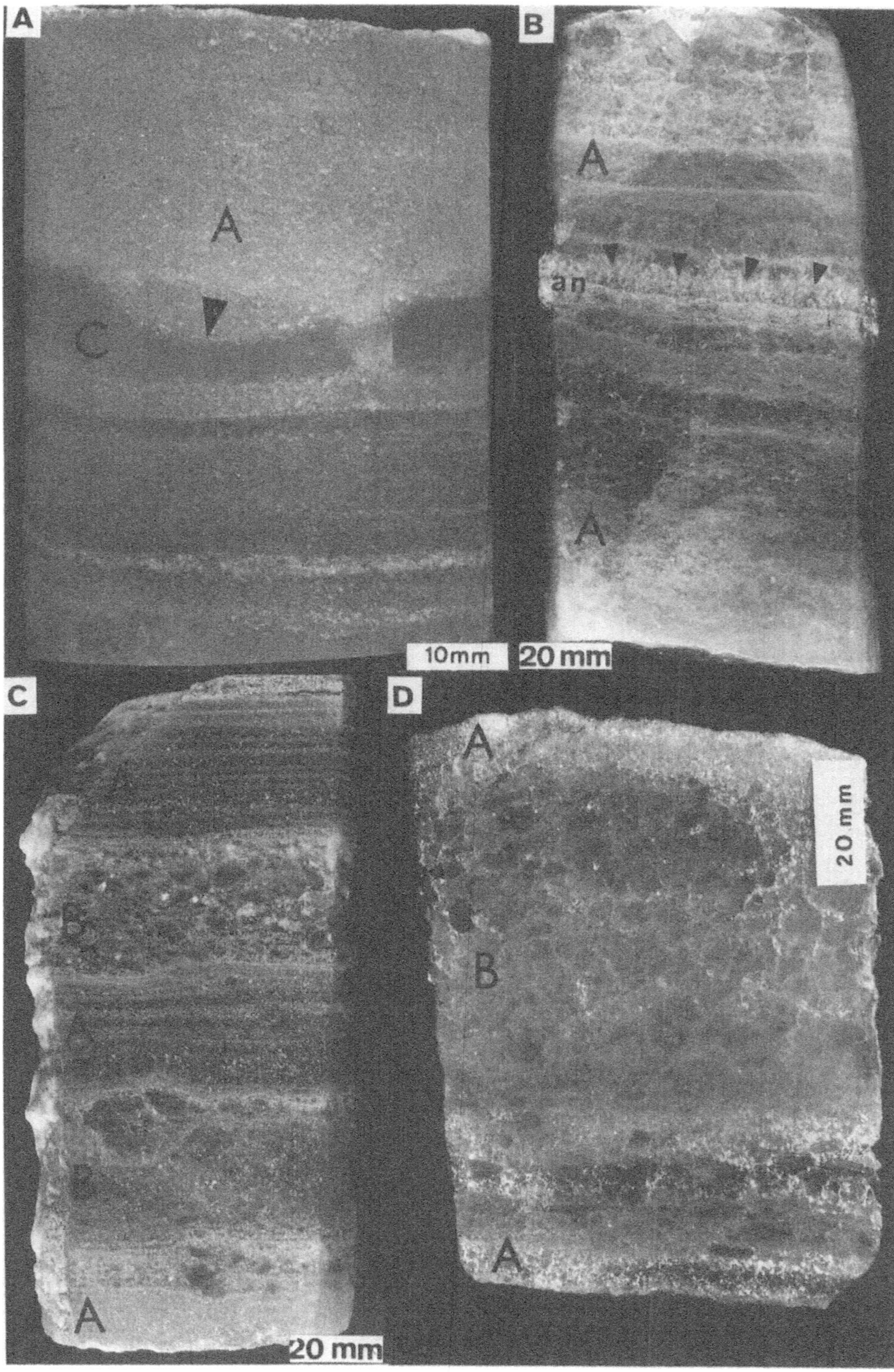
A
A
C
B
A
an
A
10mm
20 mm
C
B
B
A
20 mm
D
A
20 mm
B
A

A
B
20 mm
B
A
B
A
B
20mm
C
D
20mm
20mm

Fig. 7. Shallow salt lagoon facies (A, B) and salt pan facies (C, D). **(A)** Dissolution surface marked by fine anhydrite lamina (arrows). B-halite with cloudy/hopper crystals upward. Radoszewo IG 1 well, depth 800.0 m. **(B)** Inclined dissolution surface between A and B halite. White laminae and microaggregates of anhydrite. Chlapowo IG 1 well, depth 769.9 m. **(C)** B-halite with clay admixture (dark zones) and corroded cloudy crystals (arrows). Swarzewo IG 6 well, depth 860.9 m. **(D)** Chevron halite crystals with secondary clear halite (arrows), infilling dissolution cavities. Swarzewo IG 7 well, depth 822.0 m.

formed first at brine surface, then sunken, growing on their way to bottom (Raup, 1970; Shearman, 1970; Kendall, 1978; Sonnenfeld, 1984; Lowenstein & Hardie, 1985).

The findings of *chevron-type halite* crystals, mainly in B-halites (Fig. 7 D), inform about rapid competitive growth of halite at the bottom (Fig. 10) from highly condensed brines (Shearman, 1978; Sonnenfeld, 1984), typical of shallow parts (Fig. 11) of evaporite basins (Kendall, 1978; Orti Cabo *et al.*, 1984, Lowenstein & Hardie, 1985).

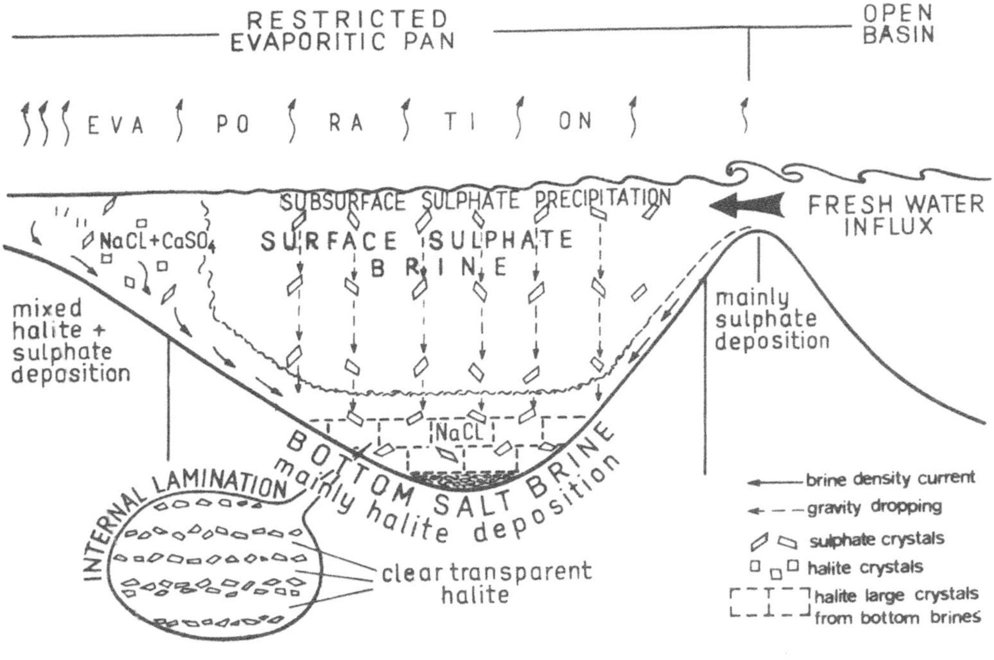

Fig. 8. Scheme of "internal lamination" development in halites.

The larger, more or less abraded halite crystals visible in the mass of smaller ones (Fig. 5 C,D), are interpreted as *halite intraclasts*, removed from primary sediment and incorporated into a new one by the erosion/mixing action of some traction currents (Figs. 10 &11).

The erosion/dissolution scours and surfaces between the halite layers (Figs. 9 A, 7 A,B &10) record the events of instant dilution of brines, probably by rapid influx of fresh water. Some surfaces are marked by fine anhydrite laminae, suggesting the strong brine dilution and post-dissolution sulphate brine precipitation. Sometimes, the dissolution cavities and vugs formed, infilled later with secondary inclusion-free halite and cloudy/hopper crystals (Figs. 7 D, 10 & 11). Such dissolution/erosion features, pointing to very unstable chemical conditions, are typical of the deposits of shallow salt lagoons and pans (Kendall, 1978; Orti Cabo *et al.*, 1984; Lowenstein & Hardie, 1984).

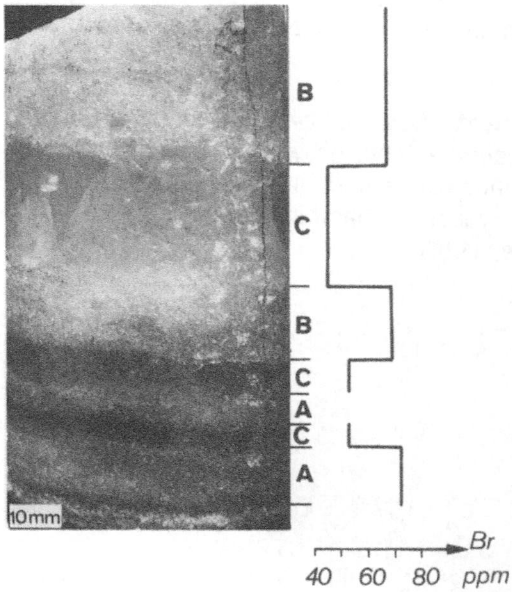

Fig. 9. Bromine content in deep-basin halite sequence. See text for explanation of symbols. Mechelinki IG 5 well, depth 1137.0 m.

The units of folded halite/anhydrite layers (Fig.12) passing gradually into undisturbed sequences, were interpreted as *slump structures*, developed on inclined slope of bottom elevations.

Some of the distinguished sedimentary structures may inform about dynamic conditions in the depositional environment. The C-halite bands with "internal lamination" and fine anhydrite lamination in A-halites suggest the generally stable dynamic conditions with continuous slow salt precipitation, regularly interrupted by dilution events, disturbing the primary brine stratification. The chevron halite crystals inform about intense competitive crystal growth on basin bottom from highly concentrated brines. Together with dissolution/erosion features as scours, surfaces, cavities and abundance of dispersed/nodular anhydrite, they record strongly fluctuating chemical conditions with high accumulation rate, interrupted by many phases of sediment erosion and dissolution.

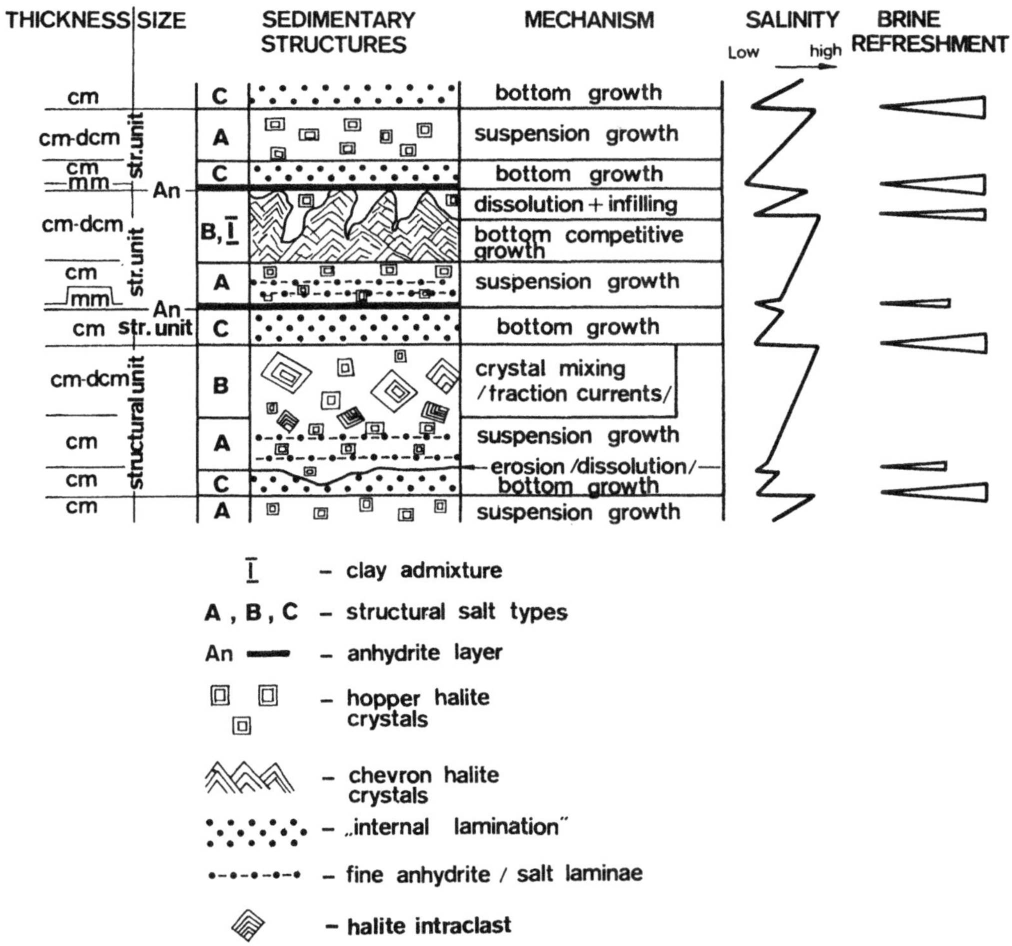

Fig. 10. Scheme of origin of halite structural units.

Structural halite units and sedimentary facies

The salt sequences are composed of many types of structural halite units. These units, composed of two to four elements (including halite structural types and anhydrite layers) contain various sedimentary structures. The units, in which all halite components have similar thicknesses, are called here "rhythmites" (Fig. 4 A).

The structural unit origin depends on the precipitation mechanism and salinity changes (Fig. 10).

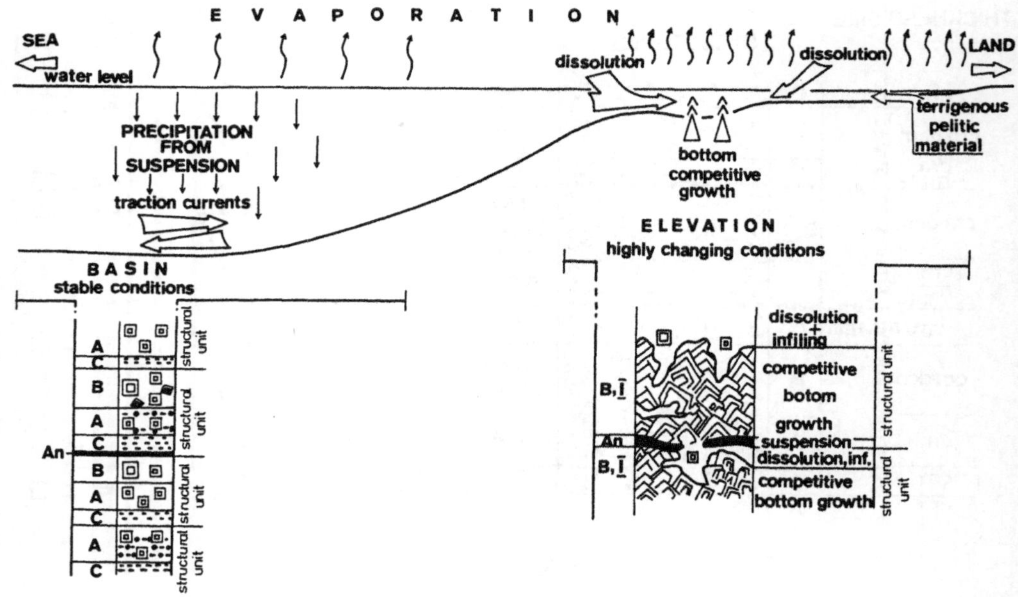

Fig. 11. Position of halite structural units within evaporitic basin.

The simple A,C couplets with "internal lamination" and hopper crystals have formed in the stable dynamic conditions with slowly growing salinity and bottom to suspension mechanism of crystal precipitation. More dynamic conditions with often dilutions and action of traction currents have produced the A,B,C units with admixture of halite intraclasts, fine anhydrite laminae and dissolution surfaces. The A,B couplets with chevron crystals, many dissolution features, and large admixture of sulphates, have formed in highly changing conditions with many events of brine refreshments and intense dissolution of primary sediment.

The described halite units are correlated in various parts of evaporitic basin (Fig. 11). The high-dynamic B,A and A,B,C units are connected with elevated, shallow-water areas where fresh water influxes constantly disturbed the brine stratification and destroyed strongly the primary sediments. Also the pelitic terrigenous material was supplied there by river input or/and wind action. Similar conditions are typical of coastal lagoons and salt pans (Shearman, 1970, 1978; Kendall, 1978; Lowenstein & Hardie, 1985) and these structural units could have formed there. The units with features of less changing conditions may represent the submerged basin shoals where brine stratification changes were better marked than in the deep parts of the basin.

The frequency of various sedimentary structures in halite sequences (Table I) confirms their environmental interpretation. The most high-dynamic features (chevron crystals, dissolution/erosion structures) are related to structural units of salt lagoons and pans. The low-dynamic structures are mainly recorded in deep basin sequences.

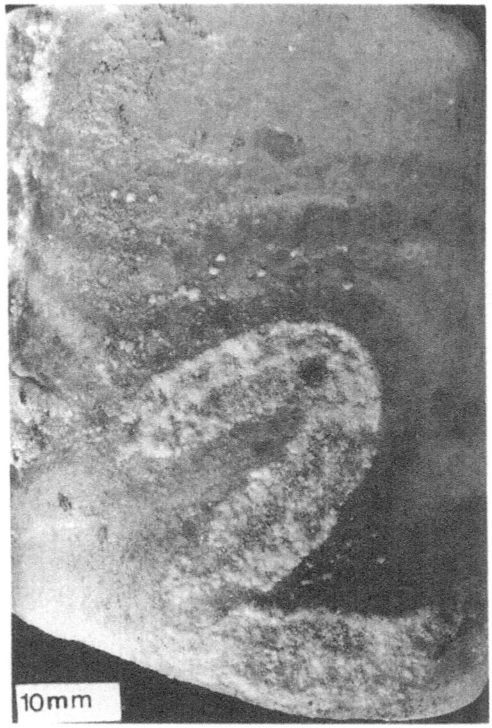

Fig. 12. Slumped AB-halite and anhydrite (white) layers. Shoal facies. Widowo ONZ 1 well, depth 680.0 m.

The salt sedimentary lithofacies consist of assemblages of various halite structural units with corresponding sedimentary structures. These lithofacies are correlated with the deposits of deep basin, shoal, coastal salt lagoon, and pan environments within evaporitic basin. The characteristic well sections are presented in Fig. 13.

Some halite units associations have been distinguished within lithofacies (Fig. 13). They include predominant structural-unit types with their characteristic assemblage of sedimentary structures. These associations representing similar or different dynamic conditions reflect the evolution of sedimentary conditions or/and position of analysed profile within environment (distal or central parts).

The characteristic sedimentary structures for the distinguished lithofacies are shown in Table I.

Table I. Occurrence of sedimentary features in the Oldest Rock Salt halites from various dynamic environments.

Sedimentary features	Distinguished dynamic environments: Stable conditions		Strongly changing conditions	
	Deep basin	Shoal	Salt lagoon	Salt pan
C-halite bands	xx	x	x	0
Internal lamination	xx	x	0	0
Fine anhydrite lamination	xx	x	x	x
Cloudy/hopper crystals	x	xx	x	x
Rafts	x	x	x	x
Chevron crystals	0	x	xx	xx
Halite intraclasts	x	xx	xx	xx
Erosion/dissolution features (scours, surfaces, cavities)	x	x	xx	xx
Slumps	0	x	x	0
Secondary clear halite infillings	0	0	x	x
Anhydrite nodules and microaggregates	x	x	xx	xx

Occurrence: xx - common, x - rare, 0 - absent

The deep basin lithofacies (Fig. 13) is composed mainly of A,C; B,C and C,A structural units (thicker element is indicated by the first position in the unit denotion), and rhythmites. Here, the low-dynamic structures are common (Fig. 4 A-D) as "internal" and fine anhydrite lamination. This facies has formed in generally low-dynamic conditions and only such phenomena as storm surges, rainfalls etc. disturbed the rhythmic precipitation from regularly refreshed brines and give the crystal redeposition and/or erosion structures. The bromine content in halites (Czapowski & Tomassi-Morawiec, 1985) increases from 30-40 ppm at the bottom of the sequence to 140-180 ppm at the top. These data suggest that the concentration of primary brines was low, probably because of dilution by influx of fresh marine waters. In some more isolated basins (II-Wladyslawowo and I-Jastrzebia Gora basins - Fig. 3) the brines became occasionally more concentrated (bromine content 300-587 ppm) and fine K-Mg salts have formed (Stepniewski, 1973). Generally, however, the salinity conditions within basin were unfavourable for such precipitation.

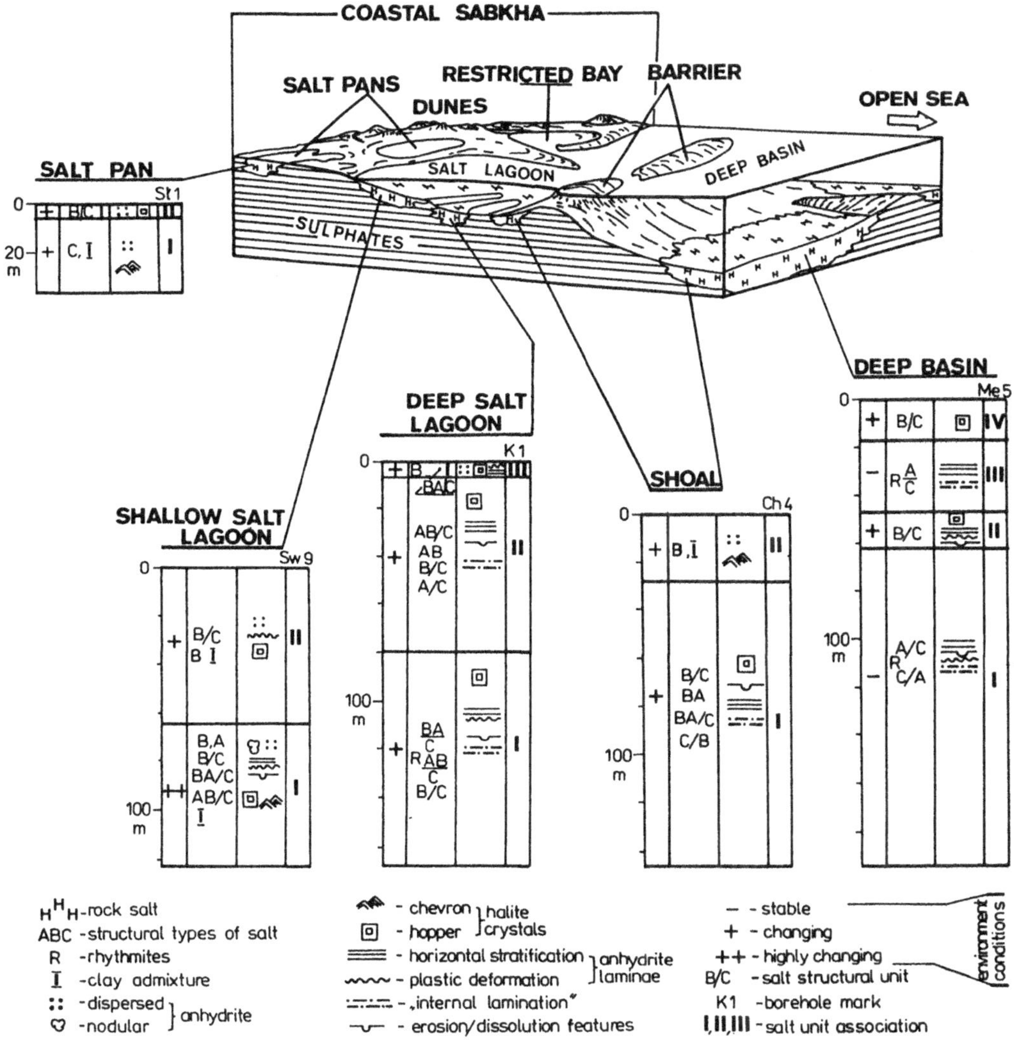

Fig. 13. Structural sequences of the Oldest Rock Salt in wells and their environmental interpretation.

The shoal lithofacies (Fig. 13) consists mainly of more dynamic structural units associations with B,C; B,A,C and B,A halite sequences. Except of "internal" and fine anhydrite laminations, the hopper crystals, halite intraclasts, dissolution scours, some slumps (Figs. 5 A-D, 9 A-B & 12) and chevron crystals are common. These features suggest the irregularly changing environmental conditions. Bromine content data indicate that salt deposits on top of shoals have formed from strongly diluted brines (bromine content of halites: 50-100 ppm) and those on the shoal base - from more condensed brines (Br contents of halites: 100-150 ppm). The accumulation of mobile water-saturated deposit at shoal margins favoured slumping.

The salt lagoon lithofacies contains various types of halite structural units with all the typical sedimentary structures (Fig. 13). The sequences with rare "internal" and fine anhydrite laminations (Fig. 6 C,D) and some erosion/dissolution features are characteristic of deeper lagoon parts where low-dynamic conditions could have existed for a long time. High-dynamic sequences with abundant chevron and hopper crystals, dissolution/erosion features and dispersed-nodular sulphates (Fig. 7 A,B) are characteristic of the shallow lagoon margins.

The salt pan lithofacies is similar to shallow lagoon type (Fig. 13) but it contains more dissolution/erosion features with secondary infillings (Fig. 7 C,D) and higher sulphate content. Various admixtures of pelitic terrigeneous material, as clay and quartz pelite, are often found in both lithofacies giving them appearance of "clayey salt". The bromine content in lagoon and pan halites is generally low, on an average 50-60 ppm, but it can range from 44 to 106 ppm. These data indicate the strong dilution of primary brines by frequent fresh of water influxes, both meteoric (as rainfall or river input) and of marine origin.

Evolution of sedimentary basin

The evaporitic basin of the Oldest Rock Salt Formation was differentiated into deeper and shallow parts from its very begining. This division was inherited after Lower Werra Anhydrite basin.

The analysis of dynamic conditions (Fig. 2) and facies distribution indicates that medium-high dynamic facies prevails in the basin. Only in central parts of larger bottom depressions, the low-dynamic sequences have been found. These data suggest predominantly shallow-water environment of deposition. The low salinity of brines and irregular changes of salinity, observations based on bromine content data, indicate according to Tucker & Cann (1986) the open shallow basin system with maximum water depth estimated at several to tens of metres. Such depths in deep basin areas allowed to maintain for a long time the undisturbed brine stratification and to form low-dynamic halite sequences.

The salt deposits have formed in two megacycles (Fig. 14). During the first megacycle (Fig. 14 A) the "clear" salt series without traces of land influence have originated. At that time, on emerged shoals and islands, the sulphate precipitation was taking place.

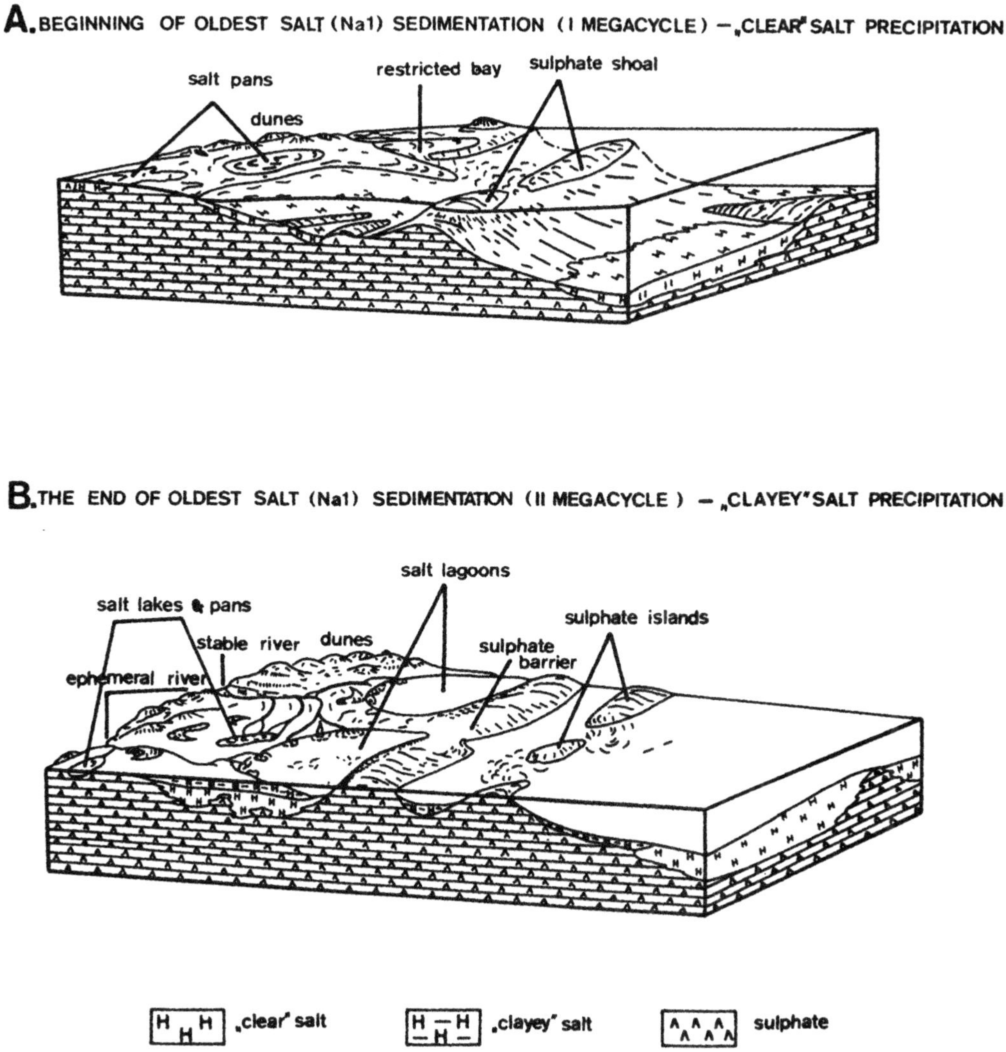

Fig. 14. Phases of the Oldest Rock Salt sedimentation in Leba elevation.

The partial infilling of bottom depressions at the end of the first megacycle caused gradual shallowing and predominance of more dynamic environments during the second megacycle (Fig. 14 B). Some smaller basins have evolved into salt pans. The more intensive river and/or aeolian inputs have supplied pelitic terrigenous material to the nearshore zone and "clayey salt" series accumulated. In the centres of larger basins, the "clear" halites have precipitated at that time. The brine salinity was generally low, probably due to high river input, but in some isolated basins (like I-Jastrzebia Gora and II-Wladyslawowo basins - Fig. 3) the K-Mg salts have precipitated from highly concentrated brines.

Acknowledgements

The author thanks Tadeusz Peryt for critical review of manuscript and his help in preparing the final version of the paper, as well as Marek Rybicki and Agnieszka Siara for drawing illustrations, and Romana Ufnal and Janina Modrzejewska for photographic work.

References

Czapowski, G.,1983. Zagadnienia sedymentacji soli kamiennej cyklotemu PZ1 na wschodnim sklonie wyniesienia Leby. *Prz. geol.*, 31: 278-284.

Czapowski, G., 1986. Internal lamination in the halite rocks. *Prz. geol.*, 34: 202-204.

Czapowski, G. & Tomassi-Morawiec, H., 1985. Sedymentacja i geochemia najstarszej soli kamiennej w rejonie Zatoki Puckiej. *Prz. geol.*, 33: 663-670.

Kendall, A. C., 1978. Subaqueous evaporites. Facies Models 12. *Geoscience Canada*, 5:124-139.

Lowenstein, T. K. & Hardie, A. L., 1985. Criteria for the recognition of salt pan evaporites. *Sedimentology* 32: 625-644.

Orti Cabo, F., Pueyo Mur, J. J., Geisler-Cussey, D. & Dulau, N.,1984. Evaporitic sedimentation in the coastal salinas of Santa Pola (Alicante, Spain). *Rev. Invest. Geol.*, 38/39: 169-220.

Peryt, T. M., Czapowski, G., Debski, J. & Pizon, A., 1985. Model sedymentacji ewaporatow cechsztynskich na wyniesieniu Leby. *Prz. geol.*, 33: 204-211.

Poborski, J., 1980. Discontinuity on Upper Permian (Zechstein) succession of evaporites in Eastern Pomerania, Poland. *Fifth Symp. on Salt.*, 2: 153-157. Northern Ohio Geological Society.

Pueyo Mur, J. J., 1979. La precipitation evaporitica actual en las lagunas saladas del area: Bujaraloz, Sastago, Caspe, Alcaniz y Calanda (provincias del Zaragoza y Teruel). *Rev. Inst. Invest. Geol.*, 33: 5-56.

Raup, O. B., 1970. Brine mixing and additional mechanism for formation of basin evaporites. *Am. Assoc. Petroleum Geologists Bull.*, 54: 2246-2259.

Shearman, D.J.., 1970. Recent halite rock, Baja California, Mexico. *Trans. Inst. Mining. Metall.* B 79: 155-162.

Shearman, D. J. ,1978. Evaporites of coastal sabkhas. *SEPM Short Course* 4: 6-42.

Sonnenfeld, P., 1984. *Brines and Evaporites*, 613 pp. Academic Press, Orlando.

Stepniewski, M., 1973. Niektore pierwiastki sladowe w cechsztynskich mineralach solnych z rejonu Zatoki Puckiej. *Biul. Inst. Geol.*, 272: 7-68.

Tucker, R. M. & Cann, J. R., 1986. A model to estimate the depositional brine depths of ancient halite rocks: Implications for ancients subaqueous evaporite depositional environments. *Sedimentology* 33: 401-412.

THE ZECHSTEIN (UPPER PERMIAN) MAIN DOLOMITE DEPOSITS OF THE LEBA ELEVATION, NORTHERN POLAND: DIAGENESIS

Tadeusz Marek Peryt
Instytut Geologiczny
ul. Rakowiecka 4, 00-975 Warszawa
Poland

Abstract: The Zechstein Main Dolomite of the Leba elevation has simple depositional history and, because of shallow burial with consequent relatively minimal depth-related diagenesis, it has been possible to construct a sequential model of the early diagenetic history which can be used as a standard for comparison with other similar intra-evaporitic carbonate sequences. Limited meteoric diagenesis and localized marine cementation governed compaction during subsequent shallow burial. This was followed by anhydritization, dissolution and cementation by dolomite, sulphate and minor halite; processes which locally overprinted and obliterated the textures of the initial diagenesis. The early diagenetic imprint was decisive for the later diagenetic history in the Leba elevation and this probably applies to the Main Dolomite over most of the basin. Local geological factors crucially influenced the intensity of the diagenesis in which early meteoric cementation coupled with late diagenetic leaching were crucial in the formation of productive hydrocarbon reservoirs.

Introduction

The Zechstein Main Dolomite has locally good porosity and permeability and provides commercial hydrocarbon reservoirs in the North Sea (Taylor, 1984) and on the continent (Hauke *et al.*, 1979; Maureau & van Wijhe, 1979; Depowski *et al.*, 1981). As in many reservoir rocks (Roehl & Choquette, 1985), key determinants of the likelihood and extent of potential reservoirs are their early diagentic pathways in addition to the nature of newly deposited carbonates and their depositional environments. Recently Scholle and Halley (1985) have stressed the importance of burial diagenesis in governing the actual porosity and permeability of carbonate rocks. In the Zechstein, the importance of some burial-related processes for reservoir properties is well-established (Clark, 1980a; Peryt, 1984). However, because of complex depositional settings (e.g., Clark, 1980b, p. 154) and/or complex diagenesis related to deep burial (as described by Clark, 1980a), the significance of early diagenesis can be underestimated. The Main Dolomite of the Leba elevation reveals a simple depositional history (Peryt, 1986a) and thus offers an unique

Lecture Notes in Earth Sciences, Vol. 10
T. M. Peryt (Ed.), The Zechstein Facies in Europe

opportunity to establish a model of the early diagenetic history which can be used as a standard for other areas where local phenomena resulted in more complicated diagenetic changes. Furthermore, the Main Dolomite deposits of the Leba elevation underwent only relatively shallow burial. The model presented by Clark (1980a) was based on the study of rocks which were more deeply buried and accordingly, the details of earlier phases of diagenesis were obliterated during the succeeding stages. As a consequence the early diagenesis of the Zechstein intra-evaporitic carbonates has been over-generalized. Clark (1980a) studied the Zechstein outcrops in England as well, but these deposits, in turn are thoroughly recrystallized and dedolomitized and are not suitable for diagenetic modelling. The Main Dolomite of the Leba elevation is therefore an ideal formation to study in order to establish a model of the diagenesis for the Main Dolomite basin, and also for any intra-evaporitic carbonate.

Geological setting

The Main Dolomite of the Leba elevation is less than 10 m thick in the basin and over 40 m thick in the carbonate platform area (Fig. 1) and represents a shallowing-upward sequence. The facies sequence includes argillaceous wackestones/mudstones with whole fossils passing shoreward into peloidal, peloidal-bioclastic and lump deposits, and then into fringing oolite flat facies and tidal-flat complex (Peryt, 1986a). A relatively uniform and gentle slope into the basin together with the facies sequence fit well the homoclinal ramp model of Read (1985).

The history of deposition (Peryt, 1986a) indicates that the first transgression of the sea in which the Main Dolomite was deposited reached the southern area first. Here peloidal-bioclastic wackestones to packstones and occasionally grainstones (Figs. 2D & 3B,D), accompanied by stromatolites (Fig. 2A) and ooid grainstones were deposited and formed a belt parallel to the shore. A rise of sea-level caused the shoreline to migrate towards the NE and W. Consequently, a coastal oolitic barrier system developed between Salino, Kopalino and Debki. It prograded relatively quickly to the east and south, before finally reaching the area several km SE of Slawoszynko (Fig. 1). Figure 1 shows the palaeogeography which predominated during deposition of the Main Dolomite. At the end of deposition, the sea-level dropped, and most of the area studied became exposed before the next transgression, related to the Basal Anhydrite, flooded the southern and central parts of the area studied. The area was again subaerially exposed at the PZ2/PZ3 boundary (Peryt *et al.*, 1985). Further emergence occurred after the PZ3 deposition. The Mesozoic sequence is characterized by great reduction and hiatuses due to periodic uplift (Dadlez *et al*, 1976). The Cenozoic sequence is less than 200 m thick. At present, the Main Dolomite deposits occur at a depth of between 520 m (in the NW part of the area shown in Fig. 1) to 970 m (in the SE part). It may be estimated that the maximum burial did not significantly exceed the present depth values.

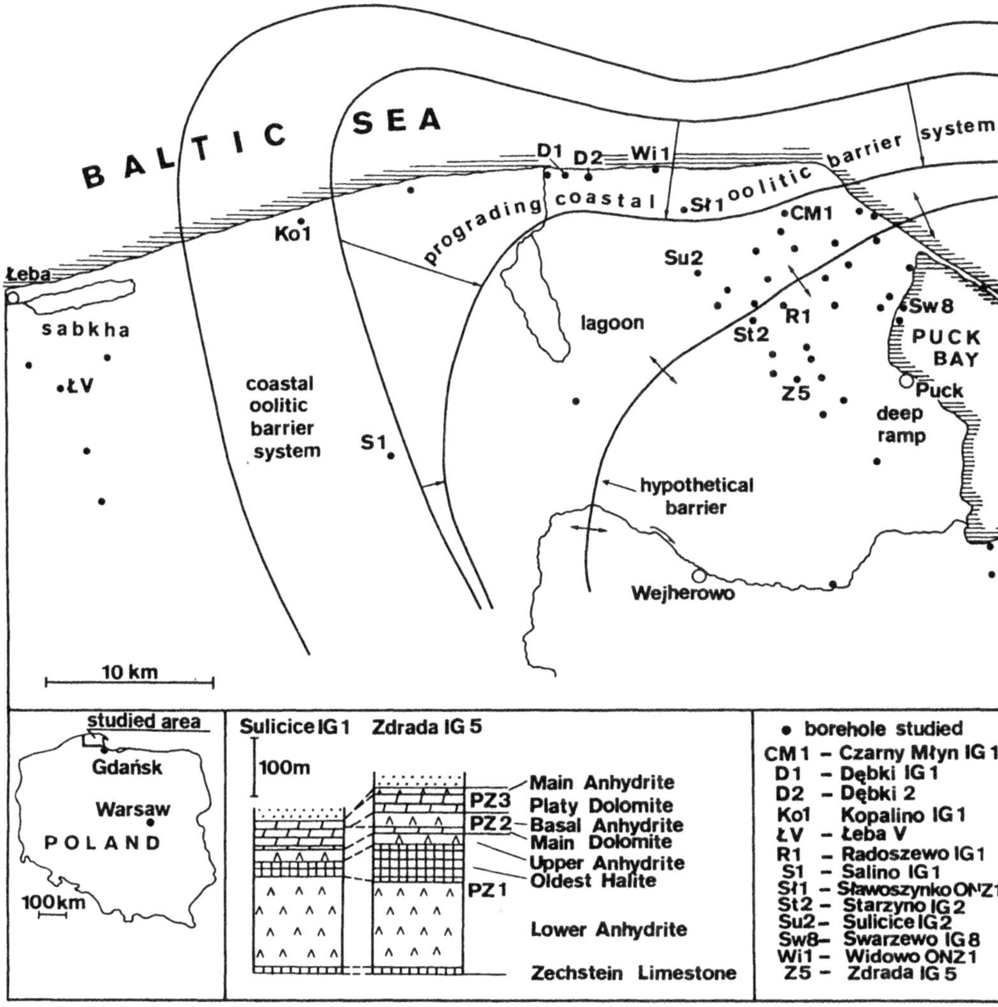

Fig. 1. Location of area studied.

The present paper is a part and parcel of the research project on the Main Dolomite of the Leba elevation, and the first paper on the subject (Peryt, 1986a) reports the facies and depositional history (conclusions are summarized in Table 1). Accordingly, the following text complements the former one.

Carbonate cementation

Several types of carbonate cements occur in the Main Dolomite and most of them are dolomitic.

Table 1. The principal features of the Main Dolomite of the Leba elevation.

Facies Zone	Thickness	Main Lithofacies
sabkha	4.7 to 12.3 m; primary thickness unknown beacuse of erosion	peloidal W & P interbedded with nodular anhydrite and accompanied by L
coastal oolitic barrier system	usually > 40 m, unless eroded	ooid G, in the southern area (prograding coastal oolitic barrier system) peloidal P & W at the bottom
lagoon	20 to 40 m	peloidal P, lump P & G, accompanied by L and vadoidal G
deeper ramp	< 20 m, diminishing toward SE	peloidal P; W & M; peloidal-bioclastic G & P, ooidal G and stromatolitic B at the bottom in the northern part

Rim cement (isopachous fringing cement) was recorded in ooid and peloidal grainstones of the bottom part of the Main Dolomite in the southern region (Fig. 3B, D) and in ooid grainstones occurring on the landward side of the coastal oolitic barrier system (Figs. 2B, 3B and 3D; Peryt , 1985a, fig. 1). It also occasionally occurs in the prograded part of the coastal oolitc barrier system (Fig. 4C, D). Such rim cements are relatively thin (15-20 µm) and probably originated syndepositionally in the active zone of the marine-phreatic environment. Similar cements are known from many recent and ancient settings where they are composed of aragonite and Mg-calcite. It is possible that the Main Dolomite rim cements were also of unstable mineralogy (*cf.* Wilkinson *et al.*, 1985, p. 182).

In the ooid grainstones of the coastal oolitic barrier system rim cement was the first cement to form, but in the southern region its formation was preceded by the precipitation of discontinuous dark crusts (Peryt, 1986a, pl. 26/3),

Fig. 2. (A) Stromatolitic boundstone, characteristic of the bottom part of the Main Dolomite of the Swarzewo-Zdrada belt. Note patchy anhydritization (arrows). Swarzewo IG8 well, depth 740.5 m. **(B, C & E)** Ooid grainstone with locally visible irregularly developed rim cement (arrow in B) and compaction features. Compaction was probably preceded by partial meteoric dissolution of the nuclei of ooids, as suggested by breaking of ooids in B & E, and may be that only these partially dissolved ooids underwent breaking while full ooids suffered pressure-solution during compaction (see C). The compaction was succeeded by carbonate B cementation which left some interparticle porosity open (x in B, C & E). Following the dissolution of some ooid nuclei and cement (arrow in E), minor dolomite cementation occurred (top and bottom left of C). Debki IG1 well, depth 588.3 m. **(D)** Peloidal-bioclastic packstone with anhydritized bioclasts (mainly gastropods) and bigger masses of the matrix (a). Starzyno IG2 well, depth 718.0 m.

which possibly originated in a splash zone environment. In the northern region only one phase of rim cementation was identified, whereas in the southern region rim cementation commonly occurred in several stages (e.g., Peryt, 1986a, pls. 26/4 & 28/6), and was sometimes accompanied by the precipitation of peloidal cement (Peryt, 1986a, pls. 26/3 & 28/5).

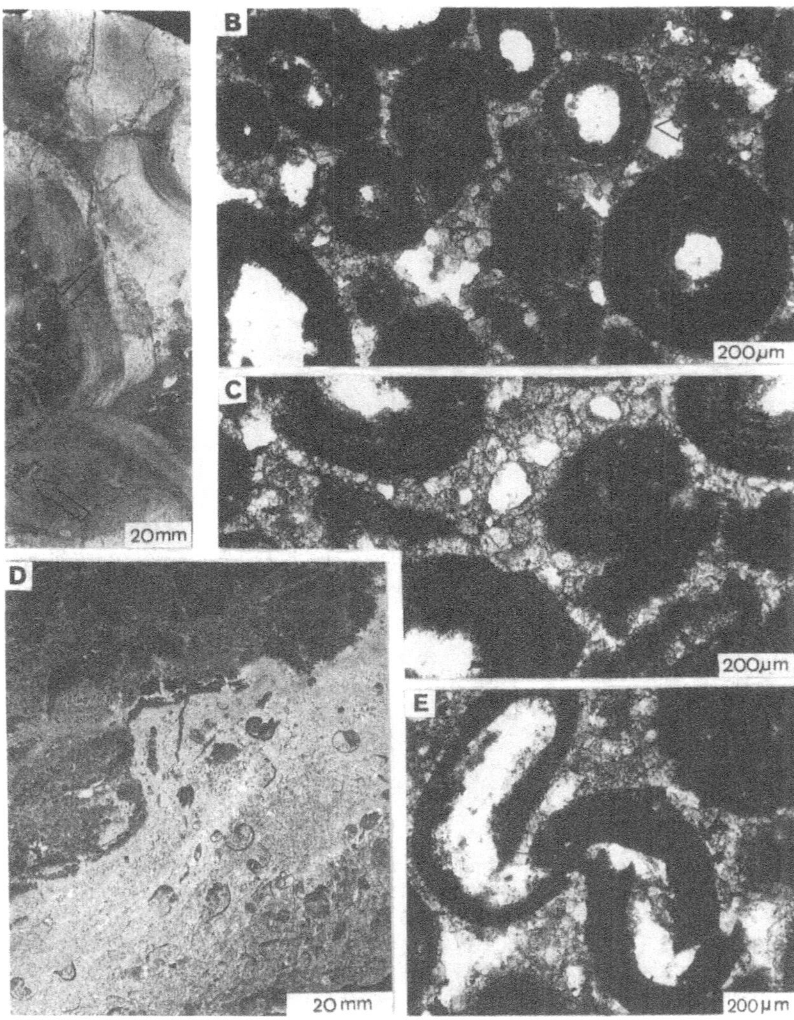

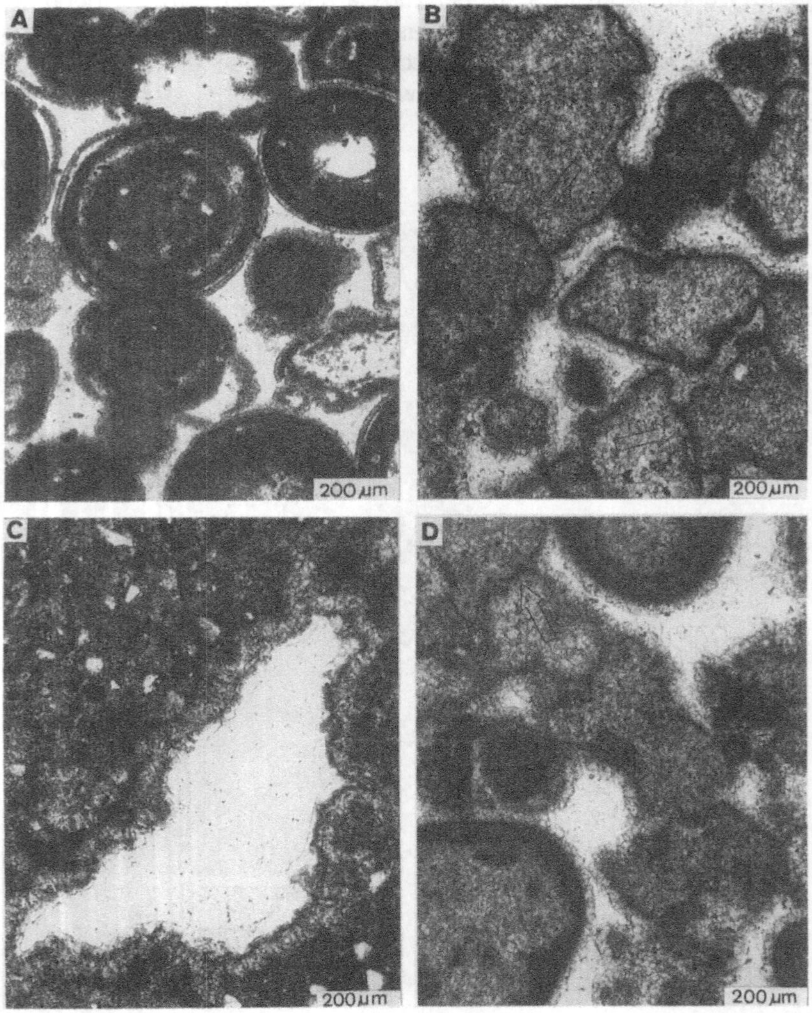

Fig. 3. (A) Spalled off cement crusts during compaction. Ooids are micritized but the lamellar structure of cortex is well visible. Most intraparticle porosity is filled by sulphate cement (mostly gypsum). Kopalino IG1 well, depth 549.3 m. **(B & D)** Peloidal grainstone. Note microstylolitic contacts between peloids (arrows) which are regarded as originated under vadose conditions, and probably some pores originated by meteoric dissolution. Pore-rimming cement postdated this meteoric diagenesis. Zdrada IG5 well, depth 754.6 m. **(C)** Pore within stromatolitic boundstone (as shown in Fig. 2A) which was filled by carbonate rim cement and then by sulphate cement. Swarzewo IG8 well, depth 740.5 m.

Peloidal infillings occurring in restricted microcavities are chemical products, probably originating, as suggested by Macintyre (1985), by repeated nucleation around centres of growth. Such a diversified association of cements in the southern region may be related to fluctuating environmental conditions which might be expected during stabilization after the PZ2 transgression.

Rim cements occurring within some pores (Fig. 3C) may also be related to such fluctuating conditions and to minor meteoric syndepositional dissolution. This rim cement succeeds the rim cement which developed around the grains and is more coarsely crystalline. However, some pores may be of later origin (Fig. 4B). To the south of the Salino-Kopalino-Debki belt (Fig. 1), an early marine cement is usually lacking (Fig. 5C) or is irregularly developed (Fig. 6C, D). This is probably related to the stagnant zone of the marine-phreatic environments. Zoning of the early marine cementation was a controlling factor during the compaction of ooid grainstones, as will be discussed below.

Cement crust is another type of early marine cement which was recorded in the bottom part of the Main Dolomite in the southern region (Peryt, 1986a, pl. 27/1). A marine origin of the cement crusts is indicated by their association with cavity-filling microbial mats (Peryt, 1986a, pl. 27/1). The cement is very similar to the primary aragonitic cements common in Permian and Triassic rocks, but dolomitization prevents any precise analysis of the original mineralogy of this and other cements of the Main Dolomite rocks.

Besides marine cementation, beach cementation also occurred during the Main Dolomite deposition. Three lines of reasoning indicate the presence of beach cements but, because of later recrystallization, direct evidence is lacking.

(1) Throughout the entire zone of ooid grainstone occurrences, in the upper part of the Main Dolomite, oomolds occur (Fig. 7C). They probably reflect the influence of mixed meteoric-marine waters (*cf.* Kaldi & Gidman, 1982), and considering the anticipated environmental sequence as proposed by these authors it is probable that cementation was related to beach environments. However, a reservation should be made that not all oomolds in the Main Dolomite are of early origin: some are late and are related to the congruent dissolution of dolomite during late diagenesis. The late oomolds differ in that they are not as uniformly developed as the early ones (Fig. 7D).

(2) In some ooid grainstones, intraclasts of ooid grainstones occur of the same composition as the ooid grainstones in which they are embedded. Both intraclasts and ooids are cemented by marine rim crusts (Peryt, 1986a, pl. 30/1).

(3) In some parts of the coastal oolitic barrier system early meteoric cementation related to solution compaction occurred (see chapter on compaction).

Sparite cements are common in the Main Dolomite. Different types of sparite may be identified related to phases of diagenesis (Figs. 2B & 6C, D; Peryt, 1986a, pl. 35/3). In the Main Dolomite, sparite cements are volumetrically important and late stage cementation was an important diagenetic event. The sparites precipitated from subsurface fluids which, in the area studied, were derived from marine waters which underwent constant modification during burial. However, the water contained no iron and accordingly, sparites which formed during burial are iron-free. It is impossible to establish a cement stratigraphy. The original mineralogy of the sparite is not known, but some may have precipitated as dolomite.

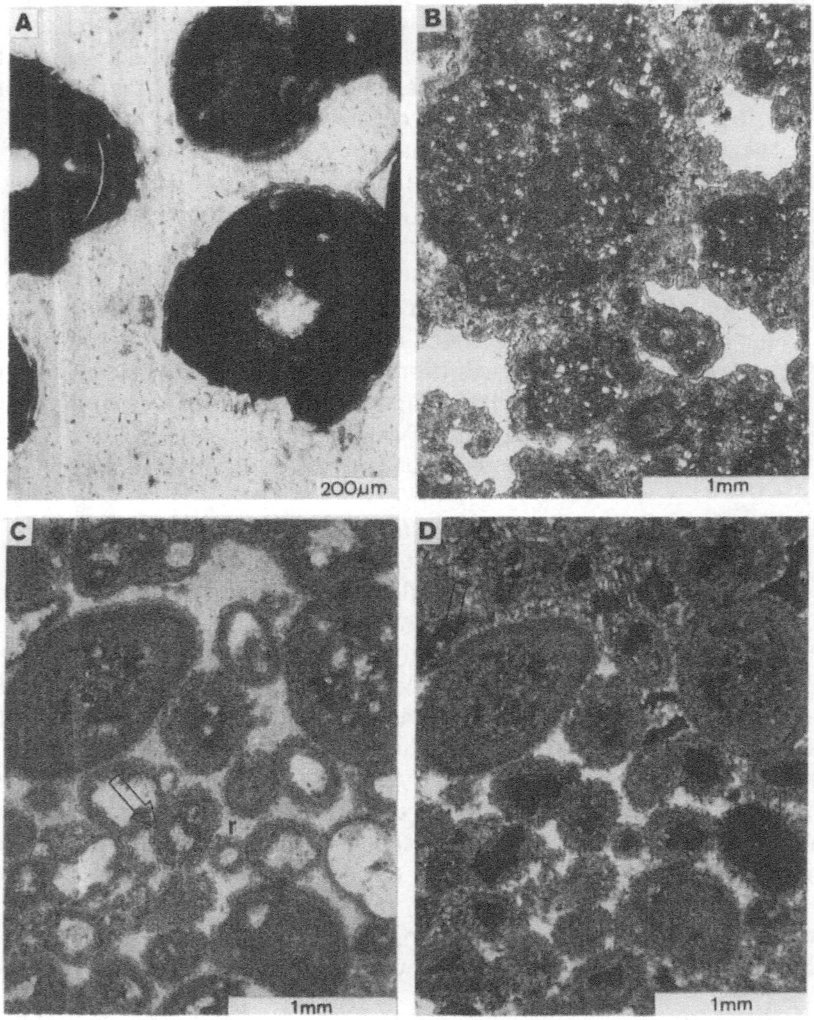

Dolomitization and dedolomitization

The lithology of the Main Dolomite is usually dolomite. Limestones, which occur locally, are clearly secondary in origin (Peryt, 1986a). The following line of reasoning seems to indicate that the first and important phase of dolomitization occurred very early in the diagenetic history.

Both in the Main Dolomite and the overlying Platy Dolomite deposits (which are separated by the sulphates of the Basal Anhydrite only in the southern part of the area studied) dedolomites occur, but the patterns of their distribution in both carbonate formations are different (*cf.* Peryt, 1986a; Gasiewicz, 1986).

In the Main Dolomite, dedolomites commonly occur in the upper part of the succession, although the uppermost part of the dedolomitized sequences are commonly redolomitized. It was concluded, therefore, that the Main Dolomite deposits underwent dedolomitization during exposure prior to deposition of the Platy Dolomite, and that the (re)dolomitization which was syndepositional with the Platy Dolomite sabkha systems (Gasiewicz *et al*, 1987) may also have affected the uppermost part of the underlying deposits. This interpretation is supported by isotope data which show that $\partial^{18}O$ values of those parts of the Main Dolomite sections which are regarded as redolomites and of the Platy Dolomite sections are similar and proper to the sabkha origin (Peryt & Magaritz, in prep. supply detailed isotope data).

Most dolomites are stoichiometric and without exception are Fe-free. There is a difference between dolomites of the central part of the basin which have $\partial^{18}O$ values about +4‰ which suggest dolomitization by evaporated marine brines and those of the carbonate platform area which have $\partial^{18}O$ values indicating important meteoric influences increasing towards the land; whereas in Debki and Widowo values between -3 and -5‰ prevail, in Slawoszynko the most dolomites have $\partial^{18}O$ about 0‰. The isotope data seem to support mixing mechanism of dolomitization, and the variation of $\partial^{18}O$ may be related to different proportion of fresh water in dolomitizing solutions.

Fig. 4. (A) Ooid grainstone. The following sequence of diagenetic events may be deduced: carbonate A cementation (rim cement) --> physical compaction including spalling off cement crusts and breaking of ooids (left side) --> sulphate cementation and replacement --> dissolution of dolomite in nuclei of ooids --> minor dolomite cementation within the oomolds. Kopalino IG1 well, depth 549.3 m. **(B)** Rim cement in small dissolution vugs which are now filled with gypsum. It is thought that this dissolution was syndepositional and related to the beach position. Kopalino IG1 well, depth 561.2 m. **(C & D)** Ooid grainstone. Ooids are micritized but locally relics of lamination are seen. The following diagenetic sequence may be deduced: carbonate A cementation --> moderate compaction (arrow in C) --> sulphate cementation (white and grey in D) and local replacement (r in C) --> dolomite dissolution (which affected the nuclei of ooids) --> rare dolomite cementation in pores (arrows in D). Radoszewo IG1 well, depth 703.7 m. C: ppl, D: xp.

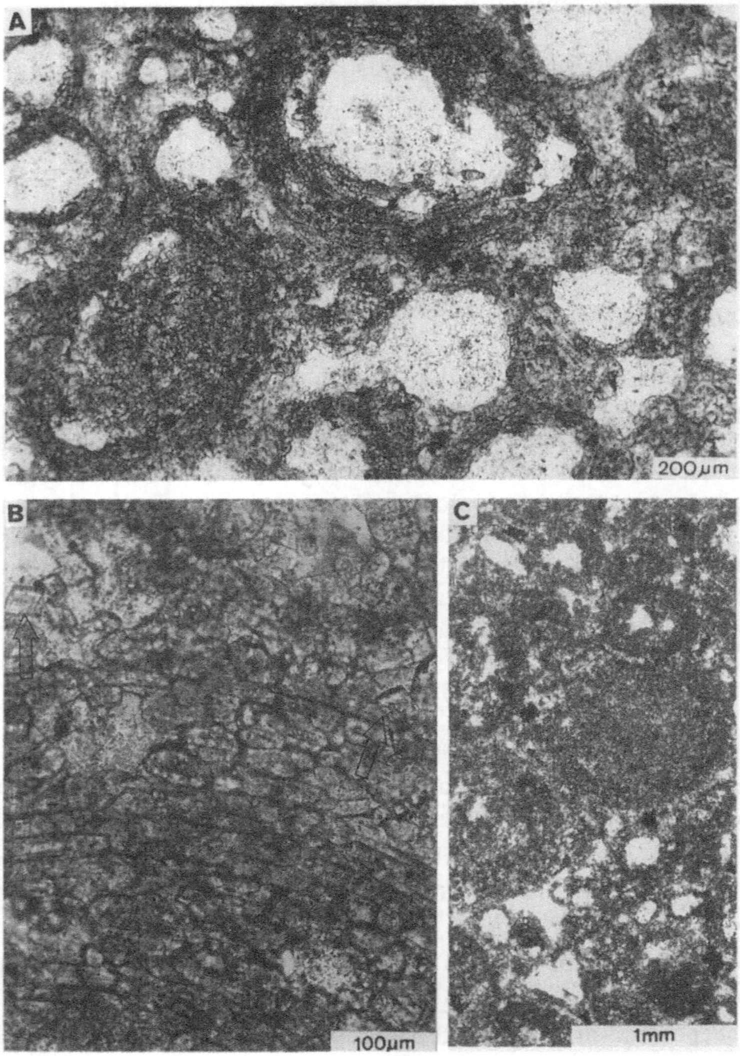

Fig. 5. (A) Ooid grainstone, dedolomitized. Oomolds are filled with gypsum cement. Sulicice IG2 well, depth 616.5 m. **(B)** A fragment of dedolomitized ooid, with occasional redolomite crystals (arrow). Widowo ONZ1 well, depth 613.7 m. **(C)** Strongly compacted, poorly sorted ooid-peloid-lump grainstone. Slawoszynko ONZ1 well, depth 612.4 m.

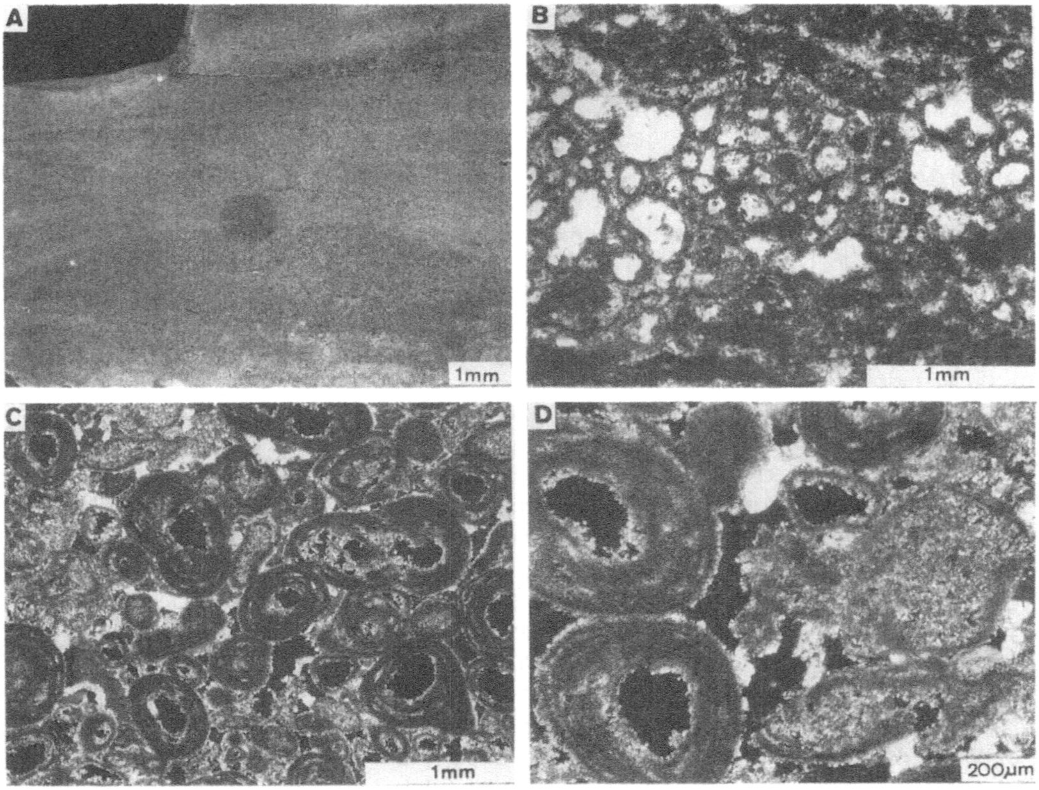

Fig. 6. **(A)** Cross-bedded recrystallized slightly calcareous dolomite, originally ooid grainstone as suggested by the presence of oomolds. Debki IG1 well, depth 562.3 m. **(B)** Recrystallized oomoldic dolomite with interlayered, compacted and dissolved ooids (magnification of photo shown in Peryt, 1985b, fig. 6B). Widowo ONZ1 well, depth 616.2 m. **(C & D)** Ooid grainstone with coated beachrock fragments (right in C). The following diagenetic sequence may be deduced: irregular carbonate A cementation --> intensive compaction --> cementation B (dolomite and next minor sulphate) --> dissolution of ooids and other grains and cement (as seen in D) --> dolomite cementation (centres of ooids). Czarny Mlyn IG1 well, depth 616.0 m. C & D: xp.

Accordingly, it has been proposed that the dolomitization was penecontemporaneous with the Main Dolomite deposition, and that the major phase of dedolomitization happened at the PZ2/PZ3 boundary, during emergence which brought about minor erosion and surficial brecciation (*cf.* Peryt *et al.*, 1985). However, is was not the only phase of dolomitization and dedolomitization.

As previously discussed, the top parts of the dedolomitized Main Dolomite sections are dolomites, and the redolomitization seems to be related to the process of dolomitization of the Platy Dolomite deposits in the sabkha system. The redolomites may be recognized on the basis of two criteria; (1) they exhibit characteristic textures which are otherwise known only from the dedolomitized sections (like the brick-like arrangement of ooid lamellae - Fig. 5A, B) - this may be due to a first texture-destructive phase of dolomitization followed by a second texture-preserving phase (*cf.* Harwood, 1984), and/or (2) their $\partial^{18}O$ is different (e.g., in Slawoszynko values of $\partial^{18}O$ oscillate between +1 and +4‰, as opposed to the other part of the section where values close to 0‰ are characteristic).

As mentioned earlier, late pore-rimming cements as well as other cements which occur in small vugs and molds (e.g., in the nuclei of ooids - Fig. 6D) are dolomites. Their origin follows anhydritization process. It is probable that they precipitated as dolomite as some burial cements. However, saddle (baroque) dolomite, which is a characteristical burial cement, was not recorded. This may be due to the absence of iron and/or relatively low temperature during diagenesis, both factors being constraints on saddle dolomite formation (Radke & Mathis, 1980).

Most of the dedolomites seem to be related to the dedolomitization event at the PZ2/PZ3 boundary, but some dedolomites originated later. This is indicated by the dedolomitization and accompanying calcitization of sulphates which postdated the sulphate replacement and cementation. As will be discussed later, the process of sulphate replacement and cementation operated during moderate burial, and the calcitization of sulphates and dolomites may have resulted from the addition of CO_2 to the pore fluids during the thermal degradation of organic matter, as suggested by Clark (1980a). The replacement of anhydrite by calcite would, in any case, indicate that the pore fluids at different times were variously saturated with anhydrite (Woronick & Land, 1985, p. 268). A cessation of CO_2 production could result in the redolomitization of calcitized dolomites and sulphates. Evidence for this process is that dolomites of similar coarsely-crystalline texture, characteristic of the late calcitization of dolomite and sulphate, occur jointly (often in the same thin section) with dedolomites.

The limestones, which are believed to be the products of burial dedolomitization are very local in occurrence and they do not form beds which can be correlated as in the case of the early dedolomites.

Compaction

Compactional phenomena have been reported in ooid grainstones occurring both in the landward and basinward parts of the coastal oolitic barrier system, but there is a difference in the fabrics of the ooid grainstones derived from each zone.

Fig. 7. (A & B) Laminated mudstone with enterolitic and nodular anhydrite - characteristic examples of the sabkha zone. Leba V well; A - depth 581.1 m, B - depth 580.8 m. **(C)** Oomoldic dolomite. Debki IG1 well, depth 559.4 m. **(D)** Oomoldic dolomite with locally occurring recrystallized ooids which have the cortices as presented in Fig. 5B. Slawoszynko ONZ1 well, depth 600.8 m.

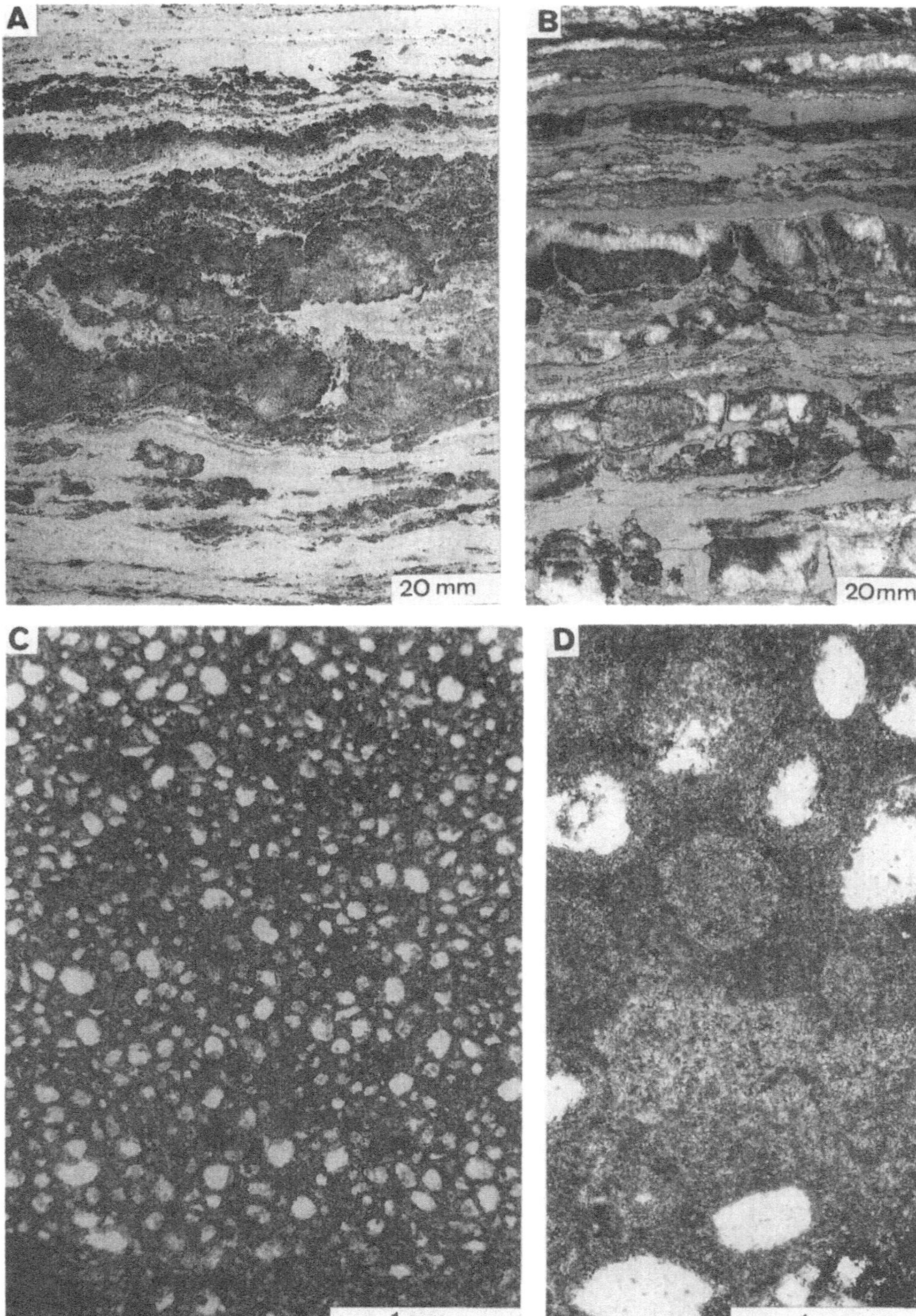
A
B
20 mm
20mm
C
D
1mm
1mm

In the landward zone, ooids have usually been cemented before compaction occurred, but the cement did not protect the ooid grainstone from compaction. Compressed ooids are the most common fabric accompanied by the spalling off cement crusts (Figs. 3A & 4B); Peryt, 1985a, fig. 1a, c, d), interpenetration of grains (Figs. 2B &3B) and occasional fractured grains (Fig. 2C). Flattened ooids, another fabric recorded in this zone, are discussed later in this chapter. In the second zone, the evidence for early cementation is lacking, and the ooids are densely packed (Fig. 5C) but no breaking of ooids was recorded, and the interpenetration of grains is rare and not as intense as in the ooids from the first zone.

The compaction of ooids results from overburden pressure, as is assumed from breaking of solid sphere models (Peryt, 1985a, fig. 1c). In the case of the non-cemented ooids, as the overburden pressure was increasing, the grains were reoriented into a more efficient packing state. This was not possible in the case of ooid grainstones with early rim cement. Cemented grains could not reorient themselves easily as overburden pressure increased. When the pressure was strong enough, the ooid grainstones underwent compaction, and the rim cement was spalled off or suffered pressure-solution.

The depth at which the compaction of grainstones occurs is a matter of debate (see Moore, 1985, p. 296) but in the case of the Main Dolomite, it was less than 600 m. The main phase of anhydrite cementation and replacement, which stopped the compaction of ooid grainstones, occurred under less than 600 m of burial.

Ooid grainstones from the bottom part of the Main Dolomite in the southern region of the area studied and the oomoldic grainstones show only sporadic evidence of compaction. This is interpreted to be the result of early cementation protecting the grains from compaction. The framework supplied by the cementation was strong enough to resist the overburden pressure prior to sulphate cementation, and compaction is manifested only in the breaking of shells (Peryt, 1986a, pl. 28/1, 2).

There are several pecularities in the distribution of flattened ooids, which merit a separate discussion. Firstly, they are recorded in deposits which were possibly subaerially exposed soon after the deposition, as the association with oomolds indicates (Fig. 6B). Secondly, they form interlayers or laminae in mudstones (Peryt, 1986a, pl. 32/2; see also Peryt, 1978. pl. 17/1 illustrating the same fabric from the Main Dolomite of SW Poland) and it seems difficult to explain how such a fabric could originate under overburden pressure. Thirdly, grainstone beds with flattened ooids are interbedded with grainstone beds containing thin rim carbonate cemented ooids which did not suffer compaction. Accordingly, if flattened ooids are interpreted as a compaction fabric resulting from strong overburden pressure, the associated ooids should be affected by the same compaction as well; however, they are not.

The distribution patterns of flattened ooids may be explained as resulting from solution compaction in the vadose environment. Such a mechanism was earlier applied by Clark (1980b) to features in the Main Dolomite of the eastern Netherlands, and also in the Jurassic oolitic limestones of the Paris basin (Clark, 1979). The tightly compacted fabric in the latter case was explained by Cussey and Friedman (1977, 1979) as originating from pressure solution of ooids in the vadose zone due to load compaction. During pressure-solution no cement was present between particles (Cussey & Friedman, 1979, p. 679). In the Zechstein Main Dolomite of the area studied, polygonal fitting (observed by Clark, 1980b, fig. 14, and by Cussey & Friedman, 1977, fig. 4) of ooids is rare and is associated with other compactional fabrics, thus suggesting an origin related to overburden pressure. The flattened ooids, as stated above, do not show such a relationship.

Bhattacharyya and Friedman (1984) experimentally proved that "plastic" deformation of ooids may result from deep-burial pressure. They observed that more intense plastic deformation occurs when the pore water is marine and suggested that longitudinal contacts and those related to pressure-solution are two independent, unrelated phenomena. The results of their experiments cannot be applied to all deformed ooids. As noted by Becher and Moore (1976, p. 51), "the only requirement necessary for solution-compaction to proceed is the presence of undersaturated pore fluid. Undersaturated conditions may be produced by either an increase in overburden pressure (pressure-solution) or by the introduction of undersaturated water from an outside source". They referred to the microstylolites of Pleistocene skeletal grainstone in Bermuda (Land, 1971, fig. 71) that could not have been buried more than 10-15 feet. Similar microstylolites have been found in the peloidal grainstones of the bottom part of the Main Dolomite (Fig. 3B, D).

There are also other records of vadose compaction in recent subaerial diagenetic environments, both published (Knox, 1977; Braithwaite, 1983) and unpublished (R.J. Dunham in Clark, 1979). It is assumed (Braithwaite, 1983, p. 354) that long concavo-convex contacts may develop when fresh water films surrounding grains promote solution at grain contacts, with little or no overburden pressure. Accordingly, the association of oomolds and distorted ooids, both in the Main Dolomite and in other cases (e.g., Lower Muschelkalk of SW Germany - Richter, 1983, fig. 3H and table 2, sample 17; Mississippian of Dakota - Elliott, 1982, fig. 15D), may be explained by the compaction solution of ooids and the cementation of ooids below; later, in the diagenetic history, when the deposits were flushed with fresh water, oomolds could originate in those underlying cemented layers.

The early beach-related cementation prevented compaction of ooids (Fig. 8). As discussed earlier, the marine rim cementation did not protect ooids from compaction, but prevented grain adjustment to the increasing overburden pressure, while the lack of marine cement made possible grain reorientation which led to the formation of an overpacked texture. Accordingly, the Zechstein grainstones demonstrate an increased packing both seawards (which is related to the absence of marine cement in the stagnant phreatic environment) and upwards. The latter trend, which has an irregular pattern, is probably related to solution compaction. Compaction of the Zechstein ooids was thus controlled by the depositional setting . The same sedimentary control over compaction of ooids has been recorded in other grainstone sequences.

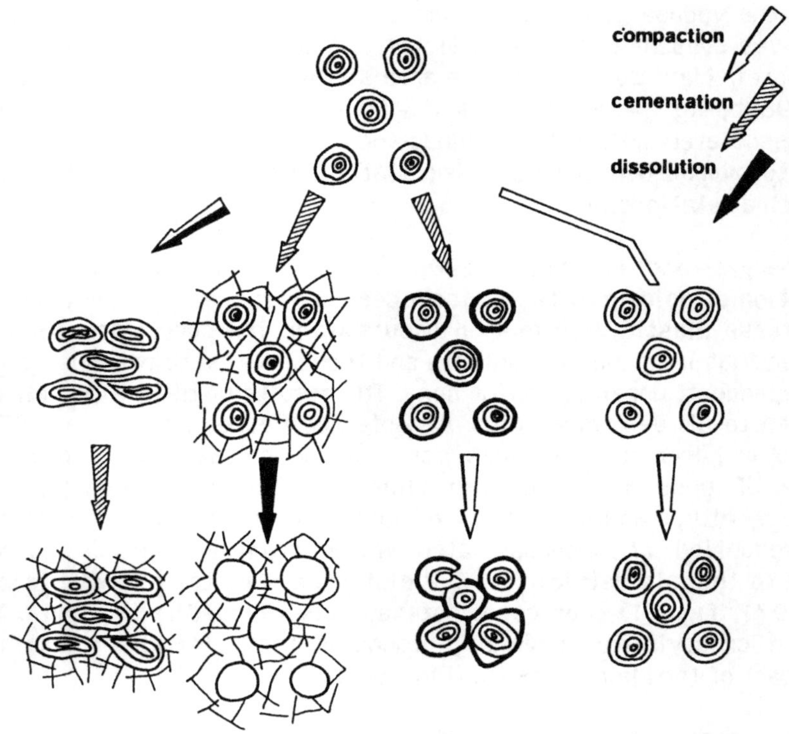

Fig. 8. Evolution of ooid grainstones in the Main Dolomite of the Leba elevation.

Moore and Druckman (1981) observed the basinward increasing compaction in the Upper Jurassic Smackover, and no compaction in the northern diagenetic zone where oomolds occur. These authors related the compaction to the burial pressure.

Cussey and Friedman (1977, 1979) also observed seaward increased compaction in the Dogger of the Paris Basin, and related it to the pressure-solution of uncemented ooids. An opposite trend - landward and upward increasing compaction - was recorded by Winchester (1977) in the Lower Permian of New Mexico. This opposite landward trend was regarded as the result of solution compaction, rather than being caused by pressure solution.

Sulphate replacement and cementation

Both anhydrite and gypsum cements occur in the Main Dolomite in the entire area studied, and although anhydrite replacement is predominant, complex relations between both minerals are the reasons why they are discussed together here. Only in a few cases (Fig. 7A, B) was sulphate precipitation contemporaneous with deposition, as in the sabkha zone of the Leba region (see Peryt, 1986a). In the upper part of the sections in the coastal oolitic barrier system, sulphate nodules (usually megacrystalline gypsum after anhydrite - Orti Cabo, 1977) are also present, but these postdate the Main Dolomite deposition and might have originated during the Basal Anhydrite deposition in a sabkha environment (*cf.* Peryt *et al*, 1985, fig. 3). Most sulphate, however, clearly postdated dolomitization, cementation (both A and B) and compaction (Figs. 4A & 9A, B). The relatively late emplacement of replacing anhydrite was previously recorded by several workers in other areas (e.g., Kendall & Walters, 1978; Clark & Shearman, 1980).

Sulphates replace fossils (which is a common feature in dolomites - Murray, 1960) (Fig. 2D), matrix, grains (Figs. 2A, D, 4A & 9D) and early carbonate cement (Fig. 4A). However, they occur primarily as a pore-filling cement (Figs. 3A & 4C, D). Poikilotopic crystals of sulphates, which are quite common, indicate that replacement and cement B formation have been, at least in some cases, genetically related.

The driving force of widespread anhydritization in the Main Dolomite of the Leba elevation (as well as in other areas - Clark, 1980a) was the conversion of gypsum into anhydrite; dehydration resulted from increased temperature and pressure related to progressively increasing depth of burial. That gypsum formed a substantial proportion of all sulphate units in the Leba elevation, is demonstrated by the presence of abundant diagnostic anhydrite crystal pseudomorphs after gypsum, including selenite fabric: these are especially abundant in the Lower Anhydrite (where gypsum mainly grew in open spaces) and in the Upper Anhydrite (where it mainly grew displacively). The conversion of gypsum to anhydrite released the crystal water, and this $CaSO_4$-rich brine was injected into the Main Dolomite deposits. The process was probably rapid, as the changes in evaporites require little time and very little heating (Schreiber, 1986). Complex association between sulphate and dolomite suggests that the process of anhydritization continued after this main phase, with some interludes of dolomite precipitation and dissolution. Such relations which have also been recorded in other areas (Clark, 1980a, p. 183) suggest periodic fluid movement (Woronick & Land, 1985, p. 273). Because of fluid movement upward and toward the basin margin, there is a distinct increase of the sulphate content landward in the Main Dolomite rocks (Fig. 10).

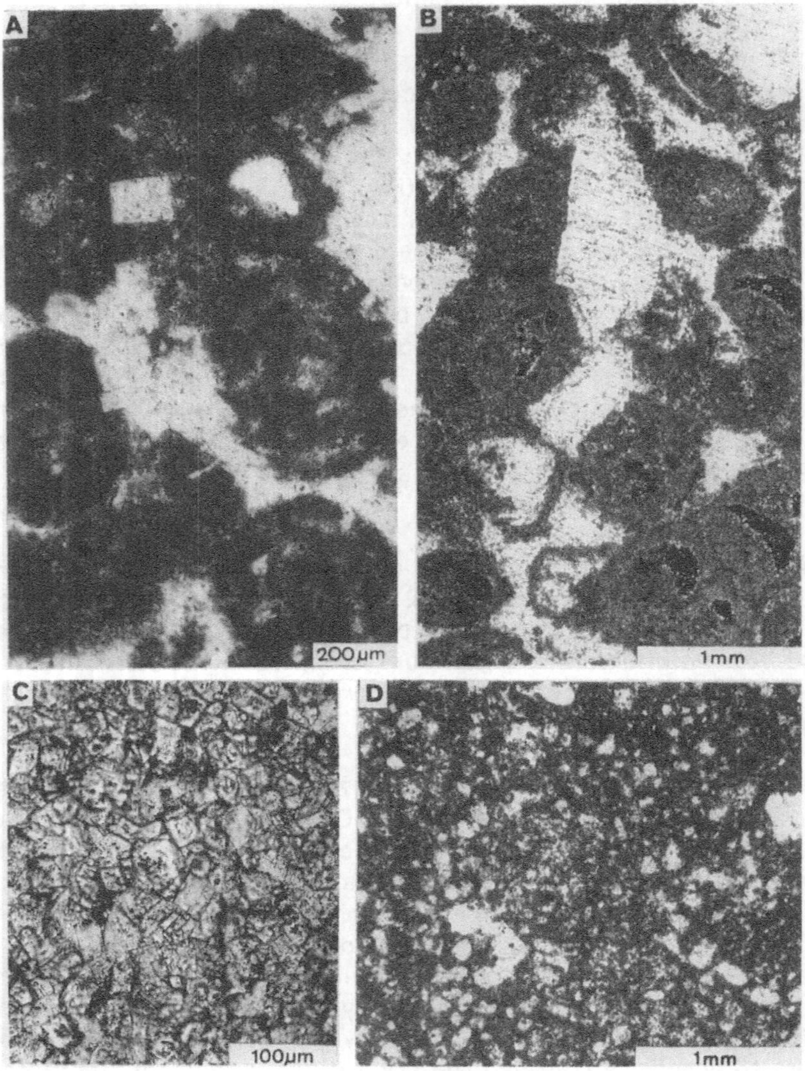

Fig. 9. (A) Ooid grainstone with sulphate replacement, ppl. Kopalino IG1 well, depth 552.3 m. **(B)** Ooid grainstone with pervasive sulphate replacement. Late dissolution of ooids (black areas) was followed by dolomite cementation, xp. Salino IG1 well, depth 821.2 m. **(C)** Recrystallized oomoldic dolomite as shown in Peryt, 1985b, fig. 6C. Debki IG1 well, depth 558.7 m. **(D)** Peloidal packstone with pervasive sulphate replacement which affected mostly peloids. Kopalino IG1 well, depth 558.5 m.

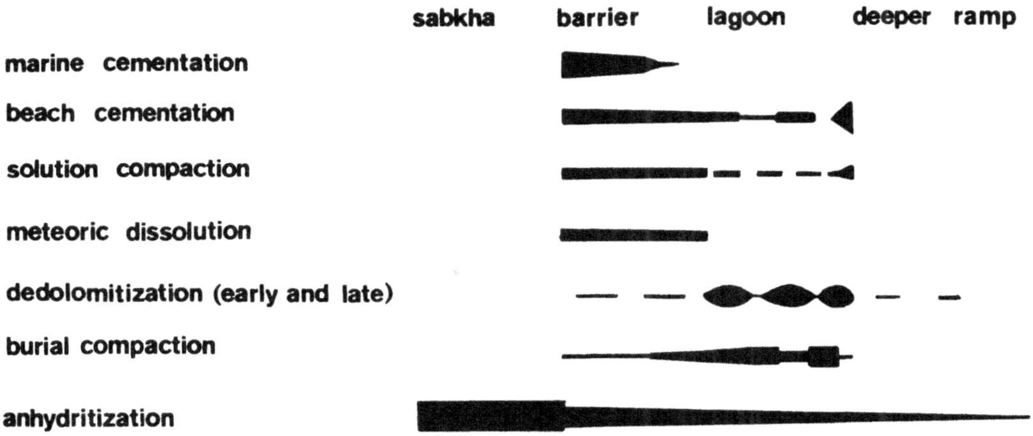

Fig. 10. Intensity of diagenetic processes in the Leba elevation.

Dissolution

Two different phases of dissolution may be distinguished. The first phase was syndepositional occurring during the stabilization of sea conditions, resulting in the origin of small vugs at the base of the Main Dolomite in the southern region (Fig. 3C), and the formation of oomoldic porosity in the upper part of the coastal oolitic barrier system (Fig. 7C). The second phase occurred periodically after compaction, as indicated by the presence of irregular vugs in the compacted deposits, and by the presence of partially dissolved nuclei of compacted ooids. These late vugs are partly cemented (Figs. 4D, 6C,D & 9B) by dolomite and sulphate and occasionally by halite.

Late dissolution affected not only dolomite but also sulphate. The driving force of dissolution in the late stage of diagenesis was the variable saturation of deep subsurface fluids with dolomite and anhydrite and eventual periodic degassing from the fluid as opposed to the first phase dissolution, related to the action of meteoric waters.

Summary of the diagenetic history

Syndepositional-early compactional phase

An active meteoric water system only slightly affected the grainstones. Most probably, lenses of meteoric water were restricted to the crests of ooid bars and because of arid conditions there is no indication of land-attached meteoric water lenses. Local lenses of meteoric water dissolved the unstable mineral suite (particularly aragonite) with creation of oomoldic porosity recorded in the upper part of the Main Dolomite in the carbonate platform area. The dissolution possibly occurred during Main Dolomite deposition and/or during the regressive phase at the PZ2/PZ3 boundary: in any case, the oomolds were formed before the second phase of diagenesis as evidenced by compaction effects in some samples (e.g., in Widowo ONZ1) in the manner shown by Kaldi and Gidman (1982, fig. 3).

Shallow marine rim cementation was only locally important. It has been recorded in settings which suggest very shallow environments of deposition, related to nearshore environments during the initial, partial submergence by the Main Dolomite transgressing sea (Zdrada area) and to the coastal oolitic barrier system which developed after the final transgression. The deposits cemented in a nearshore situation, locally suffered early leaching. This led to the origin of vuggy porosity which was soon reduced by microbial deposits and subtidal, probably aragonitic cements (Peryt, 1986a, pl. 27/1). Because the leaching did not result in oomoldic porosity in these nearshore deposits, it is probable that they were dolomitized. Deposits of the carbonate platform area were dolomitized by mixing mechanism and those at the basin by reflux mechanism.

Shallow burial-compactional phase

Increasing overburden pressure resulted in compaction affecting all grainstones in the area of the carbonate platform and to a moderate degree also the grainstones occurring locally in the area of deeper ramp, which were deposited .during the initial Ca2 transgression. The compaction in the latter case is manifested in the broken shells (Peryt, 1986a, pl. 28/1, 3).

In the area of the carbonate platform, the degree of compaction depended mainly on types of grain, grain sizes and grain sorting. In general, better sorted ooid sands (as in Kopalino) suffered grain breakage and the spalling off of the carbonate cement crusts (where present), and pressure-solution contacts are more common when compared to the poorly sorted, ooid-peloid-lump sands where the compaction resulted mainly in an overcompactional fabric. The compaction greatly reduced porosity by grain adjustment and it also supplied carbonate-rich solutions which precipitated as cement B in other Zechstein carbonates and possibly in overlying Buntsandstein silicilastics.

The B cementation clearly postdated the compactional phase in all studied sections from the coastal oolitic barrier system; although some variation in the degree of packing occurs in particular sections. This cannot be consistently ascribed as the result of B cementation which would eventually weaken the effect of overburden pressure. Accordingly, it seems more likely that the carbonates derived from the pressure-solution of ooids were not the source of carbonate cement B in the oolitic barrier complex. In some wells cement B consists of sulphate derived from Werra brines invading the connate waters. At the time of the invasion the grainstones were mostly dolomites unless they were dedolomitized during the first phase of diagenesis.

Moderate burial-Werra brine invasion phase

With progressive burial and increasing temperature, the gypsum was converted into anhydrite. This generated large quantities of water which, because of overpressing, was injected into the Main Dolomite deposits. The brines were Ca and SO_4 rich and anhydrite began to replace the dolomite grains and, occasionally, the cement. It also precipitated as cement B. As suggested by Clark (1980a), phases of anhydrite replacement in the Zechstein deposits probably occurred repeatedly, which fits a rather complex pattern of dolomite-sulphate cement relationships in the ooid grainstones. The effect of the B cementation was to occlude many pores with anhydrite, especially in the well sorted ooid grainstones.

Moderate burial-dissolution phase

Probably as a result of thermal degradation of organic matter within the carbonates, CO_2 was added to the pore fluids. This led to the congruent dissolution of dolomite. For unknown reasons, the dissolution mainly affected the interiors of ooids. The phase of dolomite dissolution was marginally preceded by a phase of local calcitization of dolomite and rarely of anhydrite. The source of CO_2 was probably again due to thermal degradation of organic matter. The dedolomitization was not complete and most of the dolomite previously preserved (e.g., the ooid interiors) was subsequently dissolved.

Late burial-cementation phase

When the production of CO_2 ceased, the pore waters again became saturated with respect to dolomite, and some dolomite cementation (especially in the ooid interiors) and occasional redolomitization occurred. This probably happened when the area was relatively uplifted at the beginning of the Tertiary where the shallower depths favoured gypsum and occasional halite cementation which passively filled some pores. This was the last major stage of the diagenetic history (Fig. 11).

Discussion

The same general succession of diagenetic processes, as established for the Main Dolomite of northern Poland, seems to be characteristic for other

Zechstein carbonates (*cf.* Clark, 1980a; Peryt, 1984) and is related to two major controls of diagenesis; overburden pressure and temperature. The Main Dolomite pattern shows many striking similarities to the diagenetic history presented by Moore and Druckman (1981) for the southern zone of the Upper Jurassic Smackover, although in their example dolomites are absent. The main difference, when compared to the Zechstein occurrences, is the intensity of preburial cementation. It is doubtful that a regional meteoric water system developed in the Main Dolomite as it is supposed to have done in the northern zone of Upper Smackover (Moore & Druckman, 1981).

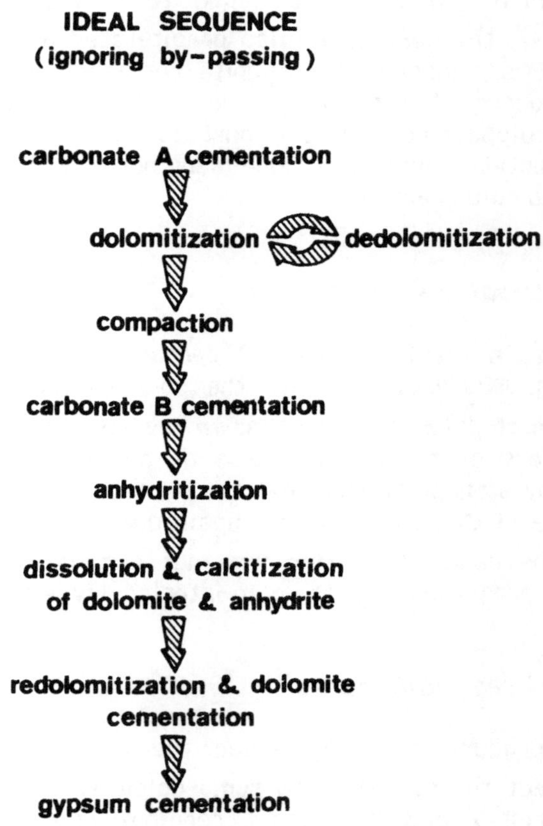

Fig. 11. Ideal diagenetic sequence in the Main Dolomite of the Leba elevation.

The Main Dolomite is the Peri-Baltic area, as opposed to the other parts of the Zechstein basin, occurs in a different palaeotectonic setting where the deposition of the PZ1 cycle rocks did not accentuate nor create the relief as commonly happened elsewhere in the basin (e.g., Sannemann *et al.*, 1978; Depowski & Peryt, 1985).

The depositional model of carbonate platform in northern Poland - a homoclinal ramp with a coastal ooid barrier system - differs from the common model of a rimmed carbonate shelf (rims consisting of barrier ooid sands) with a rapid transition from carbonate platform to deep basin environments (e.g., Sannemann *al.*, 1978; Clark, 1980b; Depowski & Peryt, 1985). The platform edge sands elsewhere show evidence of very early cementation, mostly in a beach environment (Clark, 1980b; Peryt, 1985b); and also of vadose diagenesis including dissolution of ooids, leading to the origin of oomoldic porosity, and solution compaction, which formed the flattened ooids (see Clark, 1980a, fig. 1F; 1980b, fig. 5; Peryt, 1985b, figs. 2 & 3). The latter process was more common, perhaps related to the generally arid climate.

During pre-evaporite Zechstein Limestone deposition, when the climate was more humid (Peryt, 1984), these relationships were reversed (*cf.* Kaldi & Gidman, 1982) or more balanced (Peryt, 1984). Synformational dedolomitization was of great importance during Zechstein Limestone deposition (Peryt, 1984); whereas in the case of the Main Dolomite early dedolomitization is rarely proved (as in the Leba elevation area) but late, burial dedolomitization was generally of great importance (Clark, 1980a; Depowski & Peryt, 1985).

Clark (1980a) presented a diagenetic model of Zechstein carbonate sediments in which rain-derived freshwater caused intense leaching of the platform sediments with lenses of meteoric water extending basinwards initiating dolomitization by a mixing mechanism (Clark, 1980a, p. 200). In his model, syndepositional meteoric cementation was very important (Clark, 1980a. p. 200). In the Leba elevation, syndepositional meteoric cementation was of minor importance in contrast to early surface-related dedolomitization. Clark (1980a, p. 184-185) advocated that calcitization occurred after substantial burial had taken place (with the exception of some secondary limestones that became exposed in the Cretaceous to recent time interval). In the Leba elevation late burial dedolomitization had only a local significance, however, elsewhere late burial dedolomitization could have been important. Compaction during early burial was important in the Main Dolomite of the Leba elevation but, with the exception of vadose (=solution) compaction, the significance of compaction as described here has not been considered by Clark (1980a). His definition of compaction (Clark, 1980a, p. 185-187) refers to pressure-solution. This process is not evident in the Leba elevation, perhaps because of relatively small burial.

Some conclusions by Clark (1980a, p. 201) that " the diagenesis of Zechstein carbonate sediments took place in response to the changing composition of pore fluids in the rocks as the geological history of the basin progressed" seem to be correct, although his suggested evolution of pore waters in the course of diagenesis (Clark, 1980a) differs in several important aspects from the evolution proposed here.

The main differences between the model presented here (summarized in Figs. 10 & 11) and Clark's model (1980a) are listed in Table 2. It should be noted that when applying the model proposed for the Main Dolomite in the Leba elevation to other areas in the Zechstein basin, differences arising from palaeogeography and tectonic history should be taken into consideration. This applies particularly to the intensity of early meteoric diagenesis (including dedolomitization) and of calcitization and leaching of dolomite and anhydrite.

In the Main Dolomite of the Leba elevation meteoric diagenesis was volumetrically unimportant, but in other areas such diagenesis could have been locally important, as evidenced by thick complexes of beach deposits which underwent solution compaction (e.g., Peryt, 1985b) and thick oomoldic complexes (e.g., in the Zbaszyn area, Lubuska barrier, W Poland - *cf.* Depowski & Peryt, 1985). However, the existence of widespread lenses of fresh water, as envisaged by Clark (1980a) seems doubtful, particularly when discrepancies in his interpretation are considered (*cf.* Clark, 1980a, p. 200, 1980b, p. 158). In the Leba elevation, calcitization and leaching of dolomite and anhydrite affected the platform deposits, whilst in other areas where the original elevation difference between the platform and the basin was more accentuated, these processes mainly affected the slope deposits. As deduced by Clark (1980a), important secondary porosity in carbonate of the lower slope was created by these processes. This has been confirmed in western Poland.

Most of the cementation in the Leba elevation was related to burial diagenesis, as in the case of many carbonate rocks (Scholle & Halley, 1985). However, the most comon pathways of diagenesis (Fig. 11) indicate that in fact the evolution of porosity during burial was controlled by depositional setting. Accordingly, depositional setting seems to be the overriding factor in the diagenetic behaviour of the Zechstein carbonates. Although late burial diagenesis was an important modifier, it was syndepositional and early-burial, facies-related diagenetic imprint which was mainly responsible for an irregular pattern of porosity and permeability which resulted in haphazard discoveries when prospecting for hydrocarbons in the Main Dolomite. The high unpredictability in the Main Dolomite platform edge grainstone zone (which is the most promising target for oil and gas prospection) corresponds to a similar complex pattern of porosity in the transitional zone of the Upper Jurassic Smackover (Moore & Druckman, 1981). The diagenetic histories in that zone are somewhat different when compared to those of the southern zone (Moore & Druckmann, 1981; Harwood & Moore, 1983) which show many similarities to the Main Dolomite of the Leba elevation.

To refine the model presented here, further geochemical information is needed, particularly fluid inclusion data, Sr 87/86 ratios of dolomite and sulphate cements, and trace element data of dolomite cements.

Conclusions

The history of diagenesis of the Main Dolomite in the Leba elevation indicates that the meteoric diagenesis was volumetrically of limited importance. It

Table 2. Comparison of two models of diagenesis of Zechstein deposits

Process	after Clark, 1980a	this paper
meteoric cementation	very important	locally important
fresh-water dissolution	important	locally important
surface-related dedolomitization, early	-	important
late dedolomitization	very important, especially on slope and adjacent areas	locally important
early-burial compaction	-	important
dolomitization	mostly reflux during sulphate & halite deposition	early process, not related to halite deposition
redolomitization	late burial process	mostly early burial process
leaching of dolomite	important	locally important
anhydritization	repeatedly occurring	repeatedly occurring
halitization	replacement & cementation	minor cementation
CONCLUSION	depositional and early diagenetic textures are normally completely overprinted	depositional and diagenetic imprint is decisive

included beach cementation, dissolution of ooids and solution compaction syndepositionally, and dedolomitization after deposition. There is no evidence of freshwater influence on the underlying evaporites during or after the deposition of the Main Dolomite. Marine cementation was localized and mainly limited to the landward side of the coastal oolitic barrier system and to beaches. Such patterns of cement distribution governed compaction during shallow burial. The compaction destroyed most of the interparticle porosity in the poorly sorted ooid-peloid-lump sands in the outer part of the carbonate platform area. During moderate burial $CaSO_4$ brines invaded the Main Dolomite deposits with resulting sulphate cementation and dolomite replacement. During later stages of burial evolution the Main Dolomite rocks were dissolved and cemented by dolomite, sulphate and halite. This later diagenesis locally overprinted the earlier diagenetic history but in the Leba elevation the early diagenetic imprint was preserved and governed the later diagenetic history. The same model can probably be applied to other areas of the Main Dolomite deposition as well as to other intra-evaporitic carbonate successions. The local geological factors could strongly influence the intensity of the diagenetic processes, of which early meteoric cementation, late diagenetic cementation and late diagenetic leaching are crucial in the productive areas of the Main Dolomite (Clark, 1980a; Depowski & Peryt, 1985).

Acknowledgments

The problems presented in this paper have been discussed with, and the paper has been reviewed by numerous friends and colleagues, too numerous to mention all of them. The essential part of the paper was written during the Alexander von Humboldt fellowship at Geologisches Institut, Universität Freiburg, and was presented during SEPM Research Conference held in Cancun, Mexico, in August 1985. The drawings were done by M. Rybicki and M. Bejger, and a part of photos by J. Modrzejewska.

References

Becher, J.W. & Moore, C.H., 1976. The Walker Creek field: a Smackover diagenetic trap. *Trans. Gulf. Assoc. Geol. Soc.*, 31: 43-56.

Bhattacharyya, A. & Friedman, G.M., 1984. Experimental compaction of ooids under deep-burial diagenetic temperatures and pressures. *Jour. Sed. Petrology* 54: 362-372.

Braithwaite, C.J.R., 1983. Calcrete and other soils in Quaternary limestones: structures, processes and applications. *J. geol. Soc. London* 140: 351-363.

Clark, D.N., 1979. Patterns of porosity and cement in ooid reservoirs in Dogger (Middle Jurassic) of France: Discussion. *Am. Assoc. Petroleum Geologists Bull.*, 63: 676-677.

Clark, D.N., 1980a. The diagenesis of Zechstein carbonate sediments. *Contr. Sedimentology* 9: 167-203.

Clark, D.N., 1980b. The sedimentology of the Zechstein 2 Carbonate Formation of eastern Drenthe, the Netherlands. *Contr. Sedimentology* 9: 131-165.

Clark, D.N. & Shearman, D.J., 1980. Replacement anhydrite in limestones and the recognition of moulds and pseudomorphs: a review. *Rev. Inst. Invest. Geol.*, 34: 161-186.

Cussey, R. & Friedman, G.M., 1977. Patterns of porosity and cement in ooid reservoirs in Dogger (Middle Jurassic) of France. *Am, Assoc. Petroleum Geologists Bull.*, 61: 511-518.

Cussey, R. & Friedman, G.M., 1979. Patterns of porosity and cement in ooid reservoirs in Dogger (Middle Jurassic) of France: Reply. *Am. Assoc. Petroleum Geologists Bull.*, 63: 677-679.

Dadlez, R., Dayczak-Calikowska, K., Dembowska, J., Jaskowiak-Schoeneichowa, M., Marek, S., Szyperko-Sliwczynska, A. & Wagner, R., 1976. Pokrywa permsko-mezozoiczna w zachodniej czesci syneklizy perybaltyckiej. *Biul. Inst. Geol.*, 270: 143-163.

Depowski, S. & Peryt, T.M., 1985. Carbonate petroleum reservoirs in the Permian dolomites of the Zechstein, Fore-Sudetic area, western Poland. In: Roehl, P.O. & Choquette, P.W. (Eds), *Carbonate Petroleum Reservoirs*, 253-264. Springer, N.Y.

Depowski, S., Peryt, T.M., Piatkowski, T.S. & Wagner, R., 1981. Paleogeography versus oil and gas potential of the Zechstein Main Dolomite in the Polish Lowland. *Proc. Intern. Symp. Central Europ. Permian*, 587-595.

Elliott, T.L., 1982. Carbonate facies, depositional cycles, and the development of secondary porosity during burial diagenesis: Mission Canyon Formation., Haas Field, North Dakota. *Saskatchewan Geol. Surv. Spec. Publ.*, 6: 131-151.

Gasiewicz, A., 1986. Dedolomityzacja utworow dolomitu plytowego na wyniesieniu Łeby. *Prz. geol.*, 34: 257-260.

Gasiewicz, A., Gerdes, G. & Krumbein, W.E., 1987. The peritidal sabkha type stromatolites of the Platy Dolomite (Ca3) of the Leba elevation (north Poland). *This volume.*

Harwood, G.M., 1984. Fabric preservation during dolomitization. *7th Mtg Carb. Sedim., Univ. Liverpool, Titles and Abstracts of Talks.*

Harwood, G.M. & Moore, C.H., 1983. Comparative sedimentology and diagenesis of Upper Jurassic ooid grainstone sequences, East Texas Basin. *SEPM Core Workshop* 5: 176-232.

Hauk, V.M., Petersen, H.H.F., Spoerker, H.F. & Moritz, J., 1979. Deep European H_2S is handled with special muds, cement and tubulars. *Oil Gas J.*, 77: 62-69.

Kaldi, J. & Gidman, J., 1982. Early diagenetic dolomite cements: examples from the Permian Lower Magnesian Limestone of England and the Pleistocene carbonates of the Bahamas. *Jour. Sed. Petrology* 52: 1073-1085.

Kendall, A.C. & Walters, K.L., 1978. The age of metasomatic anhydrite in Mississippian reservoir carbonates, southeastern Saskatchewan. *Can. J. Earth Sci.*, 15: 424-430.

Knox, G.J., 1977. Caliche profile formation, Saldanha Bay (South Africa). *Sedimentology* 24: 657-674.

Land, L.S., 1971. Phreatic versus meteoric diagenesis of limestones: Evidence from a fossil water table in Bermuda. In: Bricker, O.F. (Ed.), *Carbonate Cements*, 133-136. John Hopkins University Press, Baltimore.

Macintyre, I.G., 1985. Submarine cements - the peloidal problem. *SEPM Spec. Publ.*, 36: 109-116.

Maureau, G.T.F.R. & van Wijhe, D.H., 1979. The prediction of porosity in the Permian (Zechstein) carbonate of eastern Netherlands using seismic data. *Geophysics* 44: 1502-1517.

Moore, C.H., 1985. Upper Jurassic subsurface cements: a case history. *SEPM Spec. Publ*, 36: 291-308.

Moore, C.H. & Druckmann, Y., 1981. Burial diagenesis and porosity evolution., Upper Jurassic Smackover, Arkansas and Louisiana. *Am. Assoc. Petroleum Geologists Bull.*, 65: 597-628.

Murray, R.C., 1960. Origin of porosity in carbonate rocks. *Jour. Sed. Petrology* 30: 59-84.

Orti Cabo, F., 1977. Aproximación al estudio petrográfico de las microestructureas de las rocas de yeso secundario y a su origen. *Inst. Invest. Geol.*, 32: 87-152.

Peryt, T.M., 1978. Charakterystyka mikrofacjalna cechsztynskich osadów weglanowych cyklotemu pierwszego i drugiego na obszarze monokliny przedsudeckiej. *Studia Geol. Pol.*, 54, 88pp.

Peryt, T.M., 1984. Sedymentacja i wczesna diageneza utworów wapienia cechsztynskiego w Polsce zachodniej. *Prace Inst. Geol.*, 109, 80pp.

Peryt, T.M., 1985a,. Chemical control of carbonate phases: implications from Upper Pennsylvanian calcite-aragonite ooids in southeastern Kansas - Discussion. *Jour. Sed. Petrology* 55: 926-927.

Peryt, T.M., 1985b. Permian beach in the Zechstein dolomites of western Poland: influence on reservoirs. *Jour. Petroleum Geology* 8: 463-474.

Peryt, T.M., 1986a. The Zechstein (Upper Permian) Main Dolomite deposits of the Leba elevation, northern Poland: Facies and depositional history. *Facies* 14: 151-200.

Peryt, T.M., 1986b. Fossiliferous dolomites in the Upper Werra Anhydrite (Zechstein) of the Puck Bay area, northern Poland. *N. Jb. Geol. Paläont. Mh*, 1986: 193-200.

Peryt, T.M., Czapowski, G., Debski, J & Pizon, A., 1985. Model sedymentacji ewaporatow cechsztynskich na wyniesieniu Leby. *Prz. geol.*, 33: 204-211.

Radke, R.M. & Mathis, R.L., 1980. On the formation of saddle dolomite. *Jour. Sed. Petrology* 50: 1149-1168.

Read, J.F., 1985. Carbonate platform facies models. *Am. Assoc. Petroleum Geologists Bull.*, 69: 1-21.

Richter, D.K., 1983. Calcareous ooids: A synopsis. In: Peryt, T.M. (Ed.), *Coated Grains*, 71-99. Springer, Heidelberg.

Roehl, P.O. & Choquette, P.W., 1985. Introduction. In: Roehl, P.O. & Choquette, P.W. (Eds.), *Carbonate Petroleum Reservoirs*, 1-15. Springer, N.Y.

Sannemann, D., Zimdars, J. & Plein, E., 1978. Der basale Zechstein (A2-T1) zwischen Weser und Ems. *Z. dt. geol. Ges.*, 114: 461-483.

Scholle, P.A., & Halley, R.B., 1985. Burial diagenesis: out of sight, out of mind! *SEPM Spec. Publ.*, 36: 309-334.

Schreiber, B.C., 1986. Arid shorelines and evaporites. In: Reading, H.G. (Ed.), *Sedimentary Environments and Facies*, p.189-228. Blackwell, Oxford.

Taylor, J.C.M., 1984. Late Permian - Zechstein. In: Glennie, K.W. (Ed.), *Introduction to the Petroleum Geology of the North Sea*, 61-83. Blackwell, Oxford.

Wilkinson, B.H., Smith, A.L. & Lohmann, K.C., 1985. Sparry calcite marine cement in Upper Jurassic limestones of southeastern Wyoming. *SEPM Spec. Publ.*, 36: 169-184.

Winchester, P.D., 1977. Evidence for solution compaction in lime grainstones (abstract). *Am. Assoc. Petroleum Geologists Bull.*, 61: 841.

Woronick, R.E. & Land., L.S., 1985. Late burial diagenesis, Lower Cretaceous Pearsall and lower Glen Rose Formations, south Texas. *SEPM Spec. Publ.*, 36: 265-275.

THE PERITIDAL SABKHA TYPE STROMATOLITES OF THE PLATY DOLOMITE (Ca3) OF THE LEBA ELEVATION (NORTH POLAND)

Andrzej Gasiewicz
Instytut Geologiczny, ul. Rakowiecka 4 , 00-975 Warszawa, Poland
and
Gisela Gerdes and Wolfgang E. Krumbein
Geomicrobiology Division, University of Oldenburg,
P.O. Box. 2503, D-2900 Oldenburg, F.R. Germany

Abstract: The Platy Dolomite series of the Leba Elevation is predominated by the biolaminoid-nodule facies formed in supratidal and intertidal zones and it contains numerous stromatolite and stromatolite-like intercalations in the lower and upper parts of this section. Generally the stromatolite subfacies is divided in two sabkha-type lithofacies. The lower one is dominated by filamentous cyanobacteria whilst the upper one by coccoid ones. Change in character of stromatolites deposited mainly in intertidal zone and connected with alteration of biotas and character of surrounding biolaminoids indicate increase of salinity towards the end of the Platy Dolomite deposition. The general upward increase of both frequency and aggregate thickness of the stromatolite intercalations reflects the overall tendency towards slight deepening during the Platy Dolomite deposition.

Introduction

During the deposition of the Platy Dolomite (Ca3), the carbonate of the Leine, the Leba Elevation belonged to the peripheral zone of the Southern Permian Basin (Peryt *et al.*, 1985; Fig. 1).

The Platy Dolomite rocks are composed of dolomites, limestones, and anhydrites and exhibit various microfacies typical for shallow-water carbonate platform deposition. These are: (1) bioclastic wackestones and packstones which developed more shoreward, (2) peloidal wackestones and packstones, (3) oolitic packstones and grainstones which developed more seaward and built up a carbonate sandy barrier (Gasiewicz, 1985), (4) biolaminoids which clearly predominate in the facies assemblage of the Platy Dolomite carbonate platform series, (5) stromatolites, and (6) mudstones (see Gasiewicz & Peryt, in press, for detailed survey).

Lecture Notes in Earth Sciences, Vol. 10
T.M. Peryt (Ed.), The Zechstein Facies in Europe

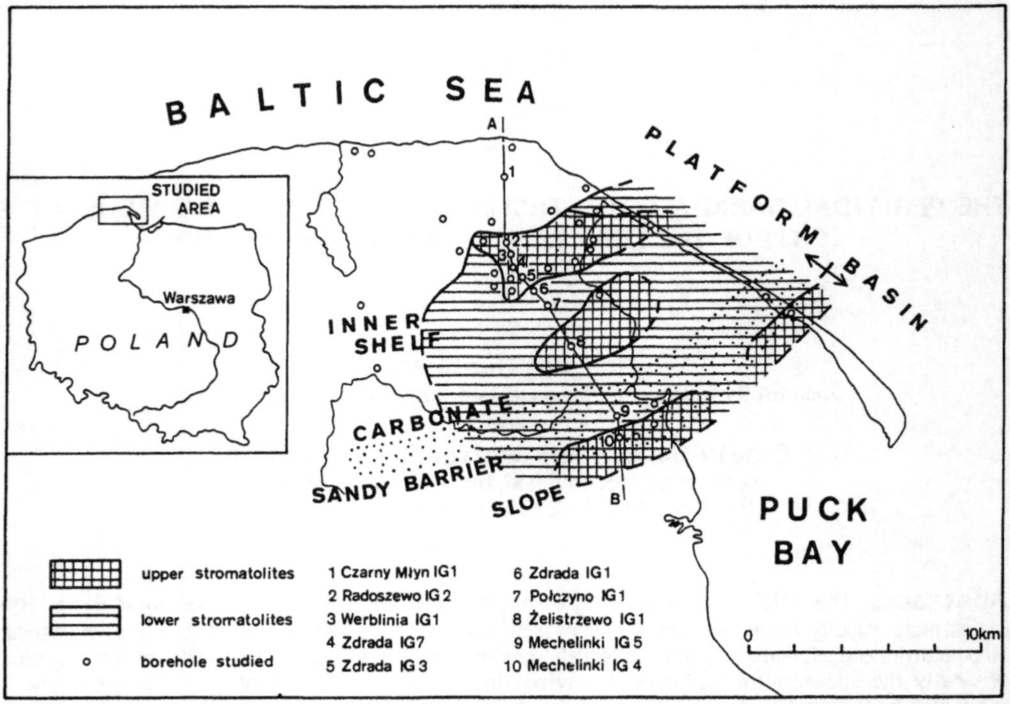

Fig. 1. Regional distribution of stromatolites on the carbonate platform area of the Leba Elevation. **A–B** - location of cross-section of Fig. 2.

The Platy Dolomite deposits exhibit distinct bipartition reflected by differences in: share of either mudstones, bioclastic and sulphate fractions; qualitative and quantitative diversity of macrofauna, microfauna and microflora; development of biolaminoids; and colour.

The stromatolites are numerous in the lower and upper parts of the Platy Dolomite section from the Leba Elevation. The presence of such stromatolite sheets associated with biolaminoids - typical coastal sabkha type deposits, as described by Gerdes *et al.* (1985) and Krumbein *et al.* (1979) from recent environments mark the different sedimentation regimes during the Platy Dolomite deposition. In addition, considerable changes in the composition and arrangement of microbial components of stromatolites and biolaminoids may indicate an evolution of the sedimentary environment at that time. Some indications are also available on the fact that the calcium distribution between carbonates and sulphates is influenced by the activity of microbial mats of the intertidal and supratidal zones. Thus, the stromatolite intercalations can reflect a general character at the beginning of marine transgression of the third Zechstein cycle.

Geological setting

The Platy Dolomite carbonate platform of the Leba Elevation developed on an almost flat surface formed by the second cycle of the Zechstein deposition. The northern part of the platform deposits is underlain (with erosional contact) by dolomites and limestones of the Main Dolomite (Ca2) belonging to the former evaporite cycle. Somewhat southward, the Platy Dolomite series rests on the anhydrites of the second cycle of Zechstein deposition. The southern part (more basinward) of the investigated platform carbonates is overlain by grey and red generally siliciclastic mudstones with admixture of sulphates belonging to the Grey Pelite (T3). A part of the Grey Pelite belongs to the third cycle of the Zechstein deposition.

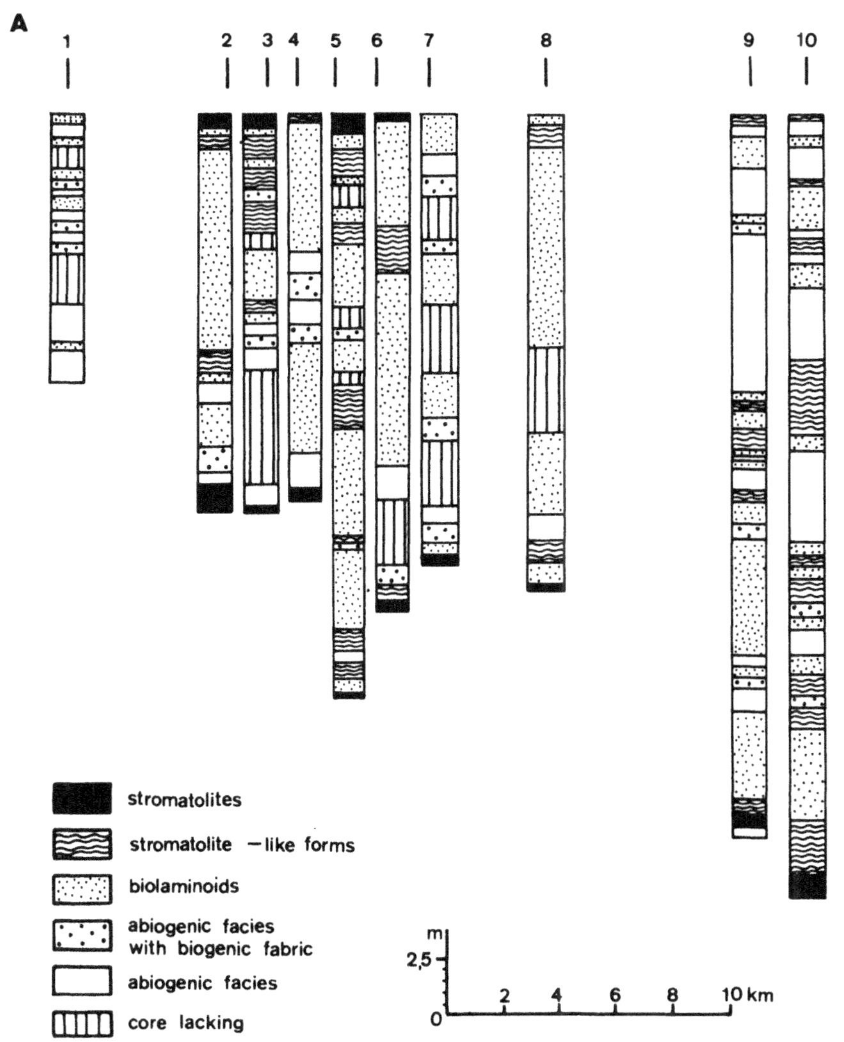

Fig. 2. Distribution of biogenic lithofacies.

In the middle part the Grey Pelite sequence is developed as a distinct intraclastic facies a few metres in thickness. This conglomeratic complex passes upwards into fine grained quartz mudstones. The Grey Pelite mudstones gradually but rapidly pass into carbonates of the Platy Dolomite.

The Platy Dolomite rocks are capped by the Triassic Buntsandstein in the northern part of the carbonate platfrom. In the southern part the Platy Dolomite deposits pass gradually into the Main Anhydrite sulphates (A3).

Beyond the carbonate platform area the Platy Dolomite sediments are strongly reduced in thickness and developed mainly as carbonate mudstones. Evaporates infill the relief which was inherited after the Platy Dolomite deposition. Accordingly, they wedge out gradually towards the coast.

Description of the stromatolites

General features

Stromatolites occur in the southern part of the carbonate platform and are limited in their horizontal distribution to the discontinuous zones adjacent to and comprised in the carbonate platform margin (Fig. 1). The stromatolites are locally well developed and usually mark the beginning and the end of sedimentation of the Platy Dolomite section in the Leba Elevation area. In addition some stromatolite intercalations occur in the lower and especially in the upper parts of the examined Zechstein sections (Fig. 2). Quantitative occurrence of stromatolitic intercalations in the Platy Dolomite section is given in Table 1.

Table 1. Quantitative comparison of the inner-shelf zone biogenic lithofacies (data from 10 boreholes). S - stromatolites, SLF - stromatolite-like forms.

Platy Dolomite section	number of biogenic intercalations		aggregate thickness (m)			thickness percentage	
	S	SLF	Platy Dolomite	S	SLF	S	SLF
upper part	16	33	148.7	9.50	26.50	6.4	17.8
lower part	11	17	81.7	4.90	10.40	6.0	12.7

Generally, the Platy Dolomite stromatolites are characterized by subtle, regular, rhythmical, flat, mm-thick and densely packed laminae (Fig. 3). These features allow to differ the stromatolite facies from biolaminoid ones (Fig. 4).

Fig. 3. Polished slabs of stromatolites developed at the base (**A**) and at the top (**B**) of the Platy Dolomite. Note characteristic flat (**A**) and wavy (**B**) lamination and the presence of evaporites (**B**).

The stromatolites are thin and relatively uniform in thickness (usually 0.20-0.50 m, 0.45 m on average, max. 1.2 m). Apart from these typical stromatolites in the middle and especially in the upper part of the Platy Dolomite section, stromatolite-like structures commonly occur in the form of intercalations. These represent transitions between biolaminoids and stromatolites in their development. These forms are relatively varied in thickness but are usually 0.30-0.90 m thick, with a maximum thickness up to 3.0 m.

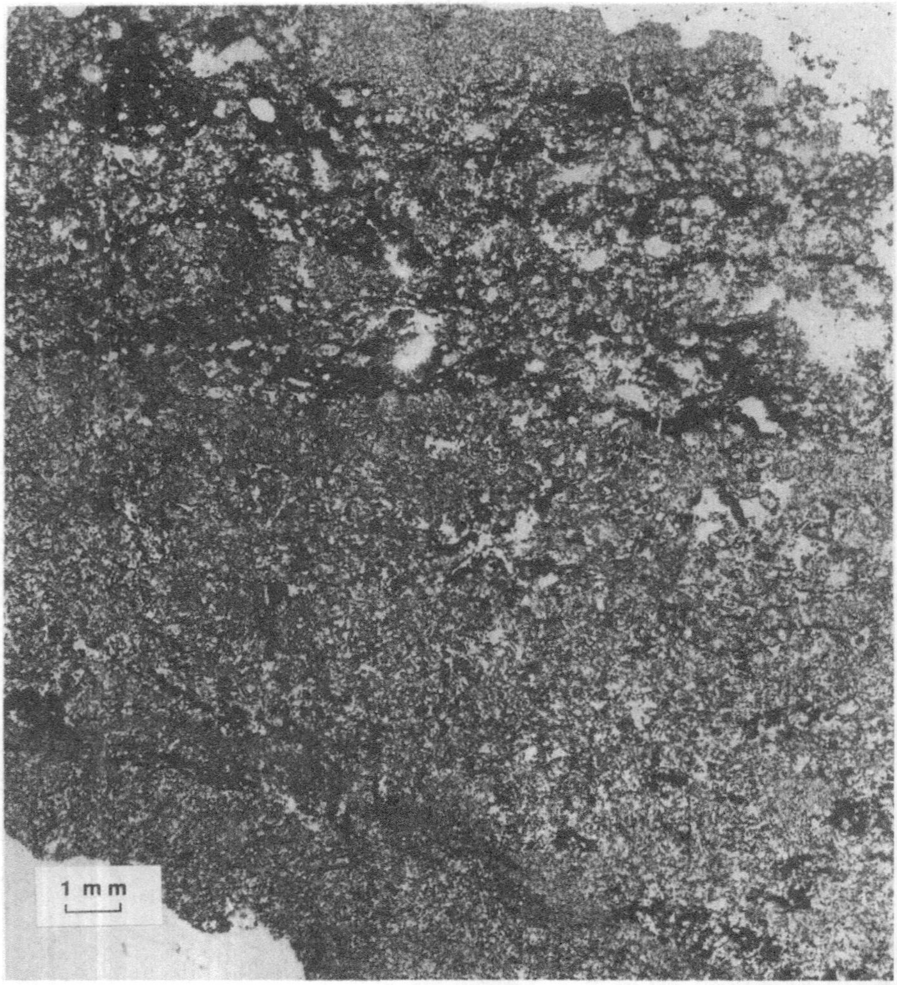

Fig. 4. Irregularly developed biolaminoid lamination. Dark organic rich and discontinuous laminae alternated with thick, somewhat cloudy muddy laminae.

Light and scanning electron microscopy studies demonstrate the presence of densely packed and abundant filamentous microfossils which built up the dark organic rich laminae of the stromatolitic subfacies. These organic remains are interpreted as filamentous cyanobacteria which may be oriented horizontally to vertically in the individual laminae (stromatoids *sensu* Kalkowsky, 1908). On the basis of our study, the stromatolites (and stromatolite-like forms) occurring in the lower part of the Platy Dolomite section may be characterized by the predominance of laminae with horizontally oriented filamentous cyanobacteria, which we will characterize as the Lh type of laminae. The laminae of the middle and the upper stromatolites and stromatolite-like structures occurring in the Platy Dolomite rocks - apart from horizontally (Lh) and diagonally to vertically (Ldv laminae) oriented filaments of cyanobacteria -

may be characterized by the presence (and alternation) of laminae enriched or impoverished in microscopically visible "cellular" structures (Figs. 5-8). These distinct "cells" may be interpreted as the remains mainly of the sheath material of coccoid and filamentous cyanobacteria of typical sabkha type phototrophic microbial mats which have been dominated by cyanobacteria.

Fig. 5. Stromatolite-like structures in the middle and upper parts of the Platy Dolomite. Condensed material alternates with looser packings. Condensed layers commonly show a lense-like microtopography within which concentric relic structures indicate the former presence of colonies of coccoid cells.

The stromatolites which occur at the base and at the top of the Platy Dolomite sequence appear in stable lithostratigraphic position. However, the stromatolites and stromatolite-like structures of the upper part of the Platy Dolomite usually occur as rather occasional, not correlatable and lense-like bodies in the biolaminoidal sedimentary matrix (Fig. 2).

No evidence of grazing, burrowing, reworking or any other type activity of Metazoa was found in the investigated stromatolitic subfacies.

The above mentioned differences in the development of stromatolites can be classified into three categories: (i) the lower stromatolites, (ii) the upper stromatolites, and (iii) the stromatolite-like forms.

The lower stromatolites (Figs. 3A & 9)

They occur as a continuous cover on parts of the carbonate platform at the base of the Platy Dolomite (Figs. 1 & 2). The lower stromatolites are relatively more diversified in thickness (up to maximum of 1.20 m, usually 0.10-0.20 m, 0.50 m on average) on the platform margin area, compared to the inner platform stromatolite area (max. 0.90 m, usually 0.20-0.50, 0.40 m on average).

Fig. 6. Close-up view of micrite matrix of lighter laminae which usually alternate with dark and more condensed layers. Light laminae are suggestive to have formed by dominance of unicellular microorganisms, their polysaccharids and decay products, dark laminae by dominance of filamentous organisms. Filamentous microfossils can be found also within the light micrite laminae. Note filamentous arrangement of larger micrite crystals which indicate precipitation of dolomite around a former sheath-encased filament bundle, the organic matrix of which had bound Mg until diagenetic dolomitization started.

The lower stromatolites directly rest on the fine-grained quartz mudstones. The transition between these two facies is continuous and rapid. Upwards, stromatolites pass gradually into the stromatolite-like packets or directly into the biolaminoid facies.

The lower stromatolites (Fig. 9) are smooth mats usually composed of distinct, regular, parallel and flat laminae up to 4 mm thick. The laminae which built the lower part of this mat usually contain small amounts of fine quartz grains. Generally the lower stromatolites are composed of three types of laminae: (1) relatively thick and lighter with microvoids and abundant thin, elongated and strongly packed, organic microlayers which are parallel to general lamination and which mark distinct secondary lamination, (2) thinner (2-5 times less than 1), darker (more micritic) and usually with less frequent and finer organic structures oriented horizontally to vertically, and (3) the lowest part of this stromatolite often contains very thin (up to 1 mm thick), flat and only micrite, single laminae which disappear upwards. The lower stromatolite is dominated by the first type of laminae so it may be defined as the Lh-type stromatolite.

Fig. 7. Micritic matrix of lighter laminae showing coccoid and filamentous microfossils. Coccoid colonies (right) show microcrystalline replacement structures; larger and smoother compounds (centre) indicate relics of the sheaths of unicellular aggregates with cast-negatives of former cells. Note also the filamentous microfossils (lower left) which are possibly fossilized in upward migration showing diagonal to vertical orientation.

The upper stromatolites (Figs. 3B & 10.)

These occur as discontinuous patches at the top of the Platy Dolomite sequence (Figs. 1 & 2). Sometimes, only in the northern part of the stromatolite area, they occur as thin and rare intercalations in the middle and especially in the upper part of the Platy Dolomite section. The uppermost stromatolites are usually thin and uniform in thickness (usually 0.20-0.30 m, 0.30 m on average and max. 0.50 m) on the platform margin area. In the northern part, however, they are thicker and. greatly varying in thickness (usually 0.30-0.50 m, 0.70 m on average and with a maximum thickness of 1.2 m).

These stromatolites (Fig. 10) are characterized by very subtle and strongly developed small-scale wavy lamination. The lamination is usually parallel, regular, continuous horizontally, sometimes more flaser-like. These stromatolites are flat mats composed of two types of laminae varying in thickness: (1) relatively thick (up to 5 mm) and light, composed of micrite and microsparite matrix, containing microvoids and sulphate granules and first of all abundant rounded and oval "cells" irregularly dispersed or connected to form elongate lenses; (2) dark, very thin, subtle, wavy and often flake-like, connected with horizons of densely packed "cells" (particularly forms under

Fig. 8. Condensed lamina of Platy Dolomite showing microcrystalline replacement structures and cast-negatives of former coccoid cells. These are far less abundant within the condensed laminae (compare their abundance in light laminae, Fig. 3).

1/3 mm in size). Both types of laminae contain filamentous cyanobacteria oriented mainly horizontally in type 2 and diagonally to vertically in type 1. Both types are firstly characterized by the dominance of cell remains, Secondly they may eventually be defined as sabkha-type flat stromatolites with low degree of undulation and domal or conical development typical for stromatolites of lower salinities, i.e., sea water salinity without evaporitic features. They may also be called evaporitic stromatolites.

The characteristic feature of the upper stromatolites is the presence of so-called augen structures, granules and nodules sometimes arranged to form distinct horizons. All these structures and the "cells" as well, are usually filled with sulphates, and their frequency increases upwards. In addition the upper part of this flat mat contains discoidal sulphate crystals more or less dispersed. It is overlain by nodular anhydrite with gradual transition between a carbonate and sulphate series. The stromatolites contain single bivalve fragments, which very probably were transported into the system from their original lower saline environment. They represent no authigenic biocoenosis.

The stromatolite-like forms (Figs. 11 & 12)

They usually occur as continuation of the lower stromatolite sequence and further as variants in thickness independent or accompanied by the upper-type

Fig. 9. Regularly developed stromatolitic laminae dominated by horizontally oriented "splices" of microfossils. Note rhythmicity of more biogenic and more muddy lamination.

stromatolite intercalation in the middle and the upper part of the Platy Dolomite section (Fig. 2) They usually occur as more or less horizontally continuous lense-like bodies in the biolaminoid series. These structures often underlie the oolite complexes of the platform margin and usually contain single coated grains. On this platform another sedimentary succession is often observed: the biolaminoid facies represented by stromatolite-like stacks of laminoids with frequently intercalated and embedded *in situ* grown coated grains among which many real ooids, sometimes merging into partly reworked oolite complexes derived from the coated grains material of the laminoid mats.

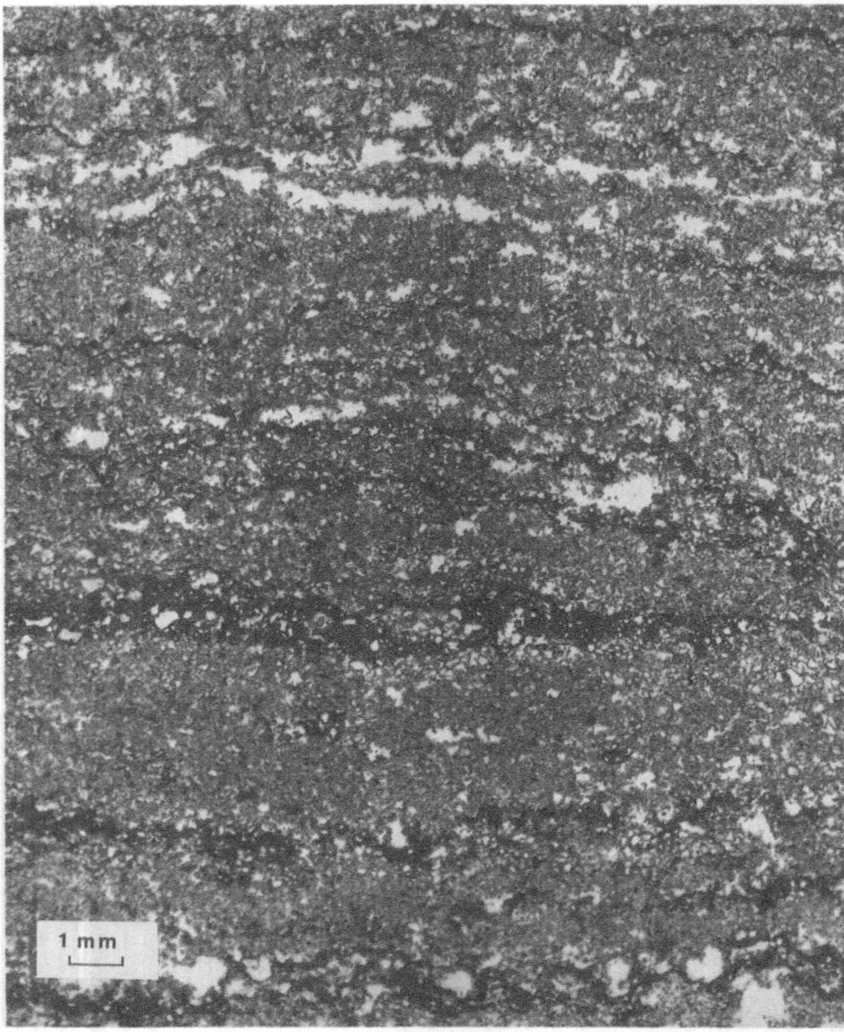

Fig. 10. Wavy laminated upper stromatolite. Note relatively thicker muddy laminae dominated by "cellular" structures inherited after the coccoid cyanobacteria colonies.

Generally these stromatolite-like forms vary in thickness (usually 0.30-0.90 m with a max. thickness of 3.0 m) but relatively less in the northern part of the stromatolite area (where they are usually thinner) compared to the southern one. This diversification in thickness increases upwards on the whole stromatolite area.

These structures, with regard to character of lamination and organic components, are similar to the lower and the upper stromatolites on one hand and to the biolaminoids on the other. They represent an intermediate between these two distinct facies. Then it is possible to distinguish two subtypes: (a) those which resemble the lower stromatolites and occur only in the lower part

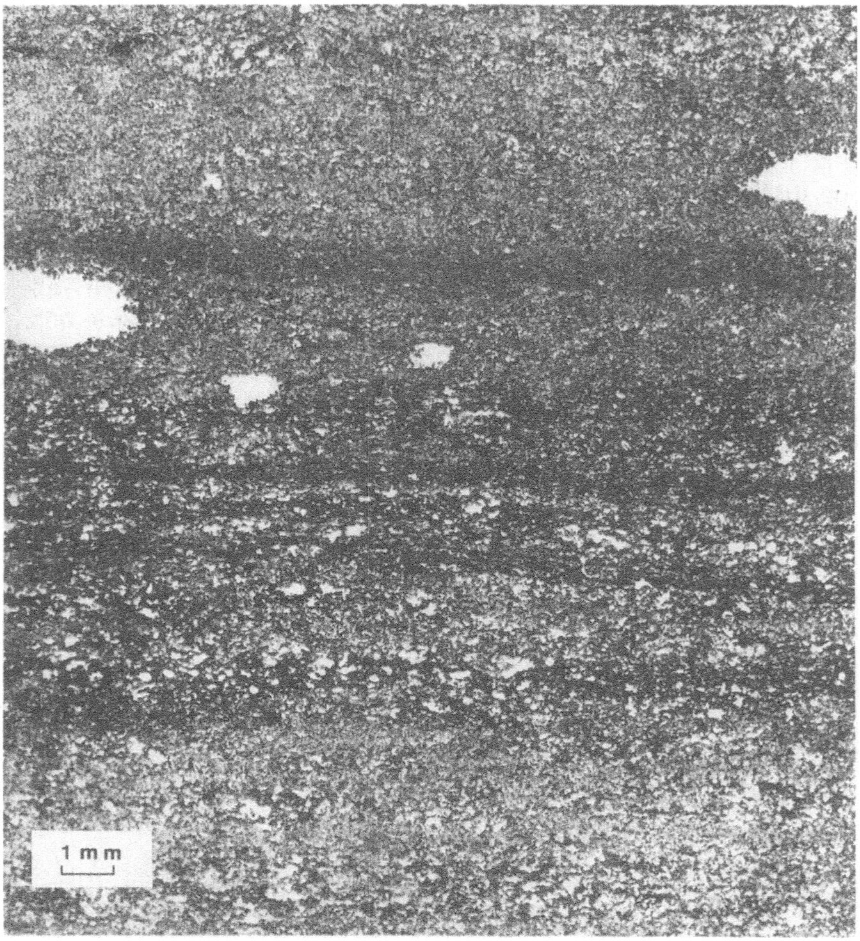

Fig. 11. Weakly developed lamination of the lower stromatolite-like form. Note distinct laminae arhtythmically alternating with thicker and muddy laminae which are biolaminoidal in character. Organic laminae are composed mainly of horizontally oriented splices of microfossils. Rare "cells" after coccoid microfossils are visible as well.

of the Platy Dolomite sequence (Fig. 11), and (b) those which resembles more the upper ones and occur in the middle and upper parts of the investigated section (Fig. 12). With regard to the presence of fossilized organic microforms they may be defined in the same way as adequate stromatolites.

Discussion

The biolaminoid-nodule facies - a major sedimentary unit of the Platy Dolomite carbonate platform - underlies, intercalates and overlies stromatolite and stromatolite-like packets (Fig. 2). The facies is firstly certainly clearly distinguishable from abiogenic sedimentary bedding and facies by its frequent content of microfossils of the cyanobacterial type and its strong dolomitization., secondly it is characterized by the authigenic formation of ooids, oncoids and pelletoidal coated grains (Dahanayake & Krumbein, 1985, 1986; Dahanayake *et al*, 1985), which sometimes are slightly moved by different degree of density and compaction in a very viscous (slimy) degradational system of mats in fine grained carbonates most of which have been precipitated within the decaying mats (Krumbein, 1979, 1983). On other place the mat has been eroded and the coated grains have been redeposited within cavities of the mats or within freshly forming biolaminoids. Finally compaction is strongly changing the microbial mats in order to generate and increase the laminoid or flasery pattern. This is certainly caused by differential compaction of the viscous cyanobacterial slimes and the more stabilized coated grains, which often get enriched in lensoid or "sedimentary augen structures".

The biolaminoid-nodule facies thus is identified as a coastal sabkha type deposit originating in the intertidal and supratidal zones (Gerdes, 1985; Dahanayake *et al.*, 1985). The stromatolites, compared to the biolaminoids reflect conditions (and periods as well) of dominance of biological productivity over sediment accretion. These are probably generated in somewhat deeper environments (in the lower intertidal or upper subtidal zone (Fig. 13). The biolaminoids, however, are forming in dryer or only perennial wetted environments where evaporative pumping and capillary water supply is more typical than tidal or seasonal seawater supply.

Thus, the facies succession:

Biolaminoids --> Stromatolite-like laminites--> Stromatolites

simply reflects an increase of water cover or the increase of the influence of water coming inshore above the sedimentary surface rather than through the pore system of the usually air exposed sabkha. At the same time, periods of stromatolite development generally reflect the tendency of the Platy Dolomite areas on the Leba Elevation to fluctuations of wetness and water cover with a general tendency to increased dryness and a topping sequence with fresh water supply. The above mentioned connected sedimentary sequence with oolite complexes in the margin zone confirms the deeper conditions for the stromatolitic subfacies as suggested by Newell *et al* (1960). They showed that the ooid growth optimum occurs at a depth of up to 2 m. Their formation, however, starts somewhat deeper and syngenetic coated grain formation proposed by Dahanayake *et al*, 1985) - usually single and simple coated grains but also multilaminated grains and grain packets (ooid bags *sensu* Kalkowsky, 1908) - are observed in mats or mat-like bodies that underly carbonate sandy beds.

The presence of densely packed smooth and flat laminated stromatolitic rocks especially at the base and at the top of the Platy Dolomite section is surprising. They reflect a permanent water condition (and deeper as well) on a larger area of the carbonate platform. On the other hand they mark the more regional events of subsidence or sea-level rise, compared to the lense-like stromatolitic packets in the lower and in the upper parts of the Platy Dolomite series.

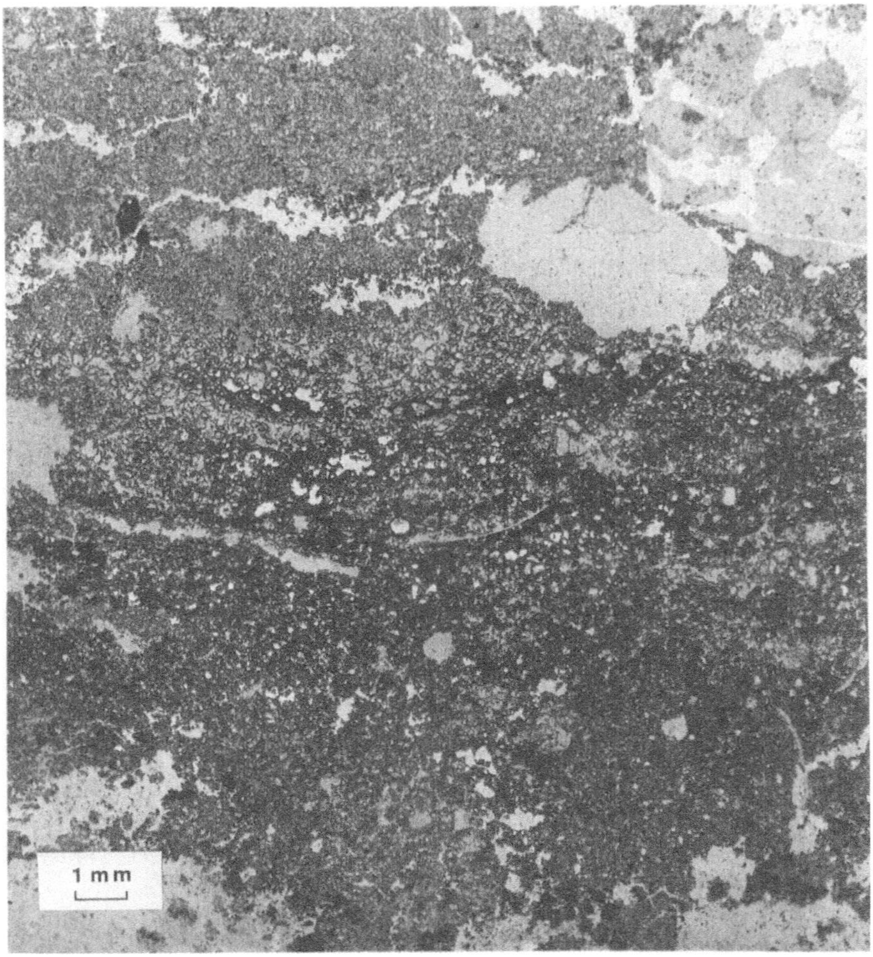

Fig. 12. The upper stromatolite-like form dominated by "cellular" structures and with badly developed lamination and frequent so-called augen-structures.

The general upward increase of stromatolitic intercalations (Table 1) simply indicates the overall tendency of the Platy Dolomite sea to deepen. The Platy Dolomite transgression started in the middle part of the Grey Pelite and is expressed by intraclastic facies. This flooding was probably rapid. The conglomeratic deposition was gradually replaced by a fine-grained one. The clastic series reflects the period of progressive transgression during the infilling of the coastal area and the appeasement of sedimentation. It caused limitation of the clastic material supply and development of normal coastal carbonate deposition in a shallow sea. Relatively stable substrates were colonized by mats to form different subfacies. An increase in water cover or subsidence was far slower than the stromatolitic accretion, making the rapid microbial mat sealing and fixing of the intertidal zone possible. It caused relatively fast decreasing of water depth and carbonate deposition above the low water level. On the entire carbonate platform area, the predominance of the biolaminoid facies in the Platy Dolomite series indicates relatively stable sedimentary conditions. However, an increase of both frequency and aggregate thickness of stromatolitic intercalations and change in the character of biolaminoids towards the top of the Platy Dolomite section indicate a slight but continual increase of water cover. It is also possible that carbonate and oolite sand shoals were exerting an additional influence on the shallow coastal strip and its hydrodynamics by hindering the sea to fully flood the shallow coastal area. Supply of water by extremely high tides, storm generated floods or seasonal changes of water level in an elongate sea under evaporation stress comparable to the situation in the recent Persian Gulf or Red Sea Gulf of Suez strips may have further influenced the alternations between stromatolites and biolaminoids. Apart from this, the rate of intertidal sedimentation was high enough to keep pace with the rate of transgression.

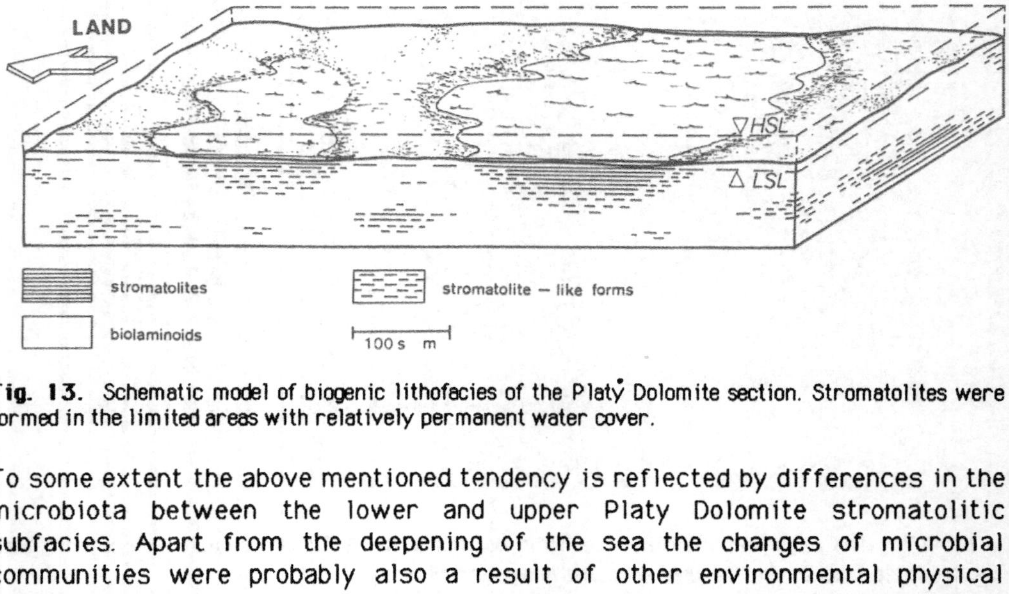

Fig. 13. Schematic model of biogenic lithofacies of the Platý Dolomite section. Stromatolites were formed in the limited areas with relatively permanent water cover.

To some extent the above mentioned tendency is reflected by differences in the microbiota between the lower and upper Platy Dolomite stromatolitic subfacies. Apart from the deepening of the sea the changes of microbial communities were probably also a result of other environmental physical factors, mainly the general increase of water salinity. Additional factors such as annual evaporation, tidal range, air temperature range, rainfall and predominant wind and wave direction probably played an insignificant role.

The more extended and pronounced stromatolitic sheets at the base and to the top of the investigated series are obviously diachronous. The finely and evenly laminated stromatolites of the lowermost and uppermost parts of the Platy Dolomite differ considerably from the biolaminoid facies in between. It seems that the densely packed layers of dark stromatolites rich in organic material have formed under the influence of conditions different from those ruling the biolaminoids. In the stromatolite zones and biolaminoid/nodule zone of the Gavish Sabkha, increasing salinity and increasing water cover are responsible for the development of ideal stromatolite mats. It has to be noted, however, that the difference in water cover are duplicate and in an astonishingly narrow range. Such narrow ranges of changing conditions with strong impacts on the sub-facies have frequently been observed in recent sabkha type stromatolitic environments (Gavish *et al.*, 1985; Gerdes *et al.*, 1985; Friedman & Krumbein, 1985; Friedman *et al*, 1985). The main differences are in the thickness of the mat overlying water column, which may be only between 50 cm and 0 cm and in the annual frequency of water cover and the question whether permanent water cover existed for extended periods or not.

The biolaminoids form rather under evaporative pumping and tidal recharge through the interstitial system. Thus, the typical sequence (salt and gypsum crusts forming thick layers in summer after a thin microbial mat formed in winter; further development of coccoid cyanobacteria around dissolving nodules of the summer crusts, later coated by microbial decay-derived carbonate deposits) will produce thick nodular layers and very thin laminae. The regime of biolaminoids is one of changing salinities and water cover, the regime of the associated "real" ideal stromatolites more of semi-permanent water cover with generally higher but stable salinities. Golubic (1973), Krumbein and Cohen (1974, 1977), Krumbein *et al* (1977) and Gerdes *et al.* (1985) have described many different types of microbial mats which may develop in an intertidal evaporitic system. The internal lamination of stromatolites will usually reflect changes from summer to winter in the cyanobacterial composition. When water cover is permanent, the coccoid layers will be thicker and with short periods of dryness the filamentous layers will become more prominent.

Precipitation of gypsum crystals connected with generation of granules and nodules in the upper part of the upper stromatolites limited and disordered the stromatolite development.

The development of stromatolites during the Platy Dolomite was not connected with any apparent metazoan grazing control so it may indicate very saline waters in the lower intertidal and shallow subtidal areas from the beginning of this Zechstein section. In addition some traces of burrowers in lower part of the Platy Dolomite deposits, relatively diversified forams and macrofauna, and the presence of locally abundant bryozoans indicate rather near-normal marine salinity.

The Platy Dolomite sabkha sediments are characterized by a horizontal and vertical zonation of several biolaminite (or stromatolitic) facies types. These comprise (a) biostromate and domed densely laminated stromatolites, (b) biolaminoids with embedded nodules, (c) evaporitic stromatolitic systems. In the recent coastal sabkha plain area it is possible to delineate several zones of biolaminites as described by Gavish *et al.* 1985) and Gerdes *et al.* (1985). In the Gavish Sabkha, densely packed, well laminated stromatolites are rarely formed and more rarely preserved (fossilized) than the biolaminoid-nodule zone sediments described above. This must depend on specific supratidal environmental conditions. The environmental setting of different types of potential stromatolites (Krumbein, 1983), cryptalgal (or cryptmicrobial) and fenestral structures, stromatoporoid structures (Monty, 1976; Kazmierczak & Krumbein, 1983) and mat generated coated grains (Friedman *et al.*, 1985; Krumbein & Cohen, 1977) has been recently reviewed (e.g., Cohen *et al.*, 1984; Friedman & Krumbein, 1985). The ideal stromatolite setting in supratidal systems with evaporative pumping and salinity changes is rather rare as compared to the less distinct microbial mat types of the biolaminoid-, nodule-, or evaporitic microbial mat-facies, in which the vertical extension of a living microbial mats embedded in gypsum and halite may reach a thickness of 8 cm and more (Krumbein, 1983; Gerdes *et al.*, 1985). Thus it is possible that the biogenicity of (i) cloudy anhydrite, (ii) chicken-wire structures, (iii) petee or enterolithic strucures (Gavish et al., 1985), (iv) carbonate nodules, (v) coated grains (oncoids and ooids) generated *in situ* within microbial mats (Friedman et al., 1985; Krumbein & Cohen, 1974), (vi) wormy dolomite, even mat generated "false cross-bedding" (Schieber, 1986), and others were not recognized previously as direct or indirect evidence for a microbial mat generated, generally stromatolitic environment with little or no direct water cover for long periods of a year and over many years. The last geological prerequisite for the development and stability of such a system is certainly that subsidence has to match the evaporative pumping and microbial mat generated evaporitic stromatolite and biolaminoid accretion rates.

Comments

The Platy Dolomite carbonate platform deposits of the Leba Elevation are dominated by stromatolitic subfacies generated in somewhat different environmental conditions (see Table 2).

The stromatolite and stromatolite-like intercalations were deposited deeper, mainly in the lower intertidal zone.

The dominance of stromatolites at the bottom and top of the series indicates a relatively fast decreasing (after the PZ3 transgression) of water cover at the beginning and increasing of water cover toward the end of the Platy Dolomite deposition.

On the basis of the development and dominated cyanobacteria the stromatolitic subfacies is diversified in two main types which reflect changes of physical conditions, mainly salinity and water depth.

Changes of water cover have been most probably associated with: stable increased salinities at the beginning, through near normal level of salinity in the lower part of the Platy Dolomite series, further increased salinity during the upper part of this section up to considerably increased at the end of the deposition of the Platy Dolomite. Largely fluctuating salinities during the deposition of thick biolaminoids intercalations have to be assumed.

Table 2. Environmental interpretation of the biogenic lithofacies of the Platy Dolomite section.

facies	tidal zone	dominant type of cyanobacteria	salinity
upper stromatolites	inter/subtidal	coccoid	considerably increased
upper stromatolite-like forms	lower intertidal	coccoid	increased
upper biolaminoids	supra/intertidal	coccoid & filamentous	increased
lower biolaminoids	lower supratidal	filamentous & coccoid	near normal (upward slightly increased)
lower stromatolite-like forms	upper intertidal	filamentous	near normal
lower stromatolites	inter/subtidal	filamentous	increased

Increase of frequency and total thickness of the biolaminoid intercalations throughout the Platy Dolomite marked increase of either the frequency or permanency of water cover and water depth. Thus it indicates the slight tendency to deepening of the Platy Dolomite sea on the Leba Elevation and in turn marks the subsidence event or sea level rise.

References

Cohen, Y., Castenholz, R.W. & Halvorson, H.O. (Eds.), 1984. *Microbial Mats: Stromatolites.* Alan Liss Publ. New York, 498 pp.

Dahanayake, K. & Krumbein, W.E., 1985. Ultrastructure of a microbial mat-generated phosphorite. *Miner. Depos.*, 20: 260-265.

Dahanayake, K. & Krumbein, W.E., 1986. Microbial structures in oolitic iron formations. *Miner. Depos.*, 21: 85-94.

Friedman, G.M. & Krumbein, W. E. (Eds.) 1985. Hypersaline Ecosystems - The Gavish Sabkha. *Ecological Studies* 53, Springer, Berlin, 484 pp.

Friedman, G.M., Sneh, A. & Owen, R.W., 1985. The Ras Muhammad Pool: Implications for the Gavish Sabkha. *Ecological Studies* 53: 218-237.

Gasiewicz, A., 1985. Krawedz platformy weglanowej dolomitu plytowego na wyniesieniu Leby. *Prz. geol.*, 33: 211-216.

Gasiewicz, A. & Peryt, T.M. (in press). Dolomit plytowy w rejonie Zatoki Puckiej: litologia i paleogeografia. *Biul. Inst Geol.*

Gavish, E., Krumbein, W.E. & Halevy, J., 1985. Geomorphology, mineralogy and groundwater geochemistry as factors of the hydrodynamic system of the Gavish Sabkha. *Ecological Studies* 53: 186-217.

Gerdes, G., 1985. Die stromatolithische Fazies marin randlicher Grenzbereiche. *Ph. D. Thesis*, Universität Oldenburg, 170 pp.

Gerdes, G. , Krumbein, W.E. & Holtkamp, E.M., 1985. Salinity and water activity related zonation of microbial communities and potential stromatolites of the Gavish Sabkha. *Ecological Studies* 53: 236-266.

Golubic, S., 1973. The relationship between blue-green algae and carbonate deposits. In: Carr N.G. & Whitton, B.A. (Eds.), *The Biology of Blue-Green Algae* 9: 434-472. Blackwell Scientific Publications, Oxford.

Kalkowsky, E., 1908. Oolith und Stromatolith im norddeutschen Buntsandstein. *Z. dt. geol. Ges.*, 60: 68-125.

Kazmierczak, J. & Krumbein, W.E., 1983. Identification of calcified coccoid cyanobacteria forming stromatoporoid stromatolites. *Lethaia* 16: 207-213.

Krumbein, W.E., 1979. Photolithotrophic and chemoorganotrophic activity of bacteria and algae as related to beachrock formation and degradation (Gulf of Aqaba, Sinai). *Geomicrobiology Jour.* 1: 139-203.

Krumbein, W.E., 1983. Stromatolites - the challenge of a term in space and time. *Precambrian Res.*, 20:493-531.

Krumbein, W.E., Buchholz, H., Franke, P., Giani, D., Giele, C. & Wonneberger, C., 1979. O_2 and H_2S coexistence in stromatolites. A model for the origin of mineralogical lamination in stromatolites and banded iron formations. *Naturwissenschaften* 66: 381-389.

Krumbein, W.E. & Cohen, Y., 1974. Biogene, klastische und evaporitische Sedimentation in einem mesothermen monomiktischen ufernahen See (Golf von Aqaba). *Geol. Rundschau* 63: 1035-1065.

Krumbein, W.E. & Cohen, Y., 1977. Primary production, mat formation and lithification: contribution of oxygenic and facultative anoxygenic cyanobacteria, In: Flügel, E. (Ed.), *Fossil Algae*, 37-56. Springer, Berlin.

Krumbein, W.E., Cohen, Y. & Shilo, M., 1977. Solar Lake (Sinai) 4. Stromatolitic cyanobacterial mats. *Limnology and Oceanography* 22: 635-656.

Monty, C.L.V., 1976. The origin and development of cryptalgal fabrics. In: Walter, M.R. (Ed.), *Stromatolites*, 193-249. Elsevier, Amsterdam.

Newell, N.D., Purdy, E.G. & Imbrie, J., 1960. Bahamian oolitic sand. *J. Geol.*, 68: 481-497.

Peryt, T.M., Czapowski, G., Debski, J. & Pizon, A., 1985. Model sedymentacji ewaporatow cechsztynskich na wyniesieniu Leby. *Prz. geol.*, 33: 204-211.

Schieber, J., 1986. The possible role of benthic microbial mats during the formation of carbonaceous shales in shallow mid-Proterozoic basins. *Sedimentology* 33: 521-536.

GPSR Compliance

The European Union's (EU) General Product Safety Regulation (GPSR) is a set of rules that requires consumer products to be safe and our obligations to ensure this.

If you have any concerns about our products, you can contact us on ProductSafety@springernature.com

In case Publisher is established outside the EU, the EU authorized representative is:

Springer Nature Customer Service Center GmbH
Europaplatz 3
69115 Heidelberg, Germany

Batch number: 10704763

Printed by Printforce, the Netherlands